How to Build Earthquake, Weather, and Solar Flare Monitors

How to Build Earthquake, Weather, and Solar Flare Monitors

Gary G. Giusti

TAB Books
Division of McGraw-Hill, Inc.
New York San Francisco Washington, D.C. Auckland Bogotá
Caracas Lisbon London Madrid Mexico City Milan
Montreal New Delhi San Juan Singapore
Sydney Tokyo Toronto

Published by TAB Books, an imprint of McGraw-Hill, Inc.

pbk 1 2 3 4 5 6 7 8 9 FGR/FGR 9 9 8 7 6 5
hc 1 2 3 4 5 6 7 8 9 FGR/FGR 9 9 8 7 6 5

Library of Congress Cataloging-in-Publication Data

Giusti, Gary G.
How to build earthquake, weather, and solar flare monitors / by Gary G. Giusti.
p. cm.
Includes bibliographical references and index.
ISBN 0-07-025209-2 (pbk.)
1. Seismometers—Design and construction. 2. Detectors—Design and construction. I. Title.
QE541.G52 1995
551.2'2'028—dc20 94-23110
CIP

Acquisitions editor: Roland S. Phelps
Editorial team: Andrew Yoder, Book Editor
Joanne Slike, Executive Editor
Production team: Katherine G. Brown, Director
Lisa M. Mellott, Coding
Brenda M. Plasterer, Coding
Toya Warner, Computer Artist
Brenda Wilhide, Computer Artist
Jeff Hall, Computer Artist
Wanda S. Ditch, Desktop Operator
Donna Gladhill, Proofreading
Joann Woy, Indexer
Designer: Jaclyn J. Boone

EL1
0252092

Contents

List of projects

Safety

Electricity is a great friend, as we use it in many ways. However, electricity does not play favorites. It will only work one way, the right way, and if you are careless, it can kill you in a second.

The following are some rules to observe when using electricity. Most are obvious, and many people have heard the rules so many times that they seem boring. But being redundant with respect to safety with electricity has its purpose: *your life depends on it!*

Because some of the circuits described in this book deal with the observation of atmospherics, lightning, and thunderstorms, remember not to make outdoor observations near high-voltage transmission lines or during an electrical storm. During a storm, make the observations indoors.

Never stand under trees, near a pole, or under wires during a thunderstorm. Do not stand near a window or electric fixtures indoors.

Do not run extension wires or cords under rugs. The insulation could rub off and cause a short circuit.

Disconnect antenna and ground lines from equipment during electrical storms. Unplugging is also a good idea, which helps prevent equipment damage from stray transient currents and voltage spikes.

When working with some of the circuits while building, repairing, or testing, remember the following points:

- Before beginning a construction project, familiarize yourself with the circuit, particularly, where applicable, in the ac and power supply sections. When repairing a circuit, turn off the power, if at all possible. If circuit testing requires the power to remain on, remember that if a circuit is defective, high voltage might be present in unexpected points in the circuit.
- Use an insulated floor material or floor mat to stand on, and an insulated work bench. Do not stand on concrete and work on equipment.
- The "one-hand-in-the-pocket" rule is good to follow when using test probes. Never touch the test leads or contact a nearby metal object during testing.

- When testing ac-powered circuits, remember that line voltage is often present at the power input points of the supply circuits—especially around switches, fuses, transformers, etc., as long as the plug is connected to an ac outlet (even if the switch in the apparatus is "off"). Unplug the device, and be sure to discharge all electrolytic capacitors present in the device; after turning the power off and disconnecting the plug, discharge the high-voltage capacitors repeatedly through a 3-MΩ resistor to ground.
- Shield and insulate high-voltage terminals to lessen the possibility of accidental shock.
- Keep low-voltage and high-voltage circuits separate when building. As well as a safety feature, most projects will be less susceptible to noise and interference this way.
- Keep jewelry and other metal objects well away from any circuit being built. Do not put these items in your pocket; put them away safely somewhere else.
- When a project can be built for both ac and battery operation, wire the circuit and test it with a battery first. If all is well and the equipment functions properly, add the ac circuit later.
- Try not to work alone. Someone, preferably with a knowledge of CPR first aid, should be around to help in case aid is necessary.
- Anytime current measurements are made, or a meter for reading current is installed, the meter must be connected in series with the load. If incorrectly connected (in parallel), the meter presents an almost direct short circuit, which is a very dangerous condition. This condition will also cause a potential total failure to the circuit in use.
- When servicing or taking measurements in high-voltage circuits, particularly in "hot chassis" equipment, always connect an isolation transformer between the ac outlet and the circuit under test. This greatly reduces the shock hazard inherent in testing these circuits.

Introduction

This book is a project book of circuits adapted to, or specifically designed for, monitoring three basic kinds of natural phenomena:

1. Atmospheric anomalies, such as lightning strikes, static discharges, and solar energy values.
2. Solar disturbances, such as sunspots and solar flares.
3. Seismicity; dealing with earthquakes and low-level seismic activity.

The book is loosely divided into three sections. The first section describes basic modular circuits designed to build a basic seismometer, its sensors, a VLF radio receiver, and, finally, an earthquake radio receiver. This section is repetitive, and contains step-by-step instructions. Some of the basic circuits, such as the power supply and the audio amplifier, are used in many of the later projects. The deliberate presentation allows the electronics beginner to "cut teeth."

The second section contains more specialized projects. Most are a bit more complex, and some serious study efforts can be made with the apparatus. Various circuitry is described, from an IC-based ULF radio receiver to several versions of seismic detectors and sensors.

The third section details some case studies and theories, and the building of monitoring equipment resulting from these studies. These "maverick" circuits are based on speculative theories, but some of the ideas are gaining acceptance, and the discussion puts the reader/user at the cutting edge in these research areas.

A portion of the book involves building and using vacuum-tube circuits. The basic circuits have existed for many years. However, many are particularly well-suited for VLF monitoring, and make excellent atmospheric and solar flare receivers. All are readily adaptable for AM broadcast shortwave reception as well. Sample tube-based power supply and audio amplifier circuits are also presented.

Most of the projects are relatively simple. Many can be built in just a few hours by even the electronics beginner. The projects are relatively inexpensive, as well. None cost over $75 to build, including the cabinet, and many can be built for less than $10.

Although many serious students and amateurs become interested in earthquakes and solar activity by observing the suddenness and the often violent consequences of these important natural phenomena, apparatus for experimentation and observation is very difficult to locate. Often, this equipment is very cumbersome to operate and is costly. Many solar flare receivers and available VLF radios require extensive antenna and ground systems. Some receivers within the $1000 to $2000 cost range do not even have an audio output, making any interface to other equipment complex, and adding cost. One excellent VLF receiver, with the features of variable tuning and an audio output, retails for over $10,000!

Commercial seismographs have the same problems, with the available gear large (often housed in trucks), complex, and costly. A complete unit, if an experimenter can find one, often runs into several thousand dollars as well.

The purpose of this book is to acquaint and inform the reader: the serious experimenter and researcher, the student, and even the homeowner who desires an earthquake alert, to cost-effective methods of monitoring and studying these interesting and important events. All of the parts used in these circuits are readily available from conventional supply sources, the circuitry is versatile as well as simple, and all of the projects have been tested—most under actual field conditions.

You should read the entire text before any of the projects are built to determine which projects best suit your needs.

Take care in building, and hopefully, you will enjoy the operation of sensitive scientific equipment in the fields of seismology, meteorology, and atmospherics.

1
PART
Background and basic projects

1 CHAPTER

Seismology

Earthquakes and their related events are among the most dramatic and disastrous of all natural phenomena, but they are feared most because they can occur without warning.

Predicting earthquakes is far from being an exact science. They can occur anywhere, and vibrations hundreds of miles away can be detected with the appropriate circuitry.

With suitable gear, you can "tune-in" to the earth and monitor and record natural vibrations. With the extreme sensitivity available in some circuits, very small wave traces (microseisms) can be observed, and you can hear the constant creaking and groaning that goes on inside the earth. Depending on sensor location and equipment adjustments, you can even listen to passing automobiles and footsteps.

The vast majority of the time, these small movements dominate the observations. The earth's louder seismic messages begin suddenly and are transmitted according to a patterned format. The radio apparatus processes these signals electronically.

Seismic waves

When graphically recorded, either by the use of a computer or with a chart recorder, quakes begin with three characteristic parts:

- P deflections, the primary waves
- S deflections, the secondary waves
- L deflections, the large waves

These deflections are the records of motions of seismic waves.

The P deflections are made by longitudinal waves, which are really sound waves, and travel about five miles per second. Vibrations in the rocks are along the direction of wave travel. This action causes the rocks to push together.

The S deflections are caused by shear waves, which influence the seismograph at the surface after moving through the earth's body. Shear waves travel about one-half as fast as longitudinal waves, and vibrations are at right angles to the wave direction. This action causes the shaking of the rocks.

The L waves, called *Rayleigh waves*, are the waves that have moved along the earth's surface. These waves are limited to the earth's surface area and travel slightly slower than shear waves. Seismic waves pass directly through the earth in about 21 minutes.

For example: refer to Figs. 1-1 through 1-4 and Table 1-1. Sudden crustal movement at epicenter A sends out a pattern of shock waves. Figure 1-1 depicts the P waves' arrival times, vibrating in the direction of wave travel, referenced by the circular percentage of the earth. Figure 1-2 shows the arrival times of the S waves, vibrating in a side-to-side manner, also referenced by the circular percentage of the earth. The L wave line of travel, in a belt of the crust around the world, is shown via the zig-zag line on Fig. 1-3. This view also diagrams the zone where no direct arrivals of S waves are observed, indicating that the earth's core is composed of materials and densities that refract, reflect, or absorb these waves. Figure 1-4 is a graphic ref-

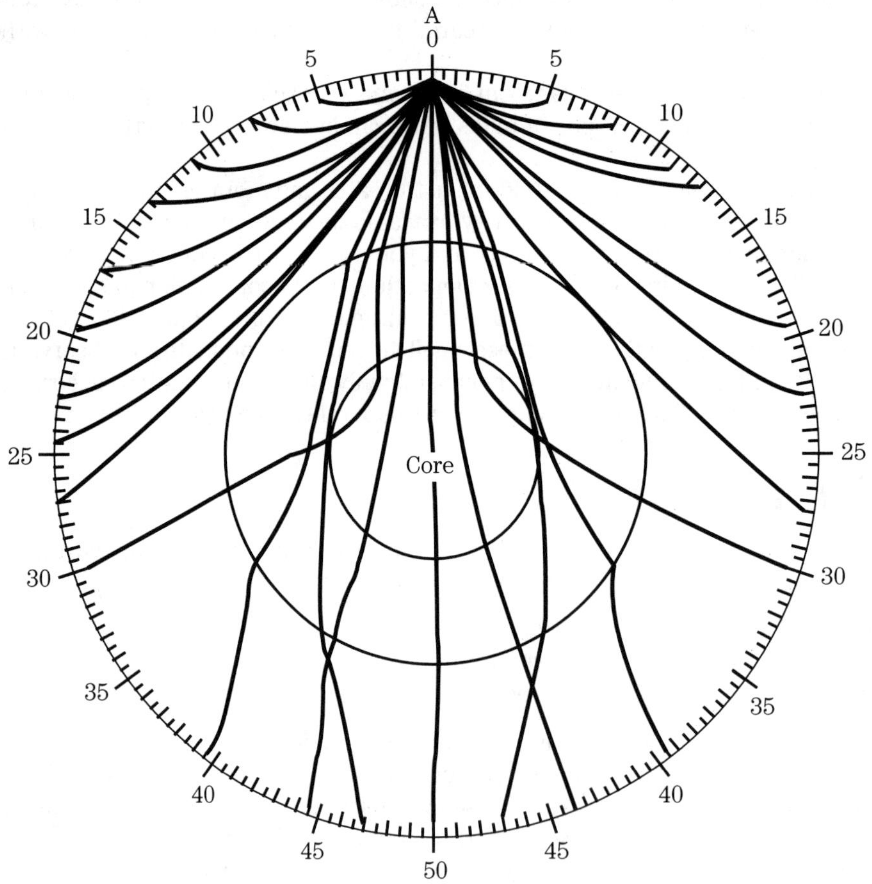

1-1 P-wave arrival time reference, with refracted traces through core area. Epicenter at A.

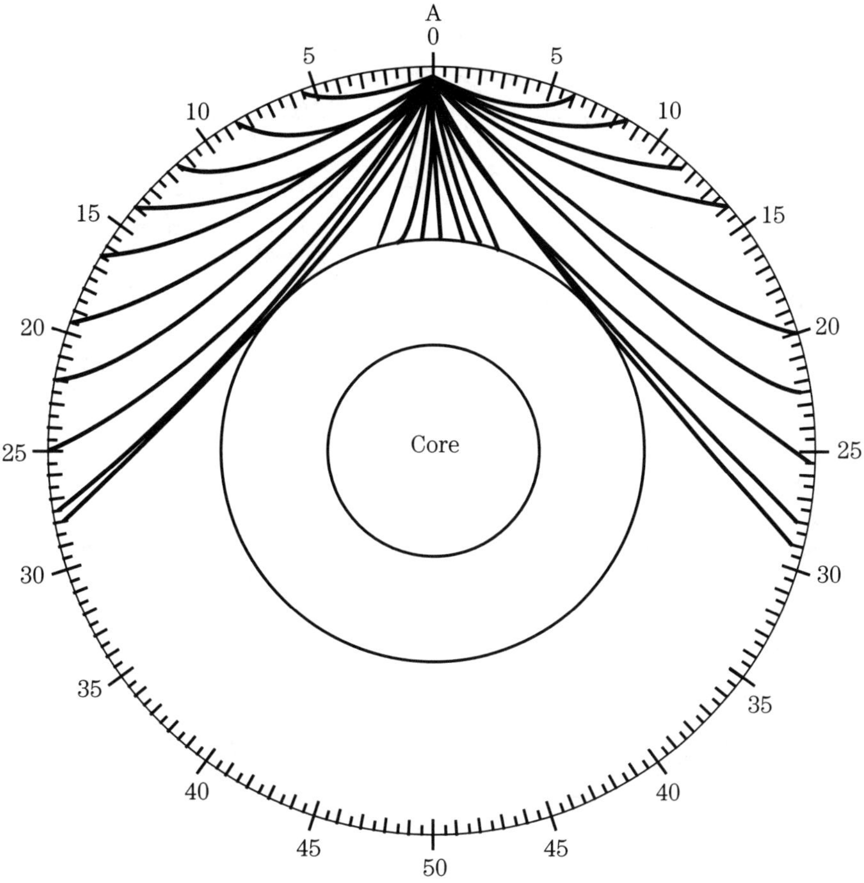

1-2 S-wave arrival reference. Earthquake epicenter at point A.

erence of the polar coordinate degrees of the earth and Table 1-1 lists the earth's circular percentage versus degrees and the corresponding wave arrival times.

Rock formations in the outer layers of the earth are continually being squeezed and stretched by the earth's internal forces. These forces result from the intense heat and pressures found toward the center of the earth. When the strain becomes greater than the rocks can stand, the rocks rupture, creating an earthquake. These ruptures are called *faults*. Most faults lie beneath the surface. Some, like the San Andreas Fault in California, can be seen at the surface. Most of the world's earthquakes occur in three main belts: The Circum-Pacific Belt; the Mediterranean and Trans-Asiatic Belt; and the Mid-Atlantic, Mid-Indian, and East African Belt.

The same forces that cause earthquakes caused mountains to be formed. This is why large earthquakes usually occur along new mountain chains and where deep sea trenches and volcanoes have formed. In contrast, the older, flat parts of continents and the sea floor are stable areas and have few earthquakes.

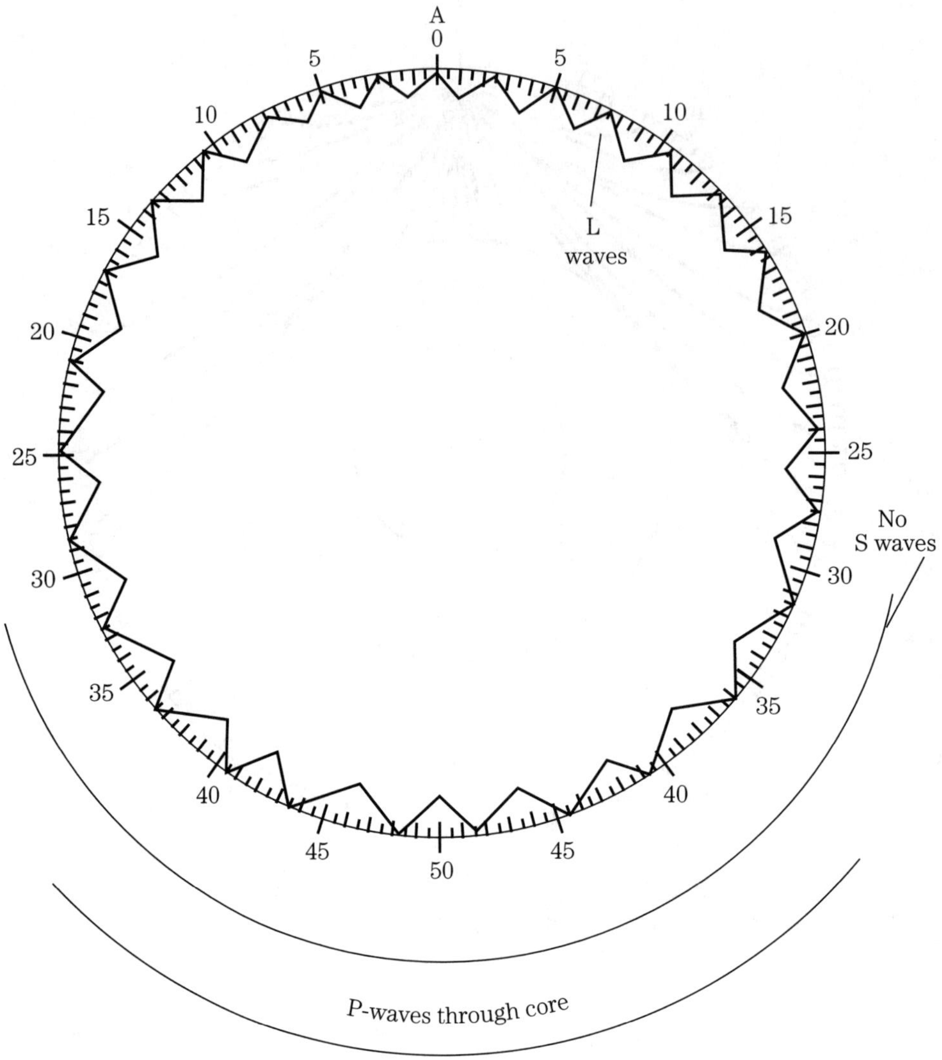

1-3 L-wave trace reference, and areas of P and S waves affected by core.

About 90% of the world's earthquakes occur in the three major belts, shown in Fig. 1-5, with the rest randomly distributed around the globe. The focus of most earthquakes is within 25 miles in depth. A few occur at the surface, and deep-focus quakes, with focal depths 250 to 400 miles in depth, are very rare, accounting for only about 3% of the world's earthquakes. Figure 1-6 illustrates average focal depths in the world's earthquake-prone areas.

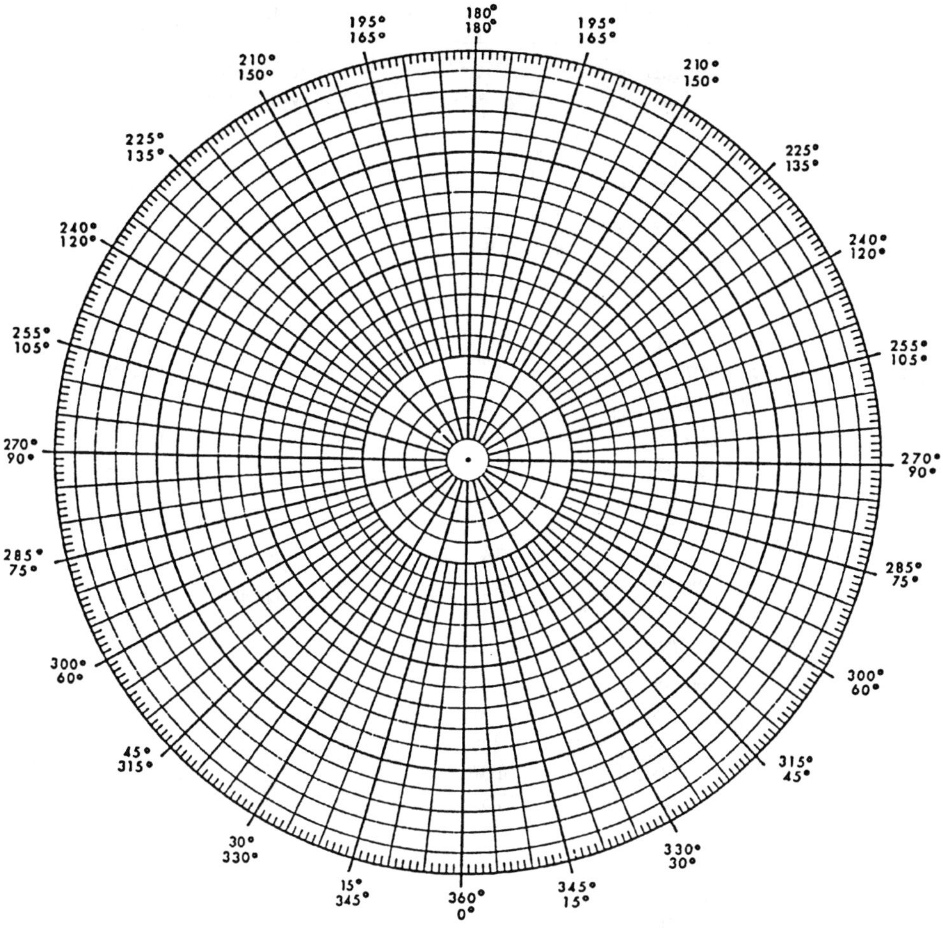

1-4 Polar coordinate degree reference.

Plate tectonics

Plate tectonics, also known as *Continental Drift*, operate according to the following conditions: the earth's lithosphere consists of a series of plates, set into motion by forces in the mantle. On the top of the plates, the lighter, continental material rides, while the plate boundaries are hot beds of geologic activity—areas where most of the world's earthquakes, mountain chains, and volcanoes are found.

Earthquakes occur at mid-ocean ridges, where spreading centers indicate plates pulling apart from each other. They also occur at *subduction zones*, where one plate descends beneath another contacting plate, and at *transform faults*, such as the San Andreas Fault, where one plate slips past another.

Table 1-1. Seismic Wave Arrival Times

Circular % of Earth	Coordinate Degrees	Arrival Times "P" Waves	"S"[a]
1.39	5		
2.78	10		5
4.17	15		
5.556	20	5	10
6.94	25		
8.33	30		
9.72	35		15
11.11	40		
12.50	45		
13.89	50		
15.28	55		
16.67	60		
18.056	65	10	20
19.44	70		
20.83	75		
22.22	80		
23.61	85		
25.00	90		
26.39	95		25
27.78	100		[b]
29.17	105[c]	15	
30.56	110		
31.94	115		
33.33	120		
34.72	125		
36.11	130		
37.50	135		
38.89	140		
40.28	145	[d]	
41.67	150		
43.056	155		
44.44	160	20	
45.83	165		
47.22	170		
48.61	175		
50.00	180		

a Arrival time in minutes.

b No arrivals of "S" waves beyond this point.

c From this point, "P" waves are sparse and diffuse, probably due to the density and composition of the material near the earth's core.

d This marks the point where primary signals pass directly through the core of the earth.

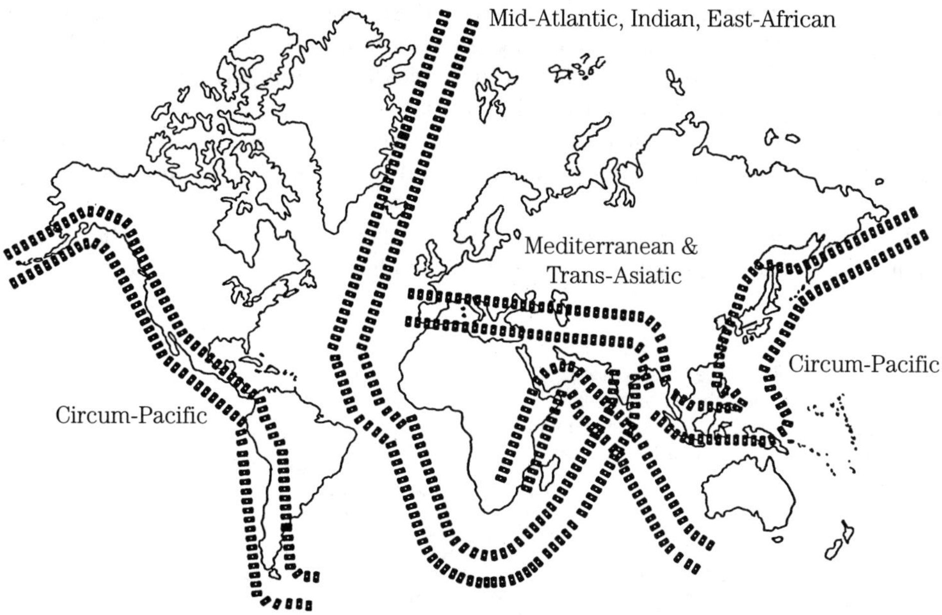

1-5 Major earthquake belts.

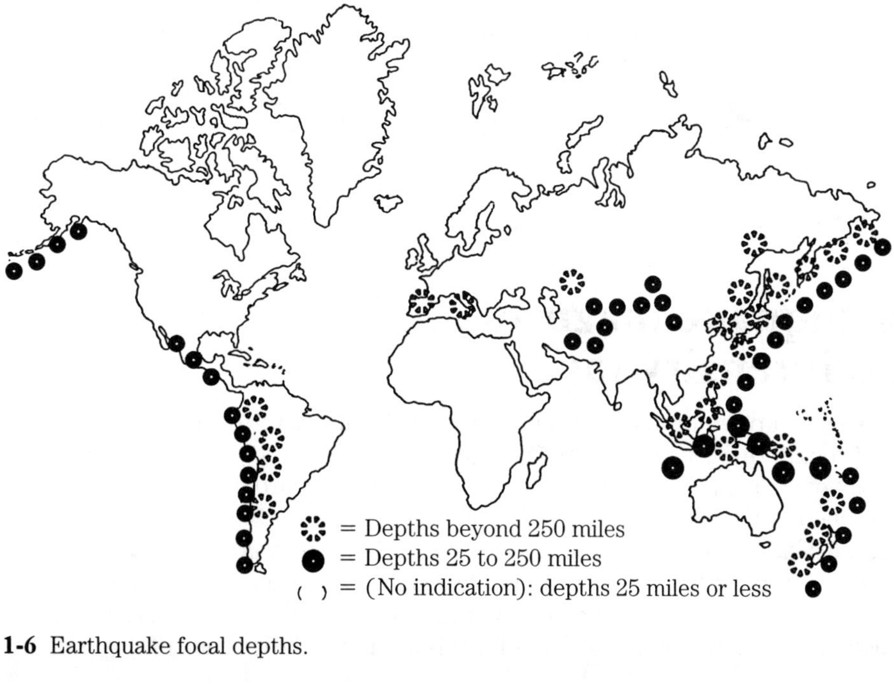

1-6 Earthquake focal depths.

Subduction zones are also associated with the deepest ocean trenches, as well as much of the world's volcanism, including the Pacific Rim's "Ring of Fire." In this region, more than 75% of the world's active volcanoes, numbering about 850, are found. Twenty percent of these are in Indonesia.

Where plates are of similar densities, collisions often cause material to "double up." Effects of this are found in the largest inland mountain ranges, such as the Alps and the Himalayan Chains.

Figure 1-7 indicates the approximate boundaries of the plates, and the type of activity associated in the area. The names of the plates are given in Table 1-2.

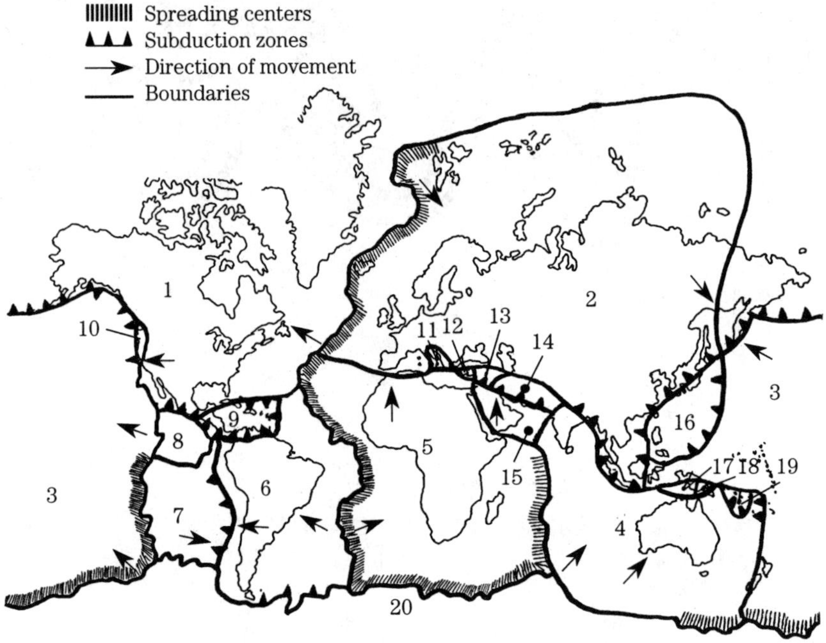

1-7 Plate boundaries and associated tectonic activity.

Earthquake magnitude and intensity ratings

Seismographs allow seismologists to make comparisons of the amplitude of motion of rock formations during an earthquake. *Amplitude* of motion depends on the quantity of energy released at the focus, and the distance from the focus. Amplitude is determined by the measurement of the heights of the deflections on seismograms. The decrease in amplitude outward from the epicenter is determined by gathering reports from many stations and plotting amplitude against the distance of the station from the epicenter. Because amplitude can vary through a large range, most stations plot the logarithm of the amplitude. Formulas for this are beyond the scope of this book.

Table 1-2. Names of the Earth's Tectonic Plates

Reference Number	Plate Name
1	North American
2	Eurasian
3	Pacific
4	Indo-Australian
5	African
6	South American
7	Nazca
8	Cocos
9	Caribbean
10	Gorda
11	Adriatic
12	Aegean
13	Turkish
14	Iranian
15	Arabian
16	Philippean
17	Bismark
18	Solomon
19	Fiji
20	Antarctic

Magnitude is a measure of the size of an earthquake, and is roughly related to the amount of energy released at the focus of an earthquake. The magnitude scale has neither "top" nor "bottom" values; the highest ever recorded was at magnitude 8.9, and the lowest at about –3. By the logarithmic scale, a magnitude 7 shallow-focus earthquake represents elastic-wave energy about 30 times that of a magnitude 6, and about 900 times greater than that of a magnitude 5 shock, and so forth. Factors entering into the determination of earthquake magnitude include earthquake focal depth, frequency content of the energy, and the earthquake's radiation pattern.

Intensity is a relative measure of the strength of an earthquake, based on destruction or disturbance and on human sensations. The two major scales used are the Mercalli and the Richter. The Mercalli scale recognizes degrees ranging from I, the weakest, to XII, the strongest. Although the Richter scale is the more precise one, it requires quantitative measurements and the use of specific formulas. The Mercalli scale serves as an adequate guide and it is presented in Table 1-3.

Actually, including the Richter and the Mercalli earthquake intensity scales, there are four magnitude scales, each appropriate for certain quakes at certain distances.

The Richter scale magnitude is also known as the *local magnitude scale*, and is usually used only to measure small to moderate intensity quakes from close proximity. The *surface-wave* and *body-wave magnitude scales* measure moderate to large quakes, and the *moment magnitude scale* is used to most accurately measure the largest quakes.

Table 1-3. Mercalli Earthquake Intensity Scale

Magnitude	Effects
I	Usually detected only by apparatus. Not felt, or rarely felt by people. Sometimes felt when the area is a great distance from a large quake.
II	Felt indoors by a few, especially on upper floors, or by sensitive persons. Delicately suspended objects may swing. Sometimes, dizziness or nausea is experienced.
III	Felt indoors by many, with usually a rapid vibration. The vibration is like the passing of light trucks. Hanging objects might swing slightly.
IV	Felt indoors by most, by a few outdoors. Vibration is like that of the passing of a heavy truck. Dishes and windows rattled, walls creak, hanging objects swing.
V	Felt indoors by all, outdoors by most. Buildings trembled, broke dishes and glassware, cracked windows in some cases. Occasional falling of unstable objects. Hanging objects swing considerably. Opened, or closed doors and shutters. Moved small objects. Spilled liquids from filled containers. Trees and bushes slightly shaken.
VI	Felt by all. Some frightened. Person's movement is unsteady. Trees and bushes shaken moderately. Small bells, like church and school bells, ring. Slight damage in poorly constructed buildings. Small amounts of plaster fall. Some chimneys have fine cracks. Broke dishes and glassware, also some windows. Overturned furniture in some instances.
VII	All frightened. Some found it difficult to stand. Noticed by persons driving cars. Trees and bushes shaken strongly. Waves on bodies of water like ponds. Turbidity of water increased from mud stirring up. Some incaving of sand stream banks. Rang large bells. Slight damage in well-constructed buildings, considerable damage in poorly built buildings. Some chimneys cracked extensively, also walls cracked. Considerable fall of plaster. Numerous windows broken.
VIII	All frightened, some near panic. Trees shaken strongly with branches and trunks sometimes broken off. Springs and well water flow changes, altered temperatures of waters. Slight damage to buildings made to withstand quakes, considerable damage to buildings of ordinary construction. Some buildings collapsed. Chimneys twisted and collapsed.
IX	Most in panic. Cracked ground. Considerable damage to most buildings. Some buildings collapsed, damage to reservoirs. Underground pipes cracked in some instances.
X	Cracked ground, especially when loose and wet, with cracks up to several inches wide. Larger fissures run parallel to canal and stream banks. Considerable landslides. Changed water level in wells. Serious damage to dams. Some wooden buildings destroyed. Bent railroad rails slightly. Torn apart pipelines underground. Open cracks in cement work and roads.
XI	Widespread ground disturbances: Fissures, cracks, land slips, etc. Sea waves, tidal waves formed. Severe damage to wood-frame houses, especially nearest the epicenter. Great damage to dams and bridges. Ruined many buried pipelines, bent railroad rails greatly.
XII	Total damage, nearly all construction works greatly damaged. Numerous shearing cracks in the ground. Landslides, rock falls are widespread. Fault slips in firm rock, geologic displacements observed. Greatly altered water channels, surface and underground, disrupting flow patterns. Dammed lakes and ponds. Waves seen on the ground, distorted levels and lines of sight. Objects thrown in the air.

Because reporting each of these measurements would add confusion, most scientists and news organizations use the term *Richter-scale magnitude*, no matter what scale the magnitude is really calculated in. Most of the so-called Richter scale measurements issued by the U.S. Geological Survey are actually surface-wave magnitudes.

As an example of the differences in official measurements, the surface-wave magnitudes of the June 28, 1992 Landers quake and the Big Bear aftershock were 7.6 and 6.7, respectively. The moment magnitude scale put the quakes at 7.3 and 6.2 respectively. In either case, it is fortunate that the quakes hit in remote areas, east of Los Angeles. Still, one person was killed, hundreds were injured, and the twin temblors caused $100 million in damages.

A sample seismogram is shown in Fig. 1-8. The quake registered 5.6 on the Richter scale, and its epicenter was 95 miles from the observation station. The apparatus used for reception is detailed in the first construction projects in this book. The chart recorder was built from a kit.

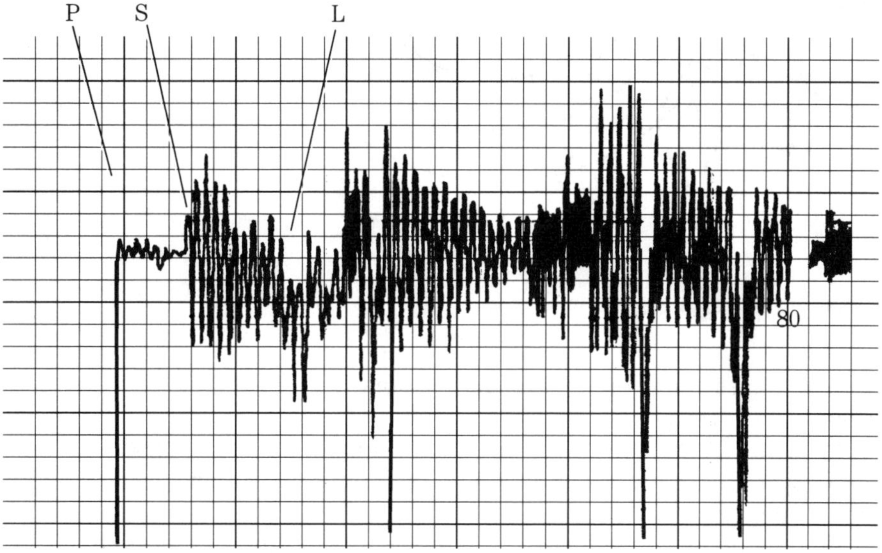

1-8 Seismogram showing various wave traces.

Tsunami

Tsunami are not "tidal waves" because they have nothing to do with lunar gravitation. They are most often set off by a severe undersea quake, but they can also be triggered by a massive submarine avalanche or even more violently by an explosive volcanic eruption.

Waves of the tsunami are gravity waves with very long lengths and periods. Some wave crests are over 150 miles apart, and wave heights can range from just a few inches to over 150 feet. Some waves can travel halfway around the world; these

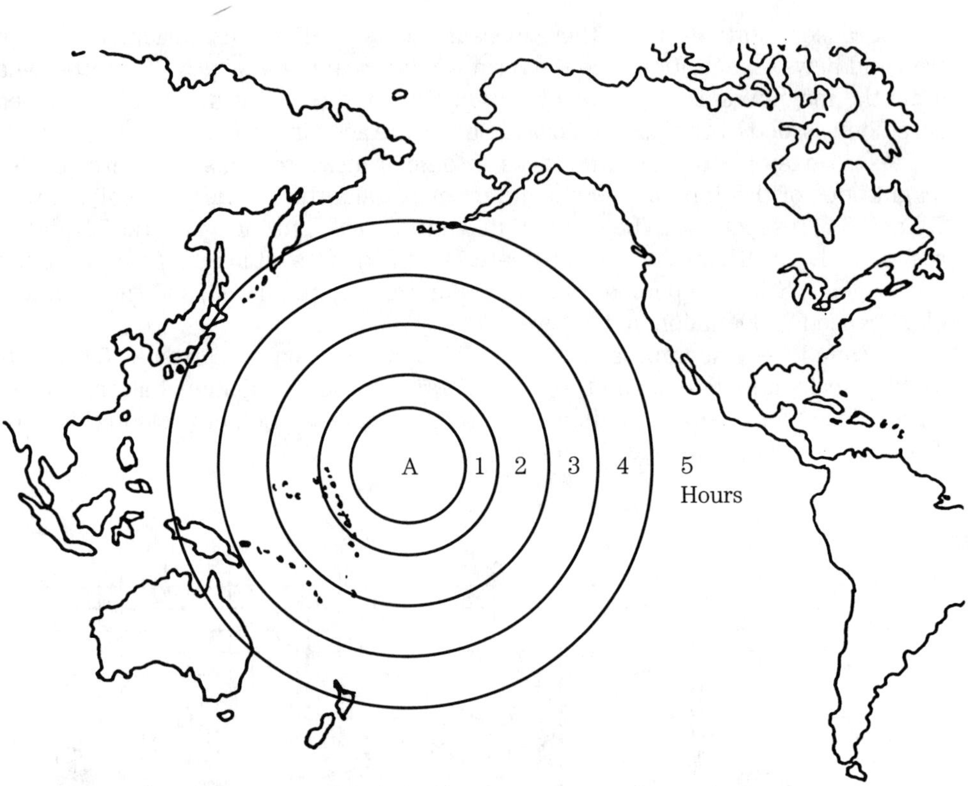

1-9 Tsunami arrival timing example.

long-range waves can even bounce off the continents that they strike and reflect back across the ocean.

A tsunami's forward speed is equal to the square root of the product of acceleration of gravity times water depth, in feet. Tsunamis can attain speeds in excess of 500 miles an hour.

The tsunami can occur anywhere in the world's oceans, but most commonly occurs when large earthquakes occur at sea around the Pacific Ocean's "Rim of Fire."

Many seismic warning stations around the Pacific Rim's earthquake belt maintain watch for tsunami. When a large sea-quake occurs, observers plot a set of concentric rings around the earthquake's epicenter and calculate the time of travel, in hours, to the possible recipients of wave damage. An example of the plotting is shown in Fig. 1-9.

2
CHAPTER

Building a basic seismometer

The seismometer circuit in this chapter is designed around the LF 356 operational-amplifier integrated circuit, which has a high input impedance. The circuit allows the minute impulses from the sensor to be displayed on a moving coil meter, reading millivolt signals and converting them so that useful display and measurements can be obtained.

The LF 356 can track input voltage references to within 0.5 V below reference ground, allowing meter deflections on even the smallest vibrations. A versatile, single-channel seismometer is shown in Fig. 2-1.

Refer to the circuit schematic of Fig. 2-2. The parts function as follows. R1 is a current-limiting resistor to protect the meter movement. R2 is a damping resistor to

2-1 Single-channel seismometer.

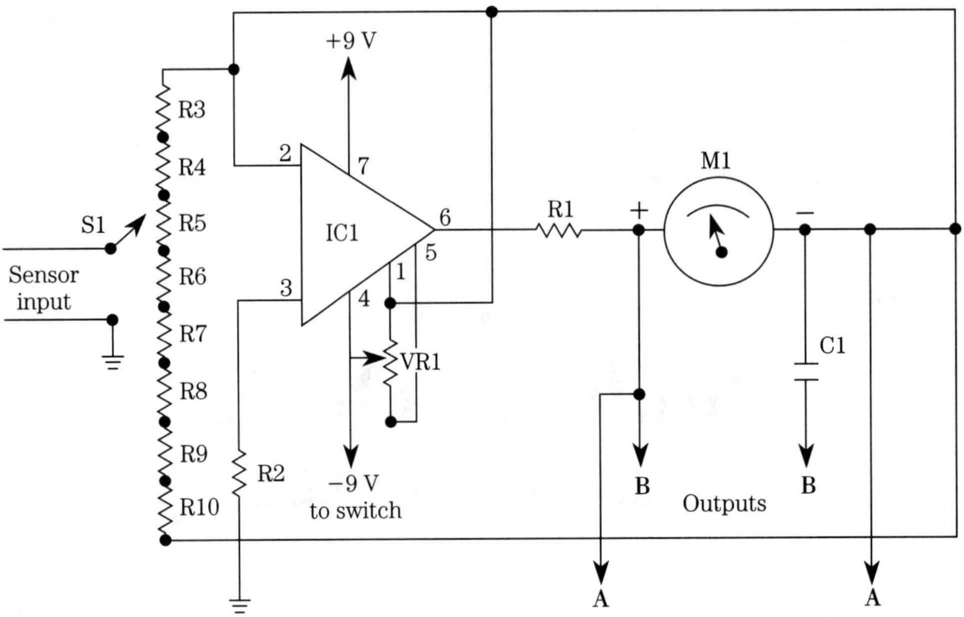

2-2 Seismometer schematic.

provide isolation to ground. VR1 functions as an offset/nulling circuit to allow the user to adapt the unit to various sensors, and, along with the attenuation circuit of R3 through R10, permits adjustment of the seismometer's overall sensitivity. Capacitor C1 protects the meter circuit and prevents drain when the seismometer's output is used to drive various devices, such as chart recorders or external amplifiers. Output B is suggested when external devices are used. Output A is tapped directly from the meter's terminals. Be sure to observe polarity.

Construct the basic circuit by following these 10 steps:

1. Connect pins 1 and 2 of the LF 356 together.
2. Connect pin 2 of the LF 356 to the switch "points" lead of the resistor attenuation circuit and to leg 3 of VR1.
3. Connect pin 3 of the LF 356 to R2. The other end of this resistor is the + input.
4. Connect pin 4 of the IC to –9 Vdc and the arm of VR1.
5. Connect pin 5 of the IC to leg 1 of VR1.
6. Connect pin 6 of the IC to R1. The other end of this resistor is the + output.
7. The ground (–) input is connected to the terminal lead of the attenuation circuit.
8. The – output to the meter is connected to pin 1 of the IC.
9. Pin 7 of the IC goes to +9 Vdc.
10. Tap meter leads and route to panel-mounted jacks for output A, and route another output, with a 10-μF electrolytic capacitor in series with one lead, for output B.

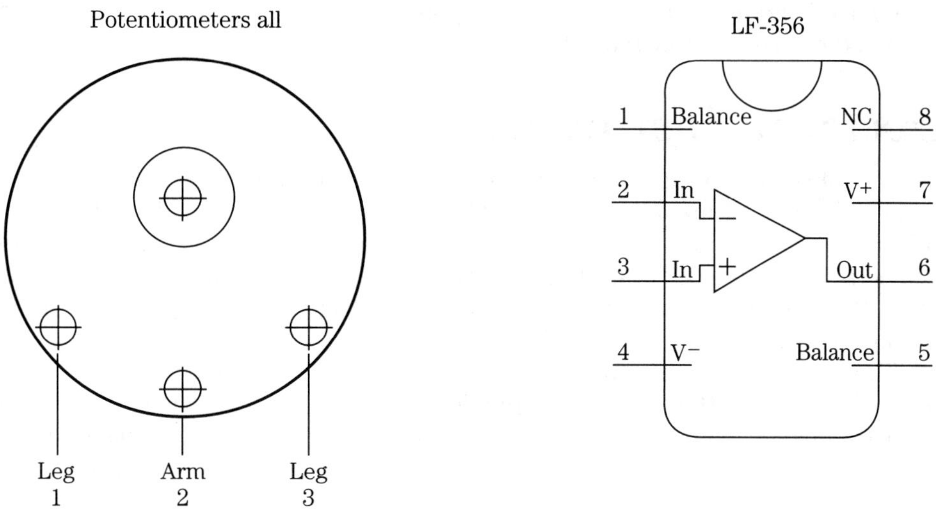

2-3 Potentiometer and LF 356 IC pinouts.

Table 2-1. Basic Seismometer Parts List

Part	Description
Switch	SPST Power Switch
S1	8-Position, Single-Pole Rotary Switch
VR1	100-kΩ ½-W Potentiometer
9 Vdc	Power Supply or Battery
R2, R6	1-kΩ, ¼-W Resistors
R1	2.7-kΩ, ¼-W Resistor
C1	10-µF, 16-V Electrolytic Capacitor
IC1	LF 356 Integrated Circuit
M1[a]	Meter, dc Milliampere
Socket	Socket for the IC
R3	1-Ω, ¼-W Resistor
R4	10-Ω, ¼-W Resistor
R5	100-Ω, ¼-W Resistor
R7	10-kΩ, ¼-W Resistor
R8	100-kΩ, ¼-W Resistor
R9	1-MΩ, ¼-W Resistor
R10	10-MΩ, ¼-W Resistor
Binding Posts	Six required: two for input and four for output
Misc.	Case, line cord, hook-up wire

a A 0- to 1-mA meter is suggested for use only when the seismometer is built for "stand-alone" use. A 0- to 2-mA FS meter is a better choice, and a meter of up to 0- to 5-mA can be used, especially when the intent for the use of the apparatus is with external components or auxiliary circuits. A 0- to 3-mA full-scale dc meter is probably the best overall choice.

Pinout diagrams of the LF 356 and the potentiometer (VR1) are provided in Fig. 2-3, and the parts list is in Table 2-1.

Operating instructions

Hook a sensor to the input terminals and switch on the power. Set the meter reading to about 1.0 mA when using a 0 to 2 or greater value full-scale meter, or set to 0.75 mA when using a 0- to 1-mA meter. Adjust VR1 only after a sensor has been connected. When using the attenuation network, set it to the position of least resistance. Place the sensor on a flat surface and tap or lightly shake its case while observing meter deflections.

Fine tune the sensitivity by adjusting for the maximum amount of meter deflection by tuning VR1, then set the attenuation switch to the resistance position providing the desired overall gain of the unit, now adapted to the sensor being used. The circuit is quite stable in operation and it can be used for continuous duty. No further adjustments are usually necessary.

3
CHAPTER
Electromagnetic seismographs

In standard electromagnetic seismographs, the case of the sensor is a large magnet. Inside the case is a core of fine wire. During an earthquake, this core of wire bobs up-and-down, inducing an electromagnetic signal through the wires. Figure 3-1 shows a typical commercial sensor and its specifications are listed in Table 3-1.

3-1 Commercial seismic sensor example.

Table 3-1. Commercial Sensor Specifications

Sensor Type: HS-10-1 Geophone, Model K	
General Specifications	
Nominal Frequency	1 Hz
Frequency Limits	0.9 Hz to 1.1 Hz
Dc Resistance @ 25° C	390 Ω
Dc Resistance Limits	371 to 409 Ω
Wire Gauge	#30
Mean Length of Turn	0.716 Foot
Length of Wire	3580 Feet
Number of Turns	5,000
Voltage Sensitivity	2.6V/Inch/Second
Open Circuit Damping	28%
Coil Form Weight	392 Grams
Coil Weight	897.5 Grams
Spring Thickness (Top)	0.016 Inch
Spring Thickness (Bottom)	0.016 Inch
Spring Constant at 1 Hz	37,000 Dynes/Cm.
Flux, bottom to top of winding	4.91 × 104 Maxwells
Total Impedance at 1 Hz	3820 Ω
Motional Impedance at 1 Hz	3430 Ω
Gap Area Per Winding	67.9 Per Sq. Cm.

This kind of sensor can be used with seismometer designs described in this book, as can several types of "generic" sensors that can be made easily. These sensors use the same principle; the inducing, by earth movements, of an electrical current in the wires by the movement of a magnet within the field of an inductor (coil of wire). Specifically, the case is a nonconductive material, the magnet serves as the weight, and the core of wire remains stationary. The movement of the magnet induces the current.

The sensor is a transducer operating in this way: a heavy mass (magnet) is suspended so that its inertia keeps it nearly at rest while the earth beneath it vibrates. It functions by converting mechanical motion into electrical impulses via a magnet that can move freely within the inductive field of a coil of wire (induction). Thus, an electrical current is induced.

The magnet hangs by a coiled spring and the records of motion are transferred by the conversion of this motion into an electrical signal. Then, it is displayed graphically or the signals are passed through an audio amplifier so that an audio signal can be recorded and analyzed.

To prevent accidental readings of surface vibrations (assuming that these are not desired), the sensor is often buried several feet below the ground or it is affixed to the bedrock. It can be secured to a concrete piling sunk 6 to 10 feet into the ground. Wires lead from the sensor to the seismometer's input terminals, and the seismometer's meters serve as galvanometers, which respond to the electrical changes induced by the magnet moving within the field of the induction coil.

Building a "generic" seismic sensor

The container is a five-gallon plastic drum with a waterproof lid. Inside is a low-loss cardboard or PVC material form. A workable dimension of the form would be about ⅛-inch-thick PVC with an overall diameter of about 3½ inches outside, about 3¼ inches inside, and about 12 inches long. The mean length of wire per turn is about 11 inches.

Around the form is wrapped 3000 to 4000 feet of #30 to #36 magnet wire. This ratio can be adjusted to give a resistance reading of 350 to 400 Ω on an ohmmeter. The higher the number of turns, the higher the resistance. About 3500 feet of wire wrapped tightly around the form should produce the desired readings.

The resonant frequency of the coil is very low. A test with 1200 feet of wire, responsive to long period seismic vibrations, has a resonant frequency of about 14 to 25 kHz and a dc resistance of about 140 Ω.

These are not absolute criteria for inductors. Coils with parameters within 20% of the figures given in the various designs will work acceptably. However, if many stations are built for research and comparison of activity, all coils should be very close in composition. Leads from the coil are attached to binding posts on the outside of the container and are connected to the seismometer's input.

From the center of the container's lid, a length of #26 wire is suspended about ¼ of the distance to the bottom of the drum. A strong (but sensitive) compression spring is attached to this wire. The other end of the spring is attached to another piece of wire, and the length is adjusted so that it can be attached to a rectangular magnet that can swing freely within the coil of wire. The form must be glued permanently to the bottom center of the drum. There is no electrical connection to the wire holding the spring and magnet, and shellac can be painted on a length of string, with the string taking the place of the wire. When these steps are complete, the transducer for the sensor is completed.

Mounting the sensor is accomplished by the use of a nonmetallic mounting plate attached to the bottom of the drum and attached to a nonmetallic post (wood or concrete, or nonferrous metal), driven into the ground several feet, or into the bedrock. Attaching the mound to a large, partially buried rock, or on an outcrop of the bedrock, also works well. The sensor, in its container, is shown in Fig. 3-2. Table 3-2 is its parts list. A level should be attached to the container's lid for adjustment. The kinds used with recreational vehicles are excellent for this.

Magnets

Many of the seismic sensors described in this book use magnets in their construction. Those used should be strong and easy to mount. The permanent magnets used in speaker construction work well, but many other kinds are available, and come in all shapes and sizes. Opportunities for some very specific building schemes exist because of this variety.

Even a magnet's shape affects its performance, often with advantageous results. A rectangular magnet, for example, exhibits some directional characteristics. Orien-

3-2 Sealed "generic" sensor; interior view.

Table 3-2. Generic Sensor Parts List

Quantity	Description
1	Plastic Five-Gallon Drum, with Lid
1	#26 Wire, or Piece of String 36 Inches Long
1	Compression Spring
3000 Feet	Magnet Wire; #30 to #36 AWG
2	Binding Posts
1	Coil Form; PVC or Cardboard
1	Magnet
1	Level Indicator for Enclosure
1	Length of Coaxial Cable for Hook-Up

tation of the container of a sensor using this shape of magnet can be used to enhance, or to null out some impulses to differentiate between north-south or east-west motion. A round magnet is omnidirectional.

The type of spring arrangement also has a bearing on how the magnet and the sensor performs. Placement of the spring at various levels on the plumb line of the magnet can affect the vertical sensitivity of the sensor. The type and length of the spring, as well as the number of springs and plumbs used can affect the type of waves the particular sensor will be most sensitive to, as well as the sensor's sensitivity to various time periods of the seismic waves.

The magnet is a critical component of the electromagnetic seismograph, regardless of the particular design. Various sensor designs are presented in this book. Choose the magnet carefully. Strength and size are important considerations. Remember, if many sensors are being built for comparative studies, the same type, size, and strength of magnets should be used in all the sensors. Various kinds of magnets are described in the following paragraphs.

Alnico

Alnico magnets are permanent-magnet alloys containing iron, nickel, cobalt, copper, and titanium. Alnico magnets have a wide range of magnetic properties. They are highly stable with respect to age and thermal influences, and are physically strong.

Barium ferrite

Barium ferrite magnets have a strong ferromagnetic permanent-magnet field. There are two forms: *isotropic*, which is oriented in many directions, and *anisotropic*, which is oriented in one specific direction. The anisotropic form is the more efficient type. It is cost-effective and very resistant to demagnetization. These magnets are ceramic-based types.

Neodymium-iron-boron

The *neodymium-iron-boron* group of magnets have the strongest energy density properties available. Their main feature is their strength relative to their physical size.

Rare earth cobalt

The *rare earth cobalt* magnets also have high-density strength. They are highly resistant to demagnetization and have a high energy output capability. They are most useful in environments where physical space is restricted. Otherwise, they are probably too expensive for general use.

Ceramic

Ceramic permanent magnets are made from a combination of iron and other chemicals. They are extremely resistant to demagnetization, which makes them very useful in seismic sensor applications. An example, shown in Fig. 3-3, uses pole pieces to increase strength. This is the type used in the prototypes of most of the sensors described in this book. Its dimensions are $1\frac{7}{8}$" × $\frac{7}{8}$" × $\frac{3}{8}$" with the lift of the magnet alone about 3 pounds. These magnets are efficient, and relatively inexpensive.

Pole pieces

By the use of pole pieces, the magnet specifications can be increased significantly. Using two pole pieces, the lift of the magnet can be increased from 3 to 40 pounds.

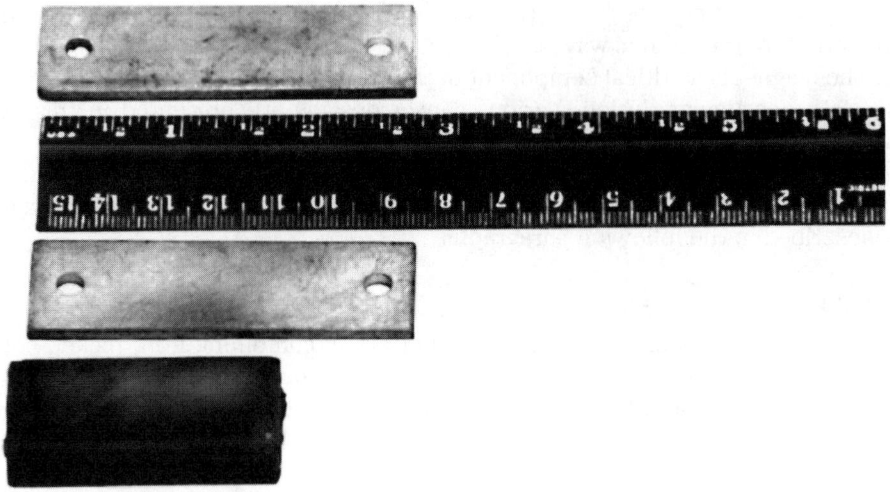

3-3 Ceramic magnet with pole pieces.

Steel plates are used as the pole pieces (Fig. 3-4). The plates move the north and south poles of the magnet closer to each other. The design and position of the pole pieces can vary depending on the size, shape, and placement of the poles on a specific magnet.

Magnets come in all shapes; bar, cylinder, disc, horseshoe, rectangular, and ring. They come in all sizes and price ranges, too. Choose according to the need and the pocketbook.

Building other induction-type sensors

Compact, efficient, and sensitive seismic sensors can be built very simply with common components. One model of sensor uses a coil form with dimensions of about four inches in diameter and eight inches in length. This will give a length-of-wire to turn ratio of about 13 inches to 1 turn. For this high resistance inductor, wrap 3000 feet of #36 magnet wire tightly around the form, making sure there are no kinks or breaks. The wire can be close-wound or wound in banks of about 750 feet each. An excellent ready-made form for this inductor is found in an 18-ounce cardboard oatmeal box (Fig. 3-5).

If bank-winding is chosen, make sure the splices, if any, are soldered and well insulated. The banks should be spread evenly from the top to the bottom of the form.

The plastic lid makes an excellent holder for the centered compression spring, directly soldered or bolted to the magnet. The magnet-spring relationship should be such that the weight of the magnet just begins to stretch the spring. A 1- to 2-inch piece of string can be suspended from the lid, and then the spring connected to the string, thus lengthening the period of reception of vibrations.

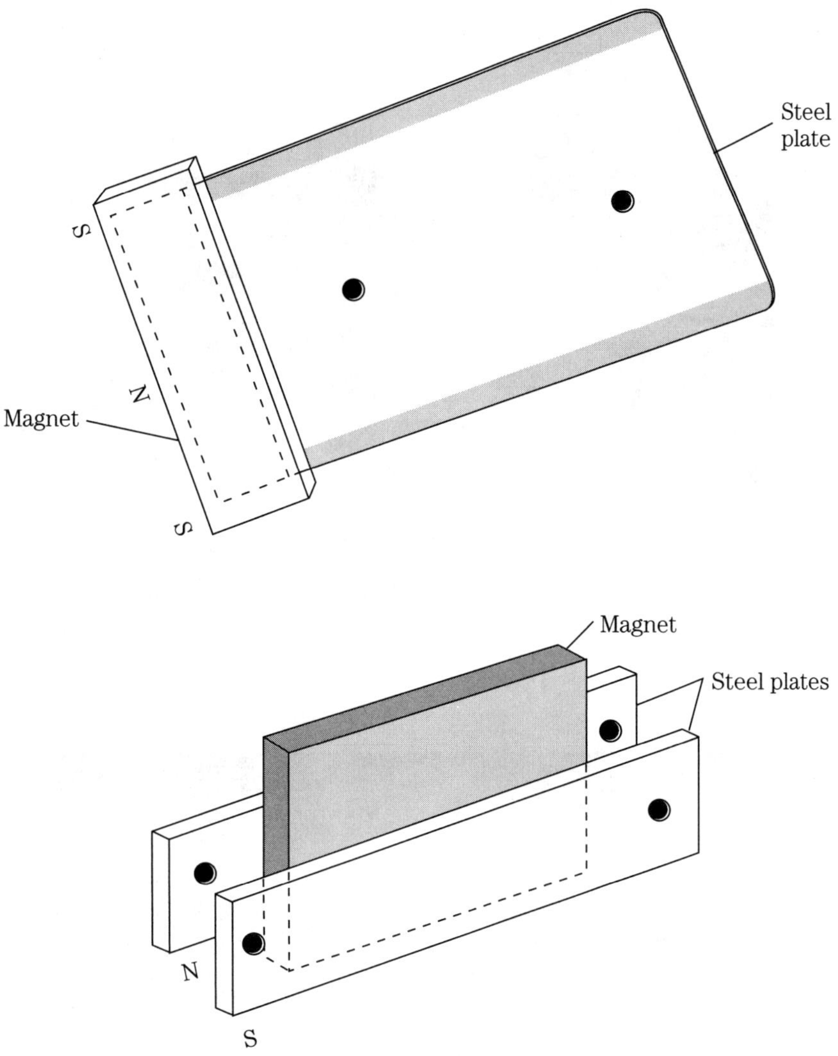

3-4 Pole pieces for magnets to increase strength and facilitate mounting.

Reinforce the lid with glue, seal it tightly, and place it in a protective container, as in the previous description (Fig. 3-6), for mounting. Again, a large rock, partially buried and away from traffic, makes an ideal mount. The cement piling method also works well.

In all installations, coaxial cable is recommended for the hook-up from the sensor to the seismometer. Use RG-58 for short runs, and RG-8 for long runs. The sensor's schematic is shown in Fig. 3-7.

3-5 Oatmeal box sensor.

3-6 Five-gallon drum as a protective container for the sensor.

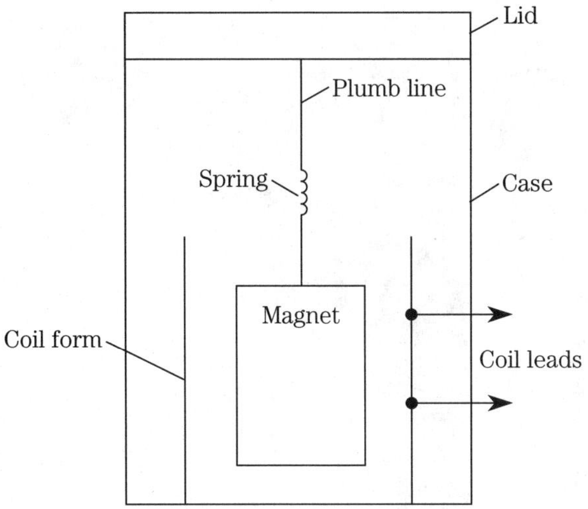

3-7
Basic sensor schematic.

Figure 3-8 shows the schematic of a sensor designed like some of the commercial sensors.

A large metal can, such as a coffee can or a bucket, is used. Iron or steel is required so that magnets will adhere to it. A magnet placed anywhere on the bucket will magnetize the entire surface. Use several magnets, spaced around the sides of the drum. Wind the coil around an appropriate form. One coil should be suspended from the plumb line that will move in the magnetic field. One coil should be stationary, wound around an insulated portion of the metal bucket, between some of the magnets. Electrical tape placed around the bucket is a good insulator for this coil. The weight should be nonmagnetic, and nonferrous. Lead is excellent. Choose a sensitive compression spring, seal it, and mount it. Various springs used in sensors are shown in Fig. 3-9.

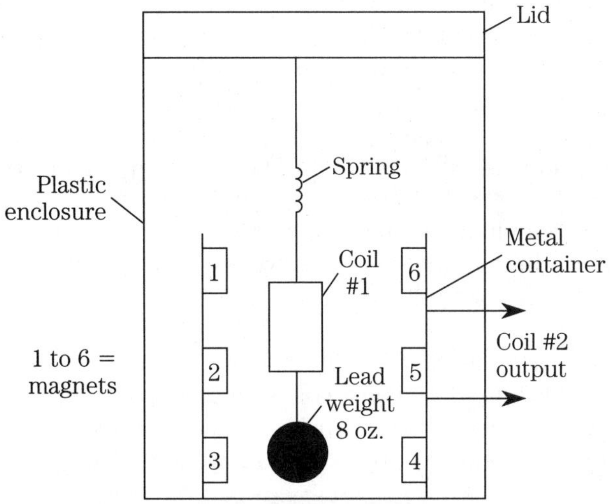

3-8
Coil-suspension sensor schematic.

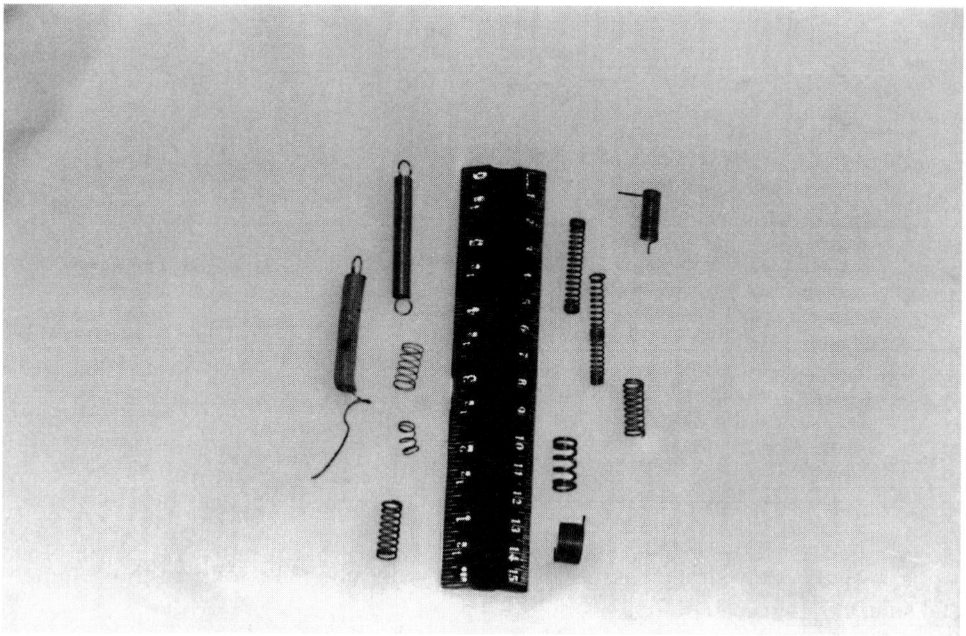

3-9 Springs used in sensors.

Figure 3-10 shows a dual-coil sensor using two coil forms and one magnet. The coils are comprised of 1500 to 2500 feet of #24 to #36 magnet wire and each coil is connected to a separate set of binding posts on the outside of the container.

This versatile hookup allows two-channel operation, if desired. One coil can be connected to a single-channel seismometer, the other to a receiver, amplifier, computer, scope, chart recorder, etc.

The coils can also be hooked in a series or a parallel fashion so that one channel can monitor a low-resistance value while the other channel can monitor a high-resistance value. The coil forms are positioned at the sides of the enclosure, and the magnet is allowed to swing freely between the two inductors.

Figure 3-11 describes an "extra-deep" sensor, which is very sensitive and is suitable for burying completely in sand or attaching to the bedrock. Notice the position of the spring in this version, which uses at least 3500 to 6000 feet of magnet wire. The size of the wire should be #28 or #30.

This sensor incorporates a pendulum type of motion. At least two plastic buckets are needed. The bottom of one bucket should be removed so that one bucket can be affixed on top of the other to make one tall bucket. The string type of plumb line should be used.

This sensor is very effective when buried in sandy soil areas at a depth of several feet. I use this type of sensor and is it mounted on an old well casing that reaches to a depth of 56 feet. It has been found to be very responsive to large earthquakes from great distances, as well as being very sensitive to small surface vibrations.

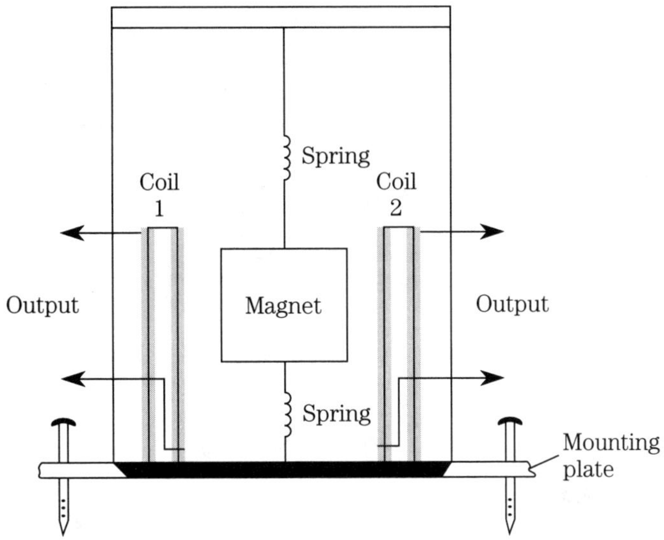

3-10
Dual-coil sensor.

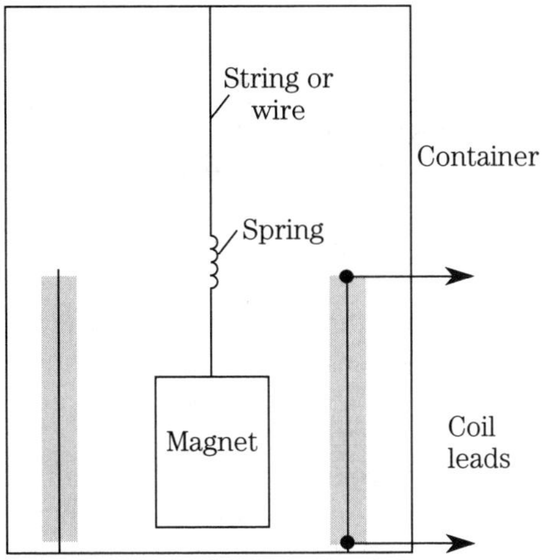

3-11
"Extra-deep" sensor.

Another "pendulum" motion-type sensor is shown in the diagram of Fig. 3-12. The suspension feature here, via two lines, is called a *bifilar system*.

First used as an earthquake detection scheme by Lord Kelvin, the bifilar pendulum is very sensitive to vibrations of the plumb line directed toward the center of the earth.

When used as a pendulum, the arc of the bifilar system shows that the earth does not rotate at a constant rate on its axis, but it speeds and slows very slightly because of the force of gravity exerted on the earth by the sun and moon.

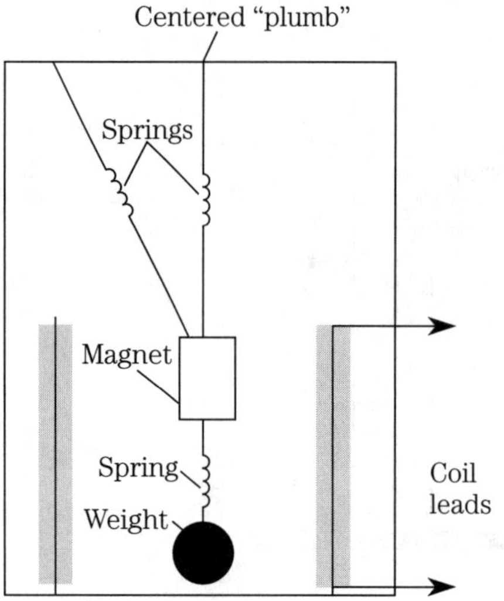

3-12
Bifilar sensor.

The round weight used here is an 8-ounce lead fishing weight. The weight is suspended from the bottom of the magnet in the same way that the magnet is suspended from the container's lid.

Especially suited to this kind of sensor, the lead weight can be used in all sensors to adjust reception period and sensitivity. With the bifilar system, use plenty of wire for the inductor. 10,000 feet of #28 magnet wire works well.

As with the use of weights, springs can also be used several ways. Springs can be placed anywhere on the suspension line(s). Generally, they work best placed near the bottom on pendulum-type sensors and near the top when used with shallow mountings.

If a spring of sufficient length is available, discard other lines and use the spring for the entire suspension line. Figure 3-13 shows the weight and some springs used successfully in the bifilar system. Other specialized sensors are detailed in later sections of this book.

When multiple electromagnetic sensors are used, such as for fault mapping or in earthquake-prediction studies, as many as 12 sensors can be hooked up in a series or parallel fashion. A geometric pattern, linear or X, is chosen, with sensors spaced 15 to 150 feet apart. Horizontal and vertical, as well as long- and short-period studies, can be made.

A handy addition for this is an inclinometer, installed on each sensor. A simple device, the inclinometer is called a *roll indicator* on ships, and when it is used with a magnetic needle, like that of a compass, it is called a *dipping needle*, which can be used to study the earth's magnetism.

The handy plastic five-gallon drum can be used as a container for a homemade inclinometer. A protractor is attached to the bottom of a wooden frame holding a suspension line with a weight and pointer attached, or an enclosed dipping needle. A hole is drilled in the drum and sealed with clear glass for observation.

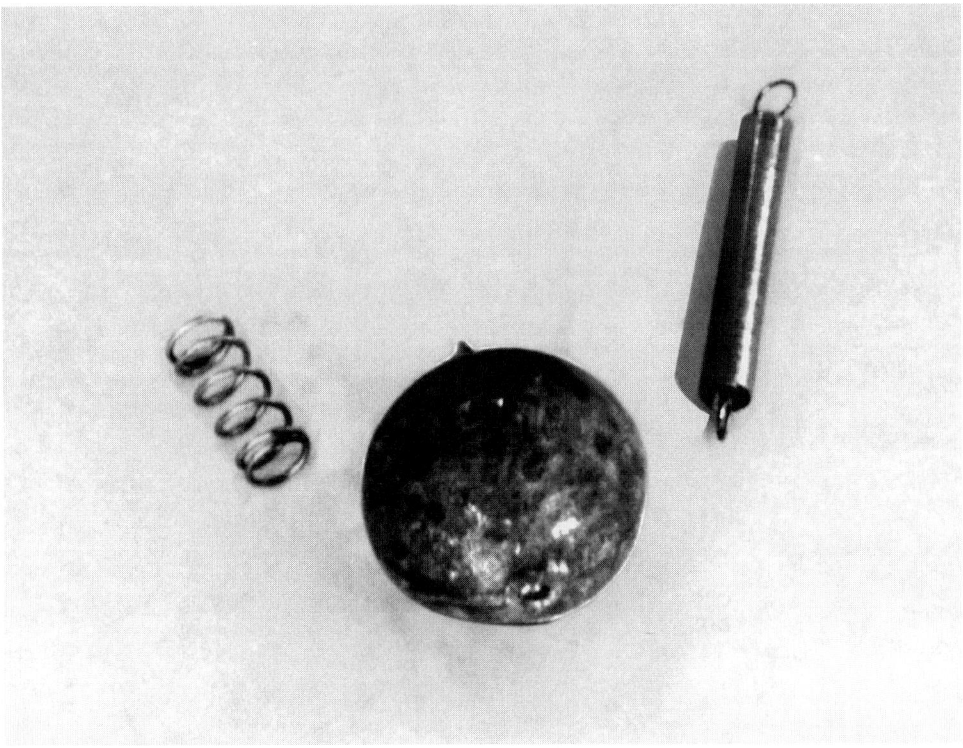

3-13 Weight and springs used in the bifilar sensor.

Ground movements will affect the inclinometer's angle readings, which are obviously of interest to the earthquake observer. When a magnetic needle is used, the needle will be affected by anomalies associated with the magnetic earth, as well as by iron and steel underground. Inclinometers are tools that can be purchased already constructed at low cost from many sources.

4
CHAPTER
Observing atmospherics

Solar flares are temporary, but violent eruptions of energy from some spot on the sun's surface. Flares emit intense radiations and affect a wide range of emissions, from the long infrared rays to the short ultraviolet rays.

Ionization of the atmosphere depends on radiation in the ultraviolet region. Therefore, conditions of the ionosphere vary with changes in solar radiation. Pronounced changes in ionization occur concurrent with the sunspot cycle, variously described as 11⅛ to 13 years in cycle.

The effect on the earth's ionosphere also affects radio propagation. High frequencies are prone to erratic signals and blackouts during times of high solar activity.

However, frequencies below 500 kHz remain effective channels of communication even during times when auroral conditions or the resulting magnetic-field disturbances are such that higher frequencies are rendered unreliable.

The National Bureau of Standards operates standard time and frequency station WWV in Ft. Collins, Colorado, as a time and frequency standards station. This station transmits hourly bulletins on solar activity and the condition of the earth's magnetic field. The bulletin begins after the time announcement at 18 minutes after each hour. Contents include the solar flux values for the previous 24 hours, the state of solar activity, the condition of the earth's magnetic field and the Boulder "K" index, and predictions for the next 24 hours.

The solar flux is a measure of solar radiation at a frequency of 2800 MHz, monitored daily. Although similar measurements are taken on many different frequencies, the 2800-MHz value is reported because it correlates well with the ionization density of the F region of the ionosphere, the Zurich sunspot number and with the maximum usable frequency for reliable long-distance communication.

The A index is a 24-hour figure for activity of the earth's geomagnetic field. When values are low and the geomagnetic field quiet, absorption of wave energy is low.

The K index is similar to the A index, except that it is nearly a current figure. It is updated every three hours. Trend is especially important when graphically charting the K index.

Geomagnetic field conditions are termed *quiet*, *unsettled*, or *active*, roughly related to the K index. Storm conditions are reported when they occur.

WWV also reports Omega signal status and VLF radio signal propagation at 16 minutes past each hour. The station operates on standard carrier frequencies of 2.5, 5.0, 10.0, 15.0, and 20.0 MHz, and gives its address for further information of services at the hour.

Application of basic circuitry

The circuitry of the basic seismometer can be used to display the effects of solar flares and atmospheric electrical activity, including lightning strikes, in display via the meter, or in graphic form with the use of a chart recorder, like in Fig. 4-1, by connecting the audio output of a low or very-low-frequency radio receiver to the seismometer's input terminals. Sensitivity is adjusted to accommodate the average signal input.

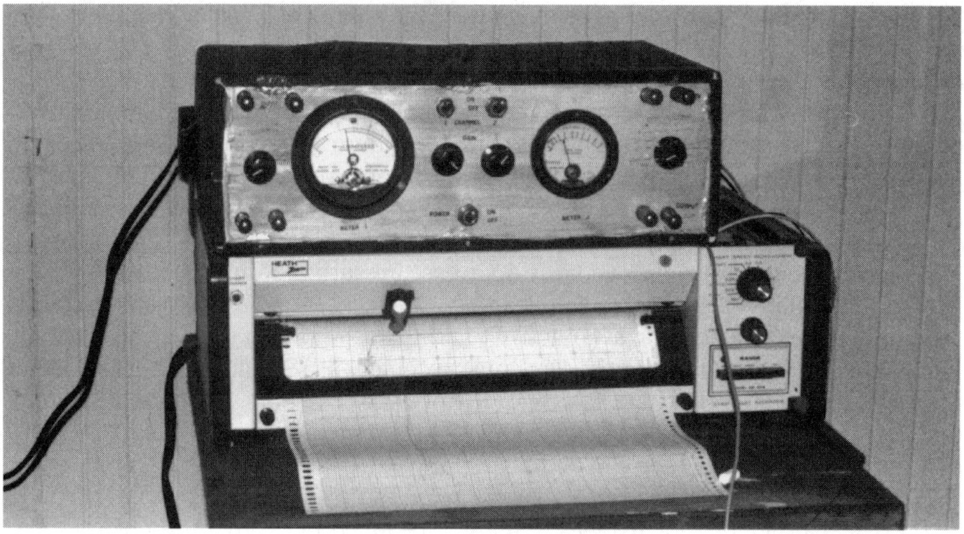

4-1 A seismometer driving a chart recorder.

The frequency range of the receiver used in the study should be from about 10 to 100 kHz. One method of solar flare detection is by monitoring the static stream most intense at 27 kHz. This method is called *monitoring the sudden enhancement of atmospherics (SEA)*.

Enhancement of atmospherics usually occurs at night. However, an SEA can occur during the daylight hours when a solar flare occurs. Whenever the radio signal strength suddenly increases toward the nighttime readings, solar activity can be suspected.

Another method of detection is by monitoring the sudden enhancement of signal strength (SES). For this, the radio receiver is tuned to a station from a known location. The output of the receiver is recorded on a strip-chart recorder or input into a computer. If the propagation path of the station is on the daylight side of the earth during a solar flare, the lowering of the ionosphere will foster the sudden enhancement of the signal, causing the characteristic fast-rise, slow-fall signal strength curve. Table 4-1 gives a partial low-frequency radio station list. There are many more stations worldwide, so consult frequency guides for further information.

The U.S. Navy operates many stations in the VLF and LF frequency spectrum for use in ship-to-shore communications. VLF stations often run very high power, some

Table 4-1. A Partial VLF Frequency List

KHz	Station	Location
0.0005 to .01 (.05 to 10 Hz)		Commercial seismic sensors, Geo-Electric Signals
0.06 (60 Hz)		Power Line Fundamental Frequency
0.1 to 3.0		Whistlers, Dawn Chorus, Tweeks, Lightning Strike Impulses, Submarines
3 to 10		Military Data, Frequency Standards Signals, Electromagnetic Anomalies
10 to 14	Omega	Worldwide Navigation
16.0	GBR	England
17.4	NTD	Japan
17.8	——	Maine
20.5	——	Several; in Russia
21.4	NSS	Maryland
22.3	NWC	Australia
23.0	——	Several; in Russia
23.4	NPM	Hawaii
24.0	NAA	Maine
24.8	NLK	Jim Creek, Washington
25.1 to 25.5	——	Several; in Russia
40.0	JJF	Japan
50.0	OMA	Czechoslovakia
60.0	MSF	England
60.0	WWVB	Colorado
66.7	RBU	Russia
75.0	HBG	Switzerland
77.5	DCF77	Germany
100 + or –	Loran	Worldwide Navigation

over one million watts, for communications with submarines. Many stations are maintained for use as precision frequency standards. VLF carriers are typically synthesized from Cesium Beam Frequency Standards, with the long-term accuracy better than one part in 10^{11} typical. Table 4-2 lists the radio frequency spectrum. When monitoring local thunderstorms, the signal patterns indicate the meteorological origin of the electrical pulses.

Table 4-2. Radio Frequency Spectrum

Abbreviation[a]	Frequency Band	Frequency Range
ULF	Ultra Low Frequency	Below 1 kHz
ELF	Extremely Low Frequency	1 to 3 kHz
VLF	Very Low Frequency	3 to 30 kHz
LF	Low Frequency	30 to 300 kHz
MF	Medium Frequency	300 to 3000 kHz
HF	High Frequency	3 to 30 MHz
VHF	Very High Frequency	30 to 300 MHz
UHF	Ultra High Frequency	300 to 3000 MHz
SHF	Super High Frequency	3 to 30 GHz
EHF	Extremely High Frequency	30 to 300 GHz

a Abbreviations, bands, and ranges used in this book.

Basic VLF receiver construction

Figure 4-2 shows the schematic for a basic transistor-based VLF receiver. It is a voltage-sensitive circuit, suitable for atmospherics study and thunderstorm monitoring, as well as for solar flare observation. The basic circuit is designed for headphone reception or for signal input into auxiliary equipment, and requires very little power to operate. Nine Vdc at as little as 10 mA can operate the basic set.

The capacitor stepping network is used as a fine-tuning for the receiver. This version is a *channel-type receiver* because no variable tuning capacitor is used. One can be substituted for the capacitor net if desired, but, for VLF monitoring of flares and atmospherics, it is desirable to have various preset frequencies to facilitate easy tuning. The fixed value inductors and capacitors accomplish this.

The ground terminal is for the connection to an extensive earth ground, a long-wire antenna, or perhaps to a ham radio antenna. Input points A and B are connected to a 1000- to 2000-μH choke coil or to an inductor stepping network, as described further on. Because the circuit is voltage sensitive, a 0- to 10-Vdc meter is used to monitor the voltage input. The voltage regulation circuit is detailed in chapter 5. In this section, construct the circuit as shown, the variable power supply can be connected later.

Sensitivity to the very lowest frequencies increases with the voltage and the antenna efficiency. The voltage should never exceed 9 V. Usually, stable reception of VLF frequencies is obtained with input voltages of 6.5 to 8.5 Vdc.

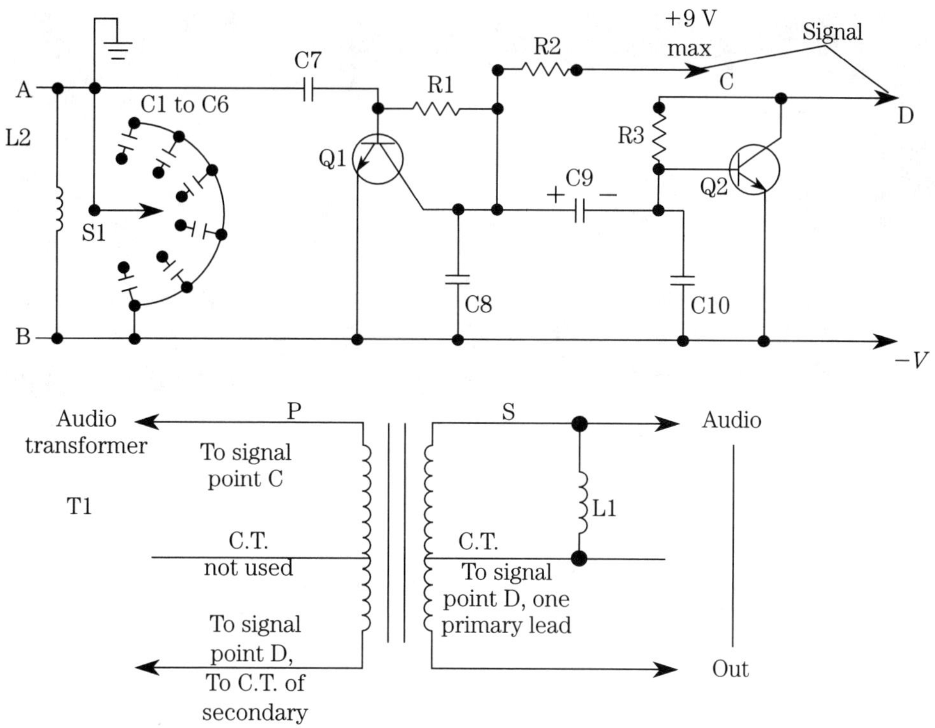

4-2 Multi-channel flare receiver.

Figure 4-3 shows a suggestion of a component substitution box that can be used for the tuning of this radio, as well as being a useful piece of test equipment.

The box contains inductors, ranging from 0.15 to 1000 μH, which would be connected to the input points of the receiver. Capacitors range from 10 pF to 0.68 μF and would be connected in place of the capacitor-stepping network. The resistors range from 1 Ω to 10 MΩ, and, although not used in the receiver, this part of the box is very helpful in building and testing other equipment.

The switches used here are single-pole, eight-position rotary type. Broadcast band frequencies can be received with inductances of 60 to 200 μH, and low frequencies with inductance values above 400 μH. With capacitance, the higher the value of capacitor selected, the lower will be the received frequency. The lower the frequency selection is, the higher the required input voltage.

Any number of switches and values can be chosen by the user/experimenter to suit a particular application. Figure 4-4 is the schematic for a single-channel receiver.

With the multi-channel receiver, an inductor selection switch can also be used to change frequencies. Try using a six- or eight-position selector, with the capacitor stepping net as well. This is quite an interesting set, and fun to experiment with because of its simplicity.

In all versions of this receiver, the connections to and from the audio transformer are the same. The receiver section is used in the next chapter's complete

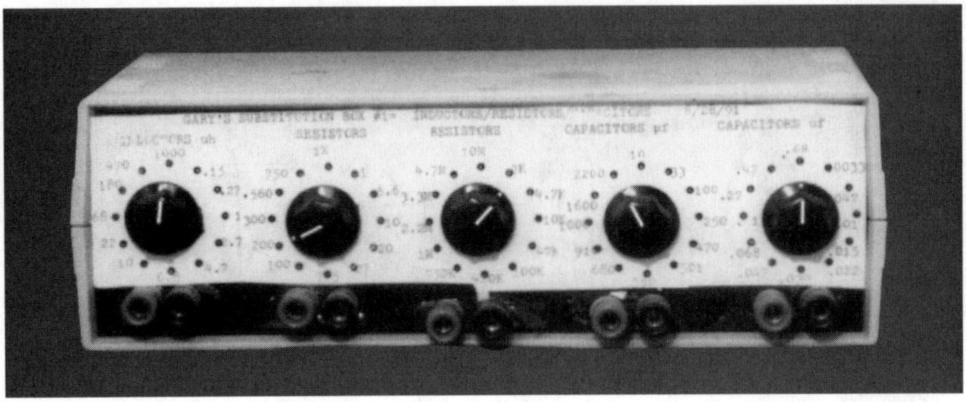

4-3 Component substitution box.

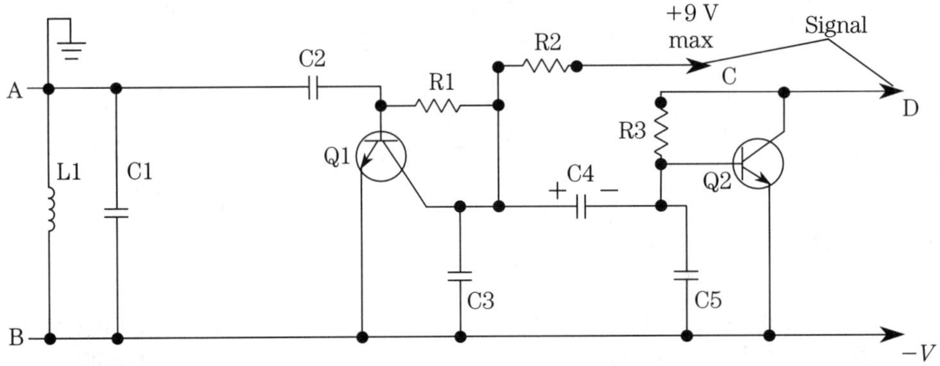

4-4 Single-channel flare receiver.

Seismic-Flare receiver project with only a power supply alteration, so construct this basic circuit carefully.

Table 4-3 is the parts list for the single-channel set, and Table 4-4 is the parts list for the multi-channel version. Experiment with other inductor and capacitor stepping values, but other parts should not be substituted for. Figure 4-5 shows a sample schematic for a multi-channel receiver with both the capacitor and inductor stepping features.

Visual observation of the sun

Much of the knowledge of the exterior features of the sun has been gained with special apparatus used when the sun is not eclipsed.

An adaptation of the spectroscope, the *spectroheliograph*, is an attachment to a telescope used to observe the chromosphere over the entire disc of the sun. It is used to observe the sun in the light of a single chemical element of the sun's atmo-

Table 4-3.
Flare Receiver Parts List, Single-Channel

Part	Description
L1	1000-μH Choke Coil
C1, C5	Capacitors; 0.022-μF Mylar
C2	Capacitor; 0.047-μF Mylar
C3	Capacitor; 0.1-μF Mylar
C4	220-μF, 50-V Electrolytic Capacitor
Q1, Q2	2N3904 NPN Transistors
R1	Resistor; 1 MΩ, ¼ W
R2	Resistor; 4.7 kΩ, ¼ W
R3	Resistor; 470 kΩ, ¼ W
T1[a]	Audio Driver Transformer; 10 KCT:2 KCT
PS	Power Supply

[a] The audio transformer diagram is omitted from the schematic in this view. Hook-up is the same in all receiver variations.

Table 4-4. Flare Receiver Parts List, Multi-Channel

Part	Description
L1	1000-μH Choke Inductor in the Audio Stage
L2[a]	Choke Inductor in the Input Stage
C1	Capacitor; 1000-pF Dipped Silver Mica
C2	Capacitor; 2200-pF Dipped Silver Mica
C3	Capacitor; 3300-pF Dipped Silver Mica
C4	Capacitor; 4700-pF Dipped Silver Mica
C5	Capacitor; 0.015 μF
C6, C10	Capacitors; 0.022 μF
C7	Capacitor; 0.047 μF
C8	Capacitor; 0.1 μF
C9	Capacitor; 220-μF; 50-V Electrolytic
Q1, Q2	Transistors, 2N3094 NPN
R1	Resistor; 1 MΩ, ¼ W
R2	Resistor; 4.7 kΩ, ¼ W
R3	Resistor; 470 kΩ, ¼ W
S1	Six-Position Rotary Switch
T1	Audio Driver Transformer; 10 KCT:2 KCT or Equiv.
PS	Power Supply

a A multi-position stepping switch of inductors can be used in this location in the circuit.

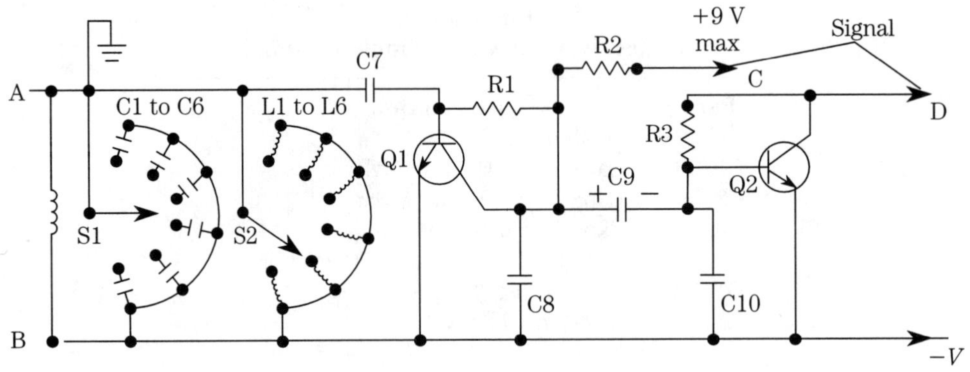

4-5 A multi-channel receiver with inductor stepping.

sphere. The spectral lines of calcium and hydrogen are generally used with ground-based apparatus.

Monochromatic filters are also used to obtain photographs in a narrow wavelength of sunlight. They can be used for viewing solar prominences and coronal features otherwise seen only during an eclipse. Polarizers are usually used with these filters to study the effects of magnetic fields on prominences.

One method of viewing the sun is to use a totally exposed and developed piece of black-and-white film (color film cannot be used). The developed silver halide crystals on black-and-white film are excellent filters, as well as excellent reflectors of heat.

Another way is to use a video camera, a monitor, and a VCR. Although pointing a camera at the sun will damage the camera as surely as looking at the sun will damage the eyes, using a piece of videotape or a piece of the black-and-white film is a simple and effective way of filtering.

Figure 4-6 shows the apparatus used to observe the sun using this method. The camera can be adapted as follows:

Obtain another lens cap for the camera. Then, cut out the center of the cap and install the photographic film in the opening. With the videotape filtering method, you have to test the tape. Some videotape will filter red or purple and be strong enough to view the sun. To test the tape, look through a piece of it at a 100-watt lightbulb. Through some tapes, you will see nothing. Through others, you will faintly see the light's filament. This is the tape to use. Cut enough to cover the hole in the lens cap and install the film. A couple of sample pictures of the sun viewed in this way are shown in Fig. 4-7.

Lunar influence

Tidal forces from the moon and the sun influence stresses and pressures on the earth, and lead to another item for observation. Tidal laws were worked out by Sir Isaac Newton after he discovered the law of gravitation. Because of gravitation, heavenly bodies (such as the moon and the sun) exert a pull on the earth. The

4-6 An apparatus used to view the sun.

strength of the pull depends on A, the mass of the heavenly body, and B, its distance from the earth. Therefore, the body exerts a gravitational pull that is in direct proportion to its mass, but this pull weakens in proportion to the square of its distance. Because of this, the sun's pull is only about 45% as strong as the moon's.

The interval of time between one apparent revolution of the moon around the earth, about 24 hours and 50 minutes, gives the tides' rise and fall a definite schedule. The gravitation of the passing moon pulls the nearest water away from the earth beneath it and simultaneously pulls the entire earth away from the water on its opposite side.

The tidal range differs each day, according to the positions of the moon and the sun. At full and new moon, the sun and moon pull along the same line, which causes unusually high tides (spring tides). At first and third quarters, the moon and the sun pull at right angles, causing unusually low levels of high tides (neap tides).

A couple of representative lunar clocks are shown in Fig. 4-8. The one-handed clock marks "Circadian" time, and has a 24-hour and 50-minute movement. This shows the time on the moon, along with tidal times on earth. The other clock is a tide timer, set to the appropriate high tide time for a particular location, then indicating hours to next tide.

Figure 4-9 shows a completed VLF receiver that also has broadcast reception. The set has a variable inductor for tuning the broadcast band, as well as the inductor and capacitor stepping networks for multi-channel reception of VLF frequencies.

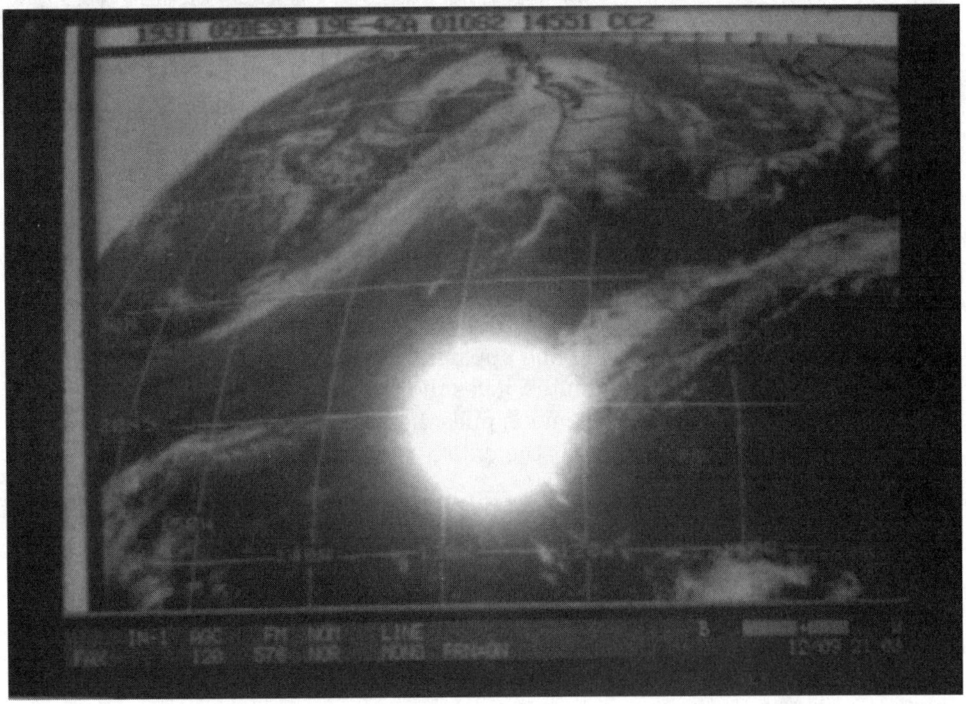

4-7 Sample pictures of the sun.

Louis Schelling; Louis Schelling Corporation

4-8 Lunar and tide clocks.

4-9
VLF receiver with broadcast band reception.

5

CHAPTER

Building a complete seismic-flare receiver

This chapter contains information necessary to build the modules for the complete receiver that will be built and interfaced with previous circuits. When completed, the apparatus is capable of visually and audibly indicating seismic or solar conditions, and atmospheric and propagation anomalies. Figures 5-1 and 5-2 show two versions of complete receivers used with a dual-channel seismometer.

Circuit description

Signals, in the form of electromagnetic impulses from the sensor, are input into the seismometer section. This is the first stage of detection and amplification of received signals. The resistor attenuation circuit is the first control over the overall sensitivity of the unit. The LF 356 integrated circuit rectifies the minute signals and the gain is determined by the setting of VR1, operating in a null/offset circuit. This control sets the meter positioning (usually between 0.75 and 1.25 mA) and the meter provides the visual display of signals.

The output tapped directly from the seismometer's meter (output A) is then fed into the detector stage, a two-transistor voltage-sensitive receiver. The capacitor stepping is a kind of fine tuning for the receiver. Q1 is used as a detector-amplifier, and Q2 further amplifies the signals for the audio output. Transformer T1 is a 10-kCT/2-kCT driver transformer, and L1 is a 1000-μH choke, a miniature epoxy-coated type, for cutting hum in the circuit.

The power requirement is 9 Vdc to the LF 356 and LM 380 integrated circuits, and a maximum of 3 Vdc to the detector stage. The 3-V supply is designed to be ad-

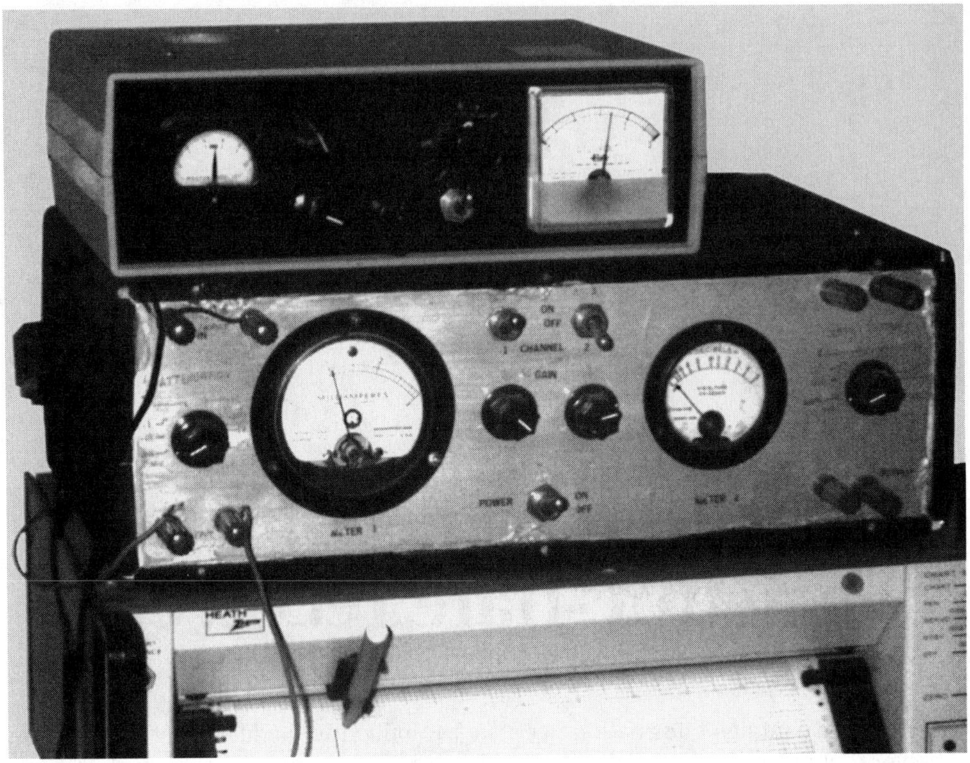

5-1 One version of a complete seismic/flare receiver, along with a dual-channel seismometer.

justable from 0 to 3 V, with sensitivity increasing with the voltage. Generally, the best reception and audio comes with about 1.7 to 2.0 V, depending on the type of sensor used. For the best results and the greatest ease in adjustment, it is advisable to use two separate 9-Vdc supplies; one driving the audio section and the other used for the seismometer and the voltage-sensitive receiver sections. Power ratings should be at least 200 mA in the audio section. Between 50 and 100 mA is plenty for the seismometer and receiver sections. The stability of the power supply to the detector stage is the most critical aspect of the unit.

The audio amp is a very simple circuit that is designed around the LM 380 IC. The output from T1 is fed directly into pins 2 and 6 of the IC. The circuit provides adequate volume for normal listening.

The sensitivity adjustment accommodates different kinds of sensors, and mid-scale positioning allows the maximum sensitivity and meter deflection to be obtained.

The attenuation-stepping network is comprised of an eight-position rotary switch with the maximum sensitivity (minimum attenuation) at the position of lowest resistance. An appropriate-value ½-W potentiometer can be used in place of the stepping network, if absolutely necessary. However, this is not advised because it makes accurate sensitivity adjustments nearly impossible and it might introduce noise into the circuit. As a stand-alone device, with no receiver or audio circuits, a

5-2 Another version of a complete receiver, also with the seismometer.

dual-channel seismometer can be used such that either channel can be used independently; to display two kinds of movements (such as north/south, east/west, or vertical) or to display one channel of seismic activity against atmospherics or solar activity. The terminals provide output for user-supplied devices (such as a chart recorder, computer, oscilloscope, etc.).

When a dual-channel seismometer is used in construction, one channel can also be used for other data acquisitions with no conversions, just a sensitivity adjustment.

Two versions are possible, one with the various sensitivity controls and one version with fixed-value components. The fixed-value type would be sufficient in installations where one type of sensor is used, or just one kind of data is being monitored and analyzed. The advantages are that fewer parts are used and that most simple adjustments are possible. Both versions work well; the user must decide on which suits best the particular requirements.

A fixed-value "complete" receiver can be adapted by following these steps:

1. Substitute a 27-Ω resistor in place of the resistance network in the seismometer section.
2. Substitute one 0.022-μF capacitor in place of the capacitor stepping network (fine tuning) in the receiver section.
3. Use separate 9-Vdc supplies for the seismometer section and the audio amplifier section.
4. When building the complete receiver, the ground lead is usually not needed.

Operating instructions

Adjust the seismometer section as was described in chapter 2. In the receiver section, begin with the receiver voltage set at about 1.0 V. Set the volume to about ¾ full and very slowly increase the power setting to about 1.6 V. Sensitivity should occur after a few seconds and any movement in the seismic sensor should also provide an audio signal along with a meter deflection. Check the polarity of the sensor input, and also check the polarity on the internal connection between the seismometer and receiver sections. Then, choose the connection that provides the best response in both the audio and meter displays. Remember to allow at least 30 seconds for the electrolytic capacitors to fully charge before making the final power setting (to somewhere between 1.6 and 2.0 V) and completing tuning adjustments. Only very minor adjustments should be necessary from here. Figures 5-3 and 5-4 show internal and external views of a finished unit. Table 5-1 is the parts list for a single-channel seismometer-based receiver (by stages).

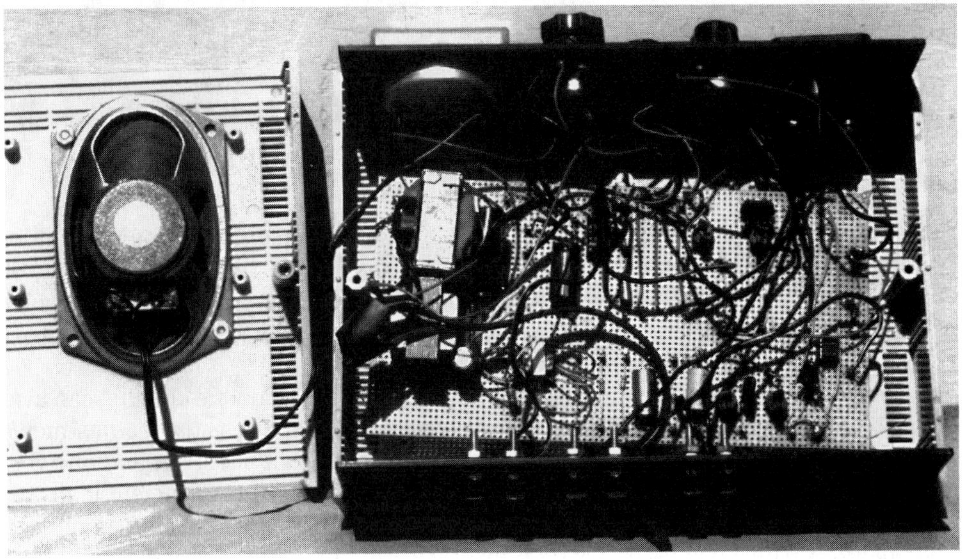

5-3 Internal view of a finished unit.

Stage descriptions

The basic seismometer section of the complete unit is covered in chapter 2. The schematic is again shown in Fig. 5-5 because this section is one of the modular units used in construction of the complete receiver.

As shown, the section functions as an unmodulated pre-amp for the reception of several types of signals. When used solely for the purpose of acquiring and converting seismic signals to audio tones, a simple additional circuit might improve reception.

5-4 External view of a finished unit.

Table 5-1.
Complete Seismic-Flare Receiver Parts List

Part	Description
	Seismometer Section
Switch	SPST Power Switch
S1	8-Position, Single-Pole Rotary Switch
VR1	100-kΩ, ½-W Potentiometer
9 Vdc	Power Supply or Battery
R2, R6	1-kΩ, ¼-W Resistors
R1	2.7-kΩ, ¼-W Resistor
C1	10-µF, 16-V Electrolytic Capacitor
IC1	LF 356
M1	Meter; 0 to 3 mA dc Full-Scale
Socket	Socket for the IC
R3	1-Ω, ¼-W Resistor
R4	10-Ω, ¼-W Resistor
R5	100-Ω, ¼-W Resistor
R7	10-kΩ, ¼-W Resistor
R8	100-kΩ, ¼-W Resistor
R9	1-MΩ, ¼-W Resistor
R10	10-MΩ, ¼-W Resistor
Misc.	Case, Line Cord, Hook-up Wire
	Detector Stage
L1	1000-µH Choke Inductor in the Audio Stage
L2	Choke Inductor in the Input Stage
C1	Capacitor; 1000-pF Dipped Silver Mica
C2	Capacitor; 2200-pF Dipped Silver Mica
C3	Capacitor; 3300-pF Dipped Silver Mica

Table 5-1. Continued

Part	Description
C4	Capacitor; 4700-pF Dipped Silver Mica
C5	Capacitor; 0.015 μF
C6, C10	Capacitors; 0.022 μF
C7	Capacitor; 0.047 μF
C8	Capacitor; 0.1 μF
C9	Capacitor; 220 μF, 50 V Electrolytic
Q1, Q2	2N3904 Transistors
R1	1-MΩ, ¼-W Resistor
R2	4.7-kΩ, ¼-W Resistor
R3	470-kΩ, ¼-W Resistor
S1	Six-Position Rotary Switch
T1	10-KCT:2-KCT Audio Driver Transformer
PS	Power Supply
	Power Supply Stage
PS	9-Vdc Power Supply
VR1	10-kΩ, ¼-W Trimpot, Internal Adjustment
Q1, Q2	2N3904 NPN Transistors
VR2	100-kΩ, ½-W Potentiometer
R1, R2	6.8-kΩ, ¼-W Resistors
Meter	0- to 3-Vdc Full Scale, with 0.1-V Divisions
	Audio Circuit
PS	9-Vdc, 200-mA (minimum) Power Supply
IC	LM 380N Integrated Circuit
Socket	Socket for the IC
C1	1000-μF, 25-V Electrolytic Capacitor
R1	1-kΩ or 2-kΩ, ½-W Audio Potentiometer
SPKR	8-Ω, at least 2-W Speaker
C or R	50-pF to 100-pF Capacitor, or 1-kΩ, ¼-W Resistor (Not used in all circuits, see text.)

For this project, a modulated audio version, the sensor is directly input into the LF 356 IC pins 2 and 3, bypassing the 1-kΩ resistor going to ground. This feature can be switch-selectable, if desired. One lone part is added; a 1-kΩ ½-W potentiometer is connected as follows: leg 1 goes to leg 1 of the potentiometer in the seismometer's null/offset circuit. The arm of the new part is connected to the – lead of the seismometer's power supply and leg 3 of the 1-kΩ pot is connected to the seismometer's input pin, #3 of the IC, through the 1-kΩ resistor that was bypassed in the direct hookup.

Adjusting the new pot will affect a tone when signals are received. This feedback sometimes improves the sensitivity to some sensors.

The detector stage, covered in chapter 4 as a VLF receiver, is shown in Fig. 5-6, changed in the following ways: input points A and B are connected to output A of the

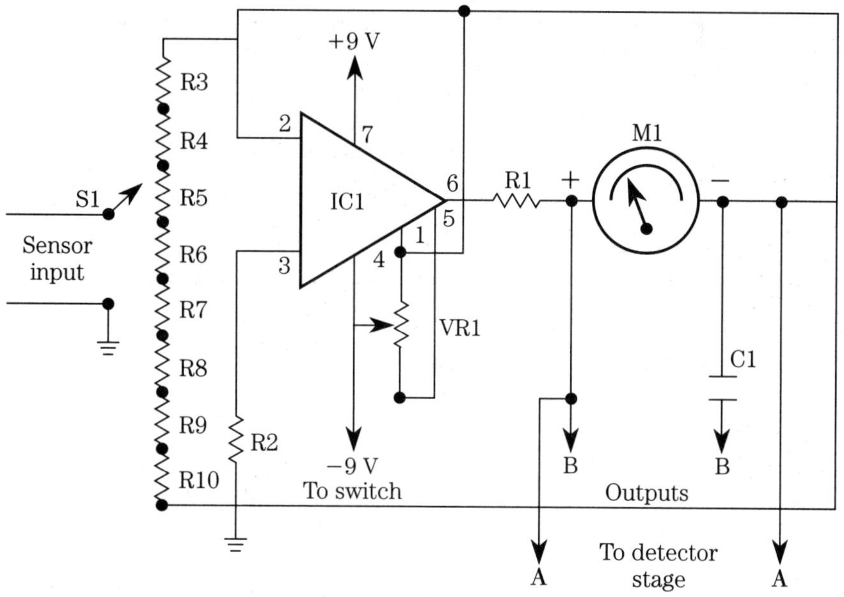

5-5 Seismometer schematic diagram.

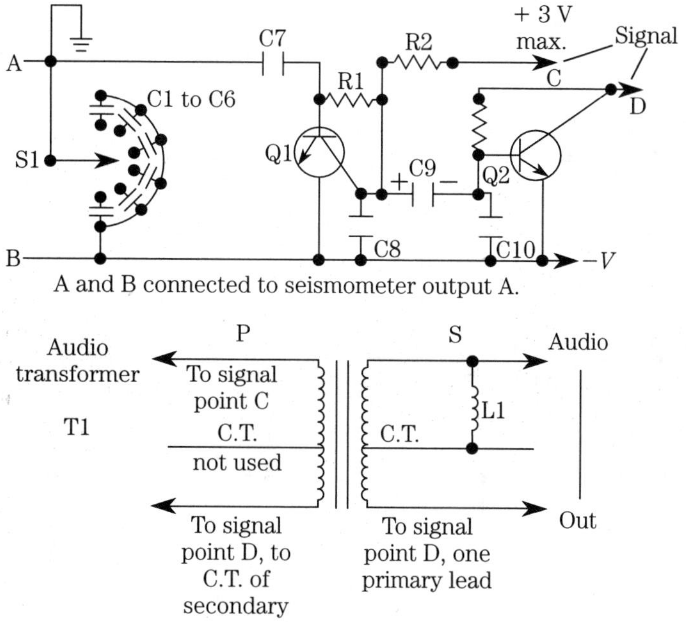

5-6 Detector stage of the receiver.

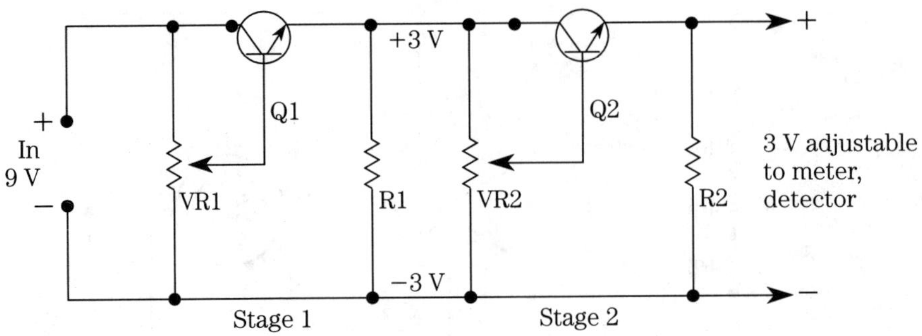

5-7 Power regulation section of the receiver.

seismometer section. Check the polarity here and connect it for the best response. The voltage to this stage is regulated to an adjustable maximum of 3 Vdc by the power circuit of Fig. 5-7. This section begins with a 9-Vdc, at least 200-mA, well-filtered supply.

Nine volts goes into the LF 356 of the seismometer section. This voltage is limited to a maximum of 3 V, adjustable, for the voltage-sensitive detector.

A two-stage supply provides regulated power to the detector-receiver as follows: one stage cuts the voltage to the 3-V maximum by using an internally adjusted trimpot or an appropriate fixed-value resistor. The second stage is user controlled by a panel-mounted 100-kΩ ½-W potentiometer, for varying the voltage into this stage from 0 to 3 V. The trimpot or fixed resistor value limits the first stage output to 3 V, and should be adjusted while the audio amplifier is set to full volume.

The audio section, utilizing a separate 9-Vdc power supply, is shown in Fig. 5-8. It is a very simple audio amplifier circuit and it provides good results. It is very easy to construct. Connect pins 3, 4, 5, 7, 10, 11, and 12 together, and always use a socket. A 1-kΩ to 2-kΩ potentiometer adequately controls the volume.

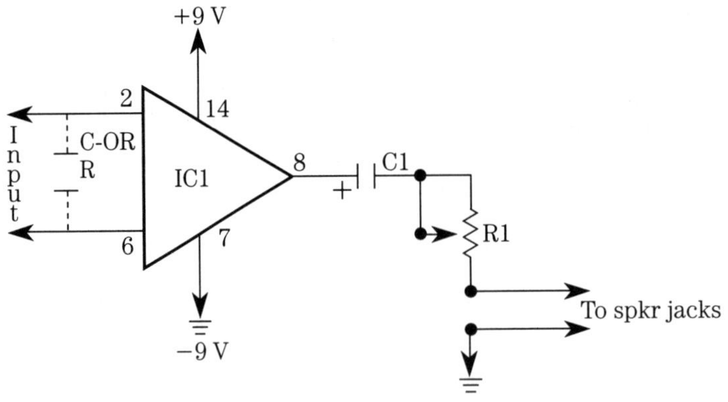

5-8 Audio section schematic of the receiver.

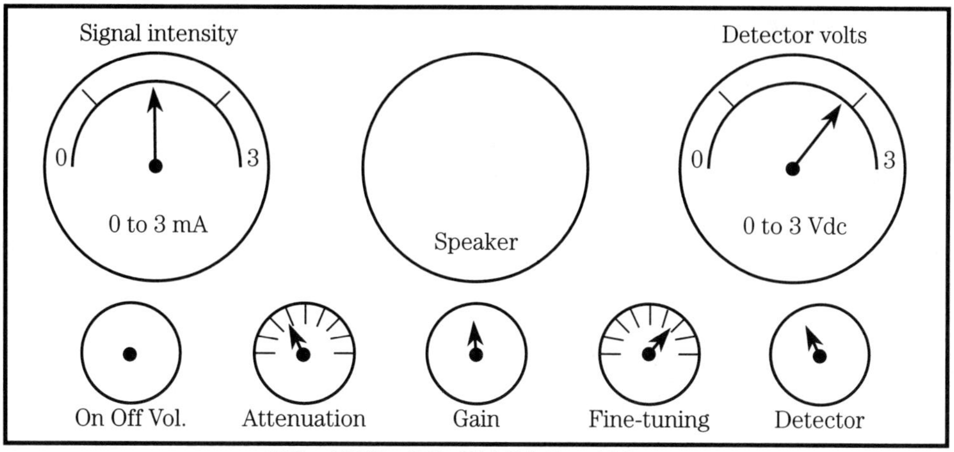

5"H x 10"W x 8"D. 2½" Rd. Panel Meters

5-9 Drawing of a sample front panel of a complete receiver.

Audio is input into this section from the output of the audio transformer. Sometimes, a 50- to 100-pF capacitor or a 1-kΩ ¼-W resistor needs to be installed between input pins 2 and 6 of the IC to get audio out. This need depends on the particular internal construction of the IC; the extra parts are usually not necessary.

Figure 5-9 shows a sample front panel look of the completed seismic-flare receiver. Many other dimensions are possible, according to your needs. Whether it is constructed to look scientific, or aesthetic, or both, the functional characteristics are the primary importance.

6
CHAPTER

Vacuum tube circuits

Simple radio circuits that use vacuum tubes are excellent low-frequency receivers. Parts can be found in old radios from the 1930s through the 1950s, or the parts can be purchased from many sources.

These radios are not toys. They can be used as dependable receivers for broadcast or aerobeacon reception, as well as for atmospheric disturbance (lightning strike) and solar flare monitoring. Some representative tube types are shown in Fig. 6-1.

You might wonder why anyone would want to use vacuum tubes; after all, tubes haven't been extensively used since the mid-1960s. Aren't they outdated anachronisms of electronics? Haven't transistors and integrated circuits taken their place?

6-1 Sample vacuum tubes.

True, transistors have certain advantages over tubes. They are smaller and cheaper, require less power, and are much more suitable for portable uses.

However, tubes also have advantages: They are durable and are less affected by temperature extremes than transistors, which will often "shut-down" or even melt in hot environments, or balk at operating in cold temperatures.

Tubes are much more forgiving of power surges than transistors are. A deviation of as much as 15 to 20% from recommended voltages to the various stages of a tube usually presents few problems. If any problems exist, the conditions are easily monitored and corrected. A transistor, on the other hand, requires more precise power regulation because they can become unstable, melt, or otherwise refuse to operate. They are particularly vulnerable to excessive voltages, and heat buildups will cause component destruction very quickly.

Tubes are also much less sensitive to static buildups or discharges than transistors. Some FETs and ICs are so sensitive that they require special handling all the way from the package to the case in which the finished apparatus is housed. Static electricity damage might not show up until a particular circuit has been in operation for quite some time, leading to the possibility of inaccurate, or otherwise spurious observations until the condition is discovered. An excellent example of this is the 40673 MOSFET, which can be used to replace a screen-grid tube, and is capable of excellent results. It is very static sensitive, however, and because ULF reception, solar flare monitoring, and atmospheric static discharges are some of the things that you want to observe, wouldn't it be prudent to use the most durable components?

In these cases, vacuum tubes and "brute-force" power supplies are very dependable. Also, when it comes time for repairs, what can be easier than replacing a tube?

Most of the circuits here are based on regeneration, which uses tube's amplification and oscillation capabilities. Often, high voltages are used, along with electrolytic capacitors in the associated power supplies. Therefore, careful construction methods and safety are musts.

See the *Selected reading list* at the end of the book for some titles that can be of help to the tube user.

One of the circuits used by the author is an old one-tube circuit called the *Pentaflex*. It is pictured in Fig. 6-2, and, as constructed receives frequencies from 74 kHz to 1600 kHz, AM and CW. The parts list for the set is in Table 6-1.

The Pentaflex

The Pentaflex is a regenerative circuit that was first used in the 1930s after the pentagrid tubes were developed.

The radio uses either a 2A7 or a 6A7 tube, and, with various configurations, can receive frequencies from nearly 0 kHz to 30 MHz, ultra-low-frequency to shortwaves. When used with an effective antenna and inductor (coil), the set makes a good wideband receiver in the Omega navigation band, below 14 kHz. Keep this in mind because the sensor for this radio is described in a later section of this book.

The various parts function as follows: R1 and C6 comprise the input circuit. R1 is the grid resistor, and values from 1 to 5 MΩ seem to work best. C6 is the grid ca-

6-2 Pentaflex radio.

pacitor, and 500- to 1000-pF values are good starting points in construction. C1 is the antenna trimmer capacitor. A 5- to 55-pF value trimmer works well. The antenna trimmer is an essential part of the set—it will not perform without it. R2 is the tube's triode stage grid bias resistor, C7 is a bypass capacitor, R5 is a resistor in the plate audio circuit, R4 is a resistor that prevents negative grid potential, and R3 is a potentiometer for regeneration control. Coils L1 and L2, along with the tuning capacitor C2, are the frequency-determining components of this receiver.

Table 6-1. Pentaflex Radio Parts List

Part	Description
C1	5- to 55-pF Antenna Trimmer Capacitor
C2	Tuning Capacitor; 365 pF to 500 pF
C3, C5	Capacitors; 0.5-µF, 400 Vdc
C4	Capacitor; 0.01-µF, 400 Vdc
C6	Capacitor; 500 pF Dipped Silver Mica
C7	Capacitor; 0.0005-µF, Dipped Silver Mica
R1	Resistor; 1 to 5 MΩ, ¼ W
R2	Resistor; 400 Ω, ½ W
R3	Potentiometer; 10 kΩ to 100 kΩ, ½ W
R4, R5	Resistors; 250 kΩ, ½ W
T1	Audio Matching Transformer; 10 kΩ to 25 kΩ (Primary), to 4 or 8 Ω (Secondary)
VT	Tube; 2A7 or 6A7
Socket	7-Prong Tube Socket
Power Supply	Filament Supply: 2.5 Vac/Vdc for the 2A7, or 6.3 Vac/Vdc for the 6A7
B+ Supply	22.5 Vdc at 50 mA, and 70 to 140 Vdc at 75 mA
L1, L2	Coils or Inductors

Coils for the AM broadcast band and low frequencies are close wound on the same form, which is about two inches in diameter, with about one-quarter of an inch separation between coils. #24 to #32 magnet wire is used and the coils can be wound on empty paper towel tubes.

A sample coil of 155 turns of #24 for the secondary coil and 52 turns on the primary uses a 500-pF tuning capacitor to allow reception down to about 300 kHz. The AM broadcast band can be received with about 80 turns on the secondary and 30 on the primary.

The Pentaflex is a special kind of tube-type receiver. It is one of the few circuits that is capable of coil-less reception. This is done by using "choke inductors." My Pentaflex uses switch-selectable 8-position values, from 56 to 6600 μH, for eight "bands." A 500-pF single-gang tuning capacitor and selectable plate inductors from 15 to 1000 μH are used for the primary coil to set its range (from 74 to 1600 kHz, in overlapping bands).

The power used in the set pictured is 6.3 Vac to a 6A7 tube, 22.5 V, and 70 V in the B circuit. To keep it simple, the audio amplifier is the same kind of circuit that is used in the seismic-flare receiver.

Table 6-2 lists the specific values for the switch-selectable circuits used in the experimental set. Here, all resistors are ¼ W, capacitors are dipped-silver mica, and inductors are miniature epoxy-coated chokes. In the rest of the set, ½-W resistors are recommended. Table 6-3 is the tube pinout for the 2A7 or 6A7. Figure 6-3 shows the basic circuit's schematic diagram.

Building other tube sets

Table 6-4 contains the parts list for the construction of various other one-tube radios. Figures 6-4 through 6-7 show the schematics of the various hook-ups.

Table 6-2. Pentaflex Circuit Parts

Grid		*Inductances*	
C6 pF	**R1 Ω**	**Band (L1)**	**μH (L2)**
10	750 kΩ	56	15
68	820 kΩ	100	22
110	910 kΩ	180	56
150	1 MΩ	330	100
500	2 MΩ	560	180
680	3.3 MΩ	680	330
820	3.9 MΩ	1000	680
910	4.7 MΩ	6600	1000
1000	5.6 MΩ		
1600	6.8 MΩ		
2000	8.2 MΩ		
2700	10 MΩ		

Table 6-3. Pentaflex Tube Pinout

Pin[a]	Function
1	Filament
2	Plate
3	Grids 3 & 5
4	Grid #2
5	Grid #1
6	K—Cathode
7	Filament
CAP	Grid #4

a Bottom view of tube, pins numbered clockwise.

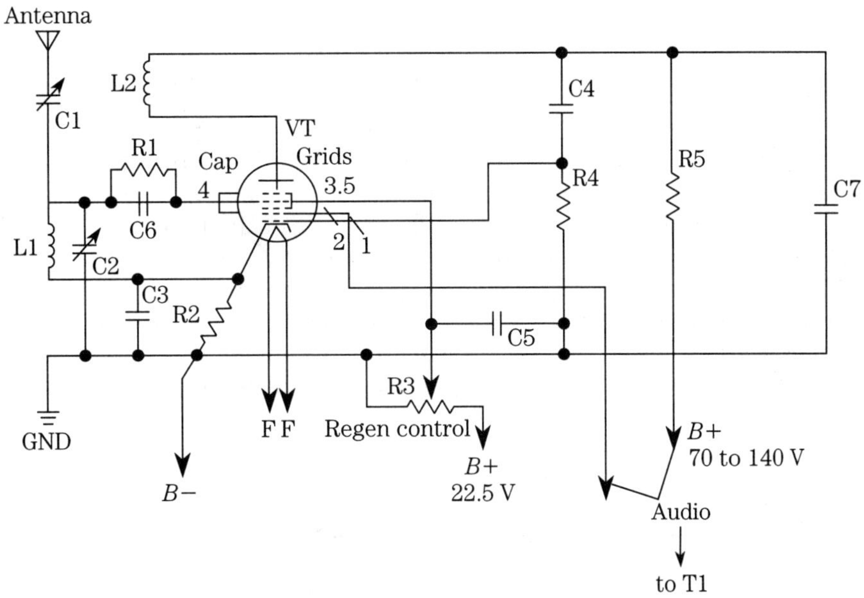

6-3 Pentaflex schematic; basic receiver design.

Testing power supplies can be constructed from 1.5-V D batteries for the filament supply and an appropriate number of 9-V transistor batteries for the B supplies, which vary from 45 to 90 V. Upon finding a suitable arrangement, a permanent ac-to-dc supply can be constructed (see chapter 7).

Any amplifier circuit can be used for speaker output from these sets, or Hi-Impedance, 2000 Ω or greater, headphones can be used. A 2.5- to 8-mH choke might be necessary, in series with the B+ voltage to the headphones, if excessive hum is experienced.

Table 6-4. Parts List for Various One-Tube Sets

Part	Description
TC1	Antenna Trimmer Capacitor
VC1	Tuning Capacitor; 365 pF to 500 pF
C1	500-pF Dipped Silver Mica Capacitor
C2	50- to 100-pF Dipped Silver Mica Capacitor
R1	10-kΩ to 100-kΩ, ½-W Potentiometer
R2[a]	1 to 10 MΩ, ¼-W Resistor
R3	1-kΩ, ¼-W Resistor
VT	Tube: 1E4, 1G4, 1P5; for 1.4-Vdc Operation 1H4 or 30 for 2.0-Vdc Operation
Power Supply	1.4 or 2.0 Vdc for Filaments 22.5 to 90 Vdc for *B*+
Phones	Hi-*Z* (Impedance) Headphones
L (Various)	Coils for Reception

[a] A 2.2-MΩ resistor is a good starting point for experimentation.

Notice that in most of the circuits, the larger secondary coil leads are hooked one end to the rotor, and one end to the stator of the tuning capacitor. The frame of the tuner is usually connected to ground, and the other lead of the tuner to the tube grid through a capacitor and/or a resistor. The smaller, primary coil, if used, is usually connected in series with the plate circuit of the tube. Also, notice that one of the audio leads is usually hooked directly to the positive terminal of the B+ supply.

Most circuits use an antenna trimmer capacitor between the antenna and the connection to the set. A 5- to 55-pF capacitor is common, although 12- to 100-pF trimmer also works well. This capacitor has a great deal to do with the overall sensi-

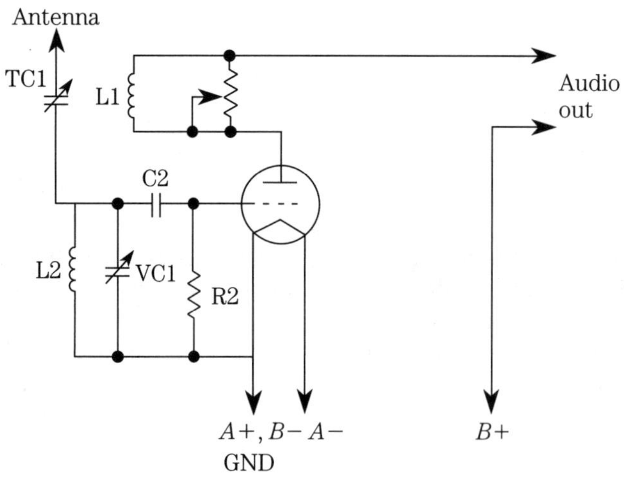

6-4
Basic radio circuit.

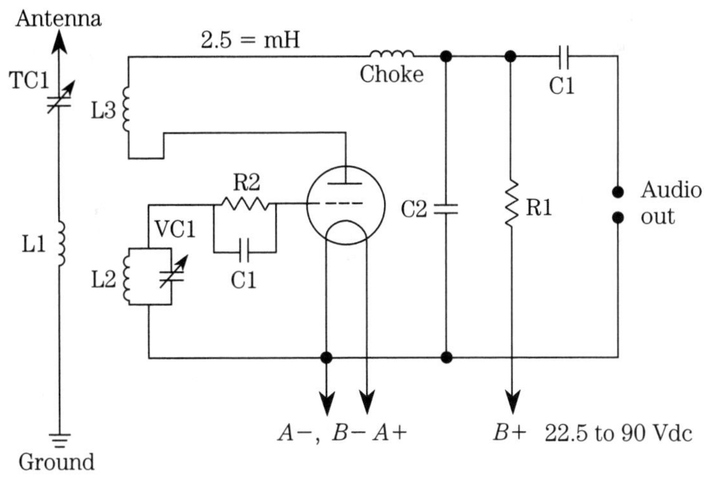

6-5 Grid-leak detector.

tivity of the set. Figure 6-4 shows a common two-coil configuration, and the circuit in Fig. 6-5 is an example of grid-leak detection.

In this circuit variation, regeneration is obtained by connecting coil L2 in series with the plate circuit. It is inductively coupled to L3, the grid coil. The antenna is connected through TC1, the trimmer capacitor, to one end of L1, with the other end of L1 connected to ground.

In all regenerative circuits, if the "rushing" sound, characteristic of the most sensitive setting of the regeneration potentiometer, is not heard or if the sensitivity is low, try reversing the coil leads in the plate circuit.

With the grid-leak circuit, the three coil windings for the AM broadcast band consist of 80 turns of #24 to #28 magnet wire in L3, and 30 turns on each of L1 and L2. Larger wire provides more volume, but smaller wire is more selective.

I use the three-coil circuit for aerobeacon and lightning-strike monitoring, with coils of #28 magnet wire; L1 = 42 turns, L2 = 42 turns, tapped at turn 15, and L3 = 135 turns, for a frequency range of 265 to 600 kHz. Coils are wound on empty paper towel tubes for forms.

The circuit in Fig. 6-6 uses only one coil and it is capable of reception all the way down to 0.5 kHz. Some of the seismic sensors can be input in place of the coil for ULF reception or seismic impulse monitoring. The set is diagrammed for a #30 type tube, but any tube designed for single-tube reception can be used.

Figure 6-7 is a three-coil variable grid-leak detector. Energy is transferred from the coil in the antenna/ground circuit to the secondary coil by inductive coupling. The coil in the plate circuit is wound on a form that is just small enough to rotate in the same plane as the coil in the grid circuit.

The grid-leak resistor in this circuit is variable from 1 to 10 MΩ, and provides variable sensitivity and selectivity. The coil windings have similar specifications to those of the other circuits described.

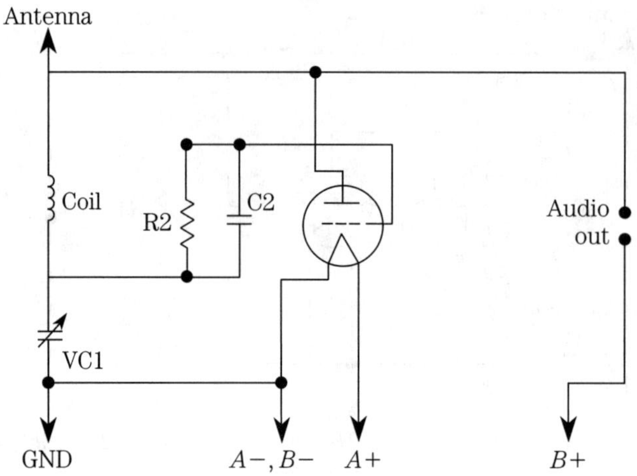

6-6 One-coil circuit.

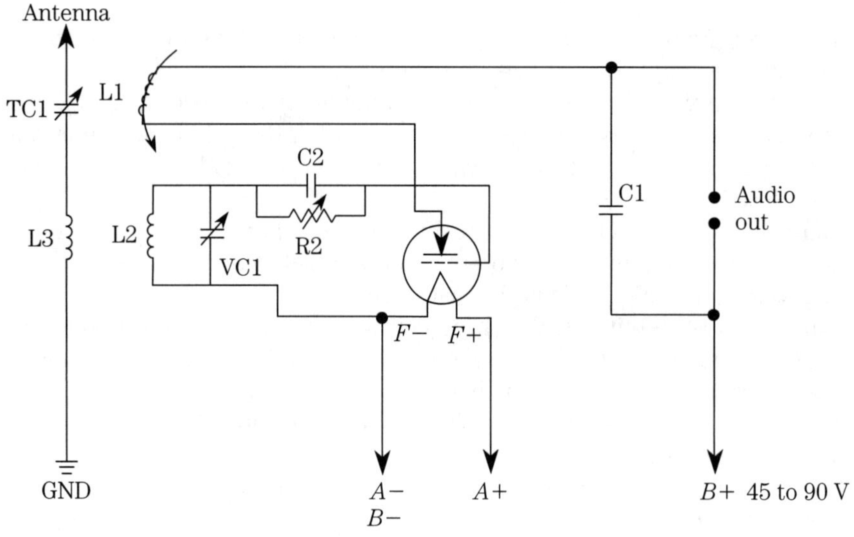

6-7 Three-coil variable grid-leak.

Building a three-tube audio amplifier

The parts list of Table 6-5 and the schematic diagram of Fig. 6-8 detail the construction of a three-tube audio amplifier that produces excellent volume to the tube radio circuits. It is a resistance-coupled amplifier, and the circuit type was widely used in the '30s and '40s.

Table 6-5. Parts List for the Audio Amplifier

Part	Description
Tubes	1 Ea; 6SL7, 6V6, 6X5
Sockets	Octal Sockets for the Tubes
L1	Filter Choke; 8 H
C1, C2	33-μF, 350-V Electrolytic Capacitors
C3	47-μF, 350-V Electrolytic Capacitor
C4, C5	10-μF, 350-V Electrolytic Capacitors
C6, C7, C8	0.02-μF, 400-Vdc Capacitors
C9	0.001-μF, 600-Vdc Capacitor
C10	0.01-μF, 1000-Vdc Capacitor
R1 (with S1)	100-kΩ, ½-W Audio Potentiometer with SPST On/Off Switch
R2	470-kΩ, ½-W Resistor
R3	3.9-kΩ, ½-W Resistor
R4, R5	120-kΩ, ½-W Resistors
R6	47-kΩ, ½-W Resistor
R7	2.2-kΩ, ½-W Resistor
R8	10-kΩ, ½-W Resistor
R9	220-kΩ, ½-W Resistor
R10	270-Ω, 1- to 5-W Resistor
R11, R12	100-Ω, ½-W Resistors
Power Xfmr	115 V Primary, 250 VCT/50 mA, and 6.3 Vac at 2 A Secondaries
Audio Xfmr	10,000: 4 to 8 Ω; Stancor A3327, or Equivalent Can Be Used; at Least 5 W
Speaker	4 or 8 Ω; Minimum 5-W Rating

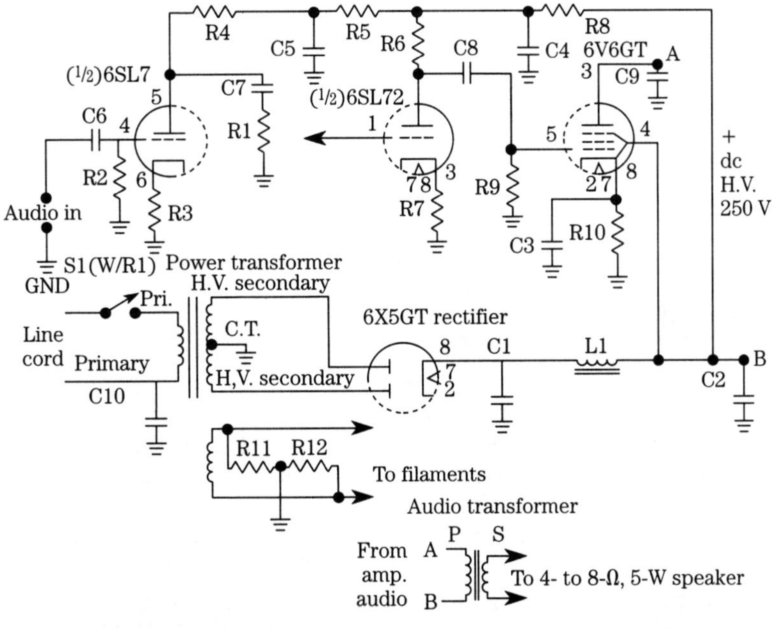

6-8 Three-tube audio amplifier schematic.

Transformer T1 provides 6.3 V to the tube filaments and 230 to 250 V, center tapped, at 50 mA for the B+ supply. Tube V1 is a 6X5 in a full-wave rectification circuit, using capacitors C1 and C2 along with choke L1 to provide the filtered dc voltage. C4 and C5, along with the resistors, provide decoupling and further filtering in the audio section.

First audio uses half of the 6SL7 tube as a resistance-coupled amplifier, which, through the volume control (R1) feeds the second audio stage using the other half of the tube. Resistance coupling provides audio from the second stage to the audio power amplifier using a 6V6 pentode beam power tube, T2 matches the output impedance of the 6V6 to the speaker. The power output is 4 W_{rms} at 4 Ω, and a 4- to 8-Ω speaker with at least a 5-W rating can be used. The input terminals are to ground and to one end of C6. The source audio should connect to the amplifier's input "ground side to ground side."

A dual-detector circuit

The schematic diagram of Fig. 6-9, along with the parts list in Table 6-6, show the construction data for a two-tube dual-detector radio. This receiver is a very efficient "lightning-strike" and atmospherics monitor; when it is coupled with the three-tube audio amplifier, it is a very stable set for aerobeacon, data, and ship-to-shore radio-telegraph signals. Details here are for a set with a frequency range from 175 to 1000 kHz.

Table 6-6. Dual-Detector Parts List

Part	Description
C1, C2	0.0002-µF Dipped Silver Mica Capacitors
C3	0.0001-µF Dipped Silver Mica Capacitor
C4	0.1-µF, 250-V Capacitor
C5	0.005-µF, 250-V Capacitor
VC1	Tuning Capacitor (See Text)
TC1	Antenna Trimmer; 5 to 55 pF
Plate Choke	20.5 mH
R1	100-kΩ, ½-W Audio Potentiometer
R2	2 MΩ, ¼-W Resistor
R3	5 MΩ, ¼-W Resistor
VT1	Tube; 1N5G
VT2	Tube; 1G4GT
Coils	Three-Coil Circuit
Power Supply	1.4 Vdc A, 65 to 75 Vdc B
Audio Xfmr	Stancor TA-33, or Equiv., or Hi-Z Phones
Misc.	2 Octal Tube Sockets, Coil Forms, Binding Posts, Hook-Up Wire, etc.

This set is efficient because separate tubes are used for detection and regeneration. A 1N5 pentode tube is used for detection, with from 65 to 90 V into the screen

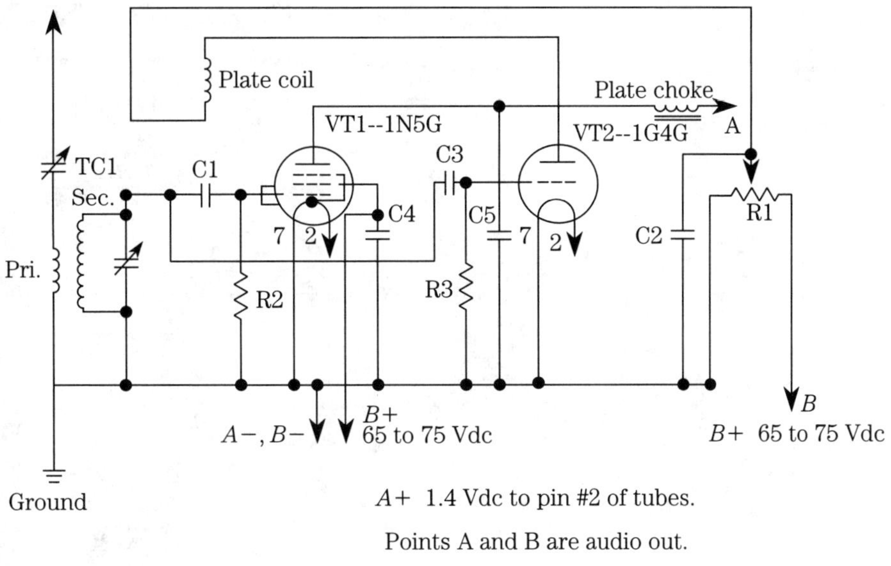

6-9 Dual-detector schematic.

grids, and a 1G4 is used for plate circuit regeneration. Power into the filaments of the tubes is 1.4 Vdc, and the current requirement is at least 100 mA. See chapter 7 for details on constructing an A power supply with a 1.5-A potential, which was used in my dual-detector set with excellent results.

The coil data for a receiver with a frequency range of 175 to 1000 kHz is as follows: the secondary coil is 160 turns of #24 with a tap at 80; the primary coil is 80 turns of #24 with a tap at 40; and the plate coil is also 80 turns, but with a tap at 20.

The tuning capacitor is a 4-gang, 500-pF-per-section variable capacitor, with one gang used in one switch-selectable setting, and the other three gangs connected in series, for a 1500-pF tuner, in the other setting. The switches used for the coil tap and tuning capacitor are SPDT types.

Figure 6-10 is the outside of the finished dual-detector, which is rather gaudy looking with the meters. The meters are not necessary for operation, but in the prototype they serve as a "test panel" for the various circuits.

From left to right: Meter 1 monitors the amplifier's B+ voltage. It is a 0- to 500-Vdc meter, with correct voltage at about 225 to 250 V. Meter 2 monitors the detector circuit's filament voltage, in operation 1.3 to 1.4 V. This is a 0- to 3-Vdc meter, same type as in the solid-state solar flare receiver circuit. Meter 3 is a 0- to 3-mA meter, for the detector circuit's current draw. The + terminal of the meter connects to the detector's B+ voltage supply, and the – lead of the meter connects to the detector's B+ input. Meter 4 monitors the detector's B+ supply of about 65 V. The + of this meter is connected to the milliampere meter's +, and the – of meter 4 connects directly to the B supply's negative lead. It is a 0- to 100-Vdc meter.

Figures 6-11A and 6-11B show two interior views of the dual-detector, which is connected to the three-tube amplifier and a 117Z6 rectifier circuit.

6-10 Outside panel of a finished dual-detector.

Figure 6-11A is an overview of the entire set, and Fig. 6-11B shows the coils and the tuning unit. The larger coil assembly on the left is constructed from two RF coils from an Atwater Kent Model 20 receiver from 1925. A tap is taken from where the two coils were joined to make one large, longwave coil, thus providing two bands for the set.

6-11A Interior of the dual-detector.

6-11B Coils and tuning unit of the dual-detector.

The homemade coil near the tuning capacitor is a wave trap (rejector) circuit used to filter out interfering AM broadcast stations. The rejector uses a 365-pF tuning capacitor and the circuit can be switched in or out of use.

The receiver itself uses the big four-gang military type tuning capacitor, switch selectable for one or three gangs use. Excellent sensitivity down to 175 kHz is obtained from this set.

The regeneration potentiometer must be carefully adjusted. This controls the B+ voltage, and an incorrect setting will cause the control to burn, while correct settings for reception will work fine. The milliampere meter, as described in the text and shown in the photo, can monitor this condition. The meter used is the same type of 0- to 3-mA meter that is recommended in the seismometer circuitry.

Meters for amperage are always connected "in series" with the current being monitored. Connect the + of the meter to the positive lead of the B+ supply, and connect the – lead of the meter to the B+ input of the radio. Correct values for receiving will be anywhere from 0.75 to 1.5 mA. Incorrect settings will pin the meter and cause a loud squeal in the audio, then burn out the potentiometer.

Audio out of this circuit is taken from points A and B. High-impedance headphones can be used or an amplifier can be used for speaker output. The TA-33 audio transformer suggested in the parts list is a very small transformer capable of handling about 0.75 W of load. A larger transformer can be used. It is best to use a transformer in any case, even if not absolutely necessary, because using one isolates the B+ voltages from the headphones, speaker, or amplifier hook-up. This thereby reduces the potential from noise or from stray electricals.

The harbor master

The harbor master, in Fig. 6-12, is a two-tube version of the BFO-type seismic signal processor, which is described a little later in this book. The set uses two miniature pentode tubes, which were common in battery sets of the '50s and '60s.

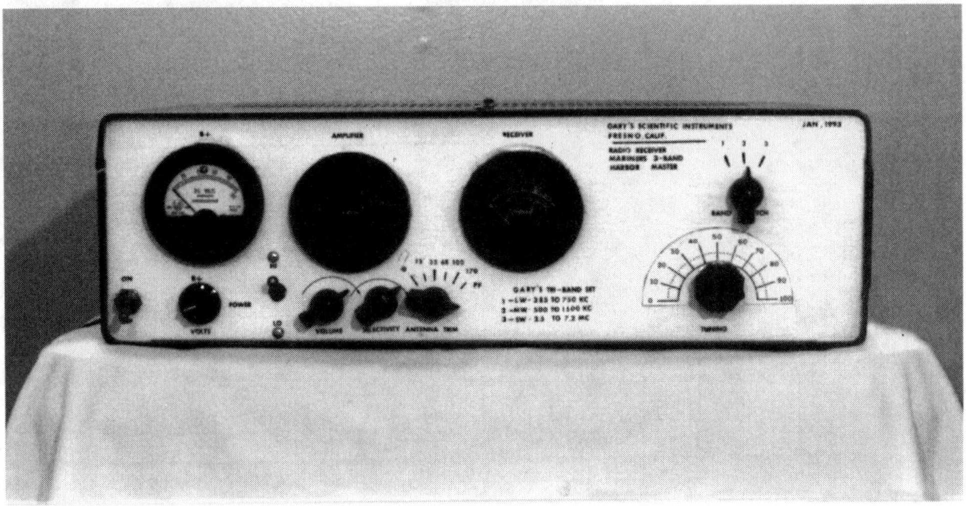

6-12 The two-tube harbor master radio.

With various three-winding coil circuits, all-wave reception, from VLF through shortwave, is possible. The set in the picture is designed to operate on three bands, including 2.5 to 7.0 MHz, and worldwide reception is possible with the circuit.

For seismic monitoring, a VLF-resonant three-winding coil is used. The coil assembly is separated from the radio's chassis and used as a sensor.

The sensor uses a dense metal plumb, positioned over the coil and connected to the radio's plate circuit so that any movement of the plumb changes the inductance of the coil. This results in a tone, or variance of tone, being generated in the receiver section. For this, the receiver is tuned to a point of oscillation or near-oscillation for monitoring. See the BFO seismic signal processors for more information.

If the set is to be used only for this kind of seismic study, only one frequency for reception is necessary. The capacitance value used for tuning to the point of oscillation can be measured, and a correct value, strict-tolerance dipped silver-mica capacitor can be substituted for the tuner. The audio output can be monitored via speaker, phones, oscilloscope, computer, chart recorder, etc. This feature can also be selected as one "band," then the receiving portion can be retained for other uses.

In this set, stability of the detector filament supply is important. When the two-tube set is constructed, a separate power supply is provided for each tube. A Stancor P-2945, 6.3-VCT/2-A transformer was used, with the center tap, providing one-half of the output. Each side is rectified by the LM 317T regulator circuit (see chapter 7), then the correct voltages for each tube were set by fixed, 5-W resistors for a stable supply. Trimpots can be used for a variable supply, if desired. The 1T4 detector tube requires 1.4 V, and the 3Q4 audio tube uses 1.4 to 2.8 V. This tube was operated with 2.2 V, which seemed to work well.

The audio amplifier hook-up is very simple. A transformer is coupled from the detector stage, using no bias resistors, as a variable B voltage to the 3Q4 (adequately controls the volume). The transformer used is equivalent to a Stancor A53C, which is similar to the kind used in antique radios. For shortwave reception, a second stage of audio would help boost the volume for improved reception. For local reception, surprising volume is obtained from this circuit, using only a 30-V power supply to the audio section. Of course, any other amplifier configuration can be used; the IC amplifier (covered earlier) provides excellent volume and is simple to hook up, which eliminates the need for the 3Q4. Again, an audio-matching transformer is used to match the plate of either the detector or the amplifier to the speaker. The transformer used in this set is the Stancor A3327.

Coil data

For the broadcast band, a replacement-type antenna coil can be used. A sample, shown in Fig. 6-13, is readily available from parts houses, and it contains two of the coils. An additional coil consisting of 25 to 30 turns of #36 magnet wire is wound around the same form, and it serves as the plate coil, L3.

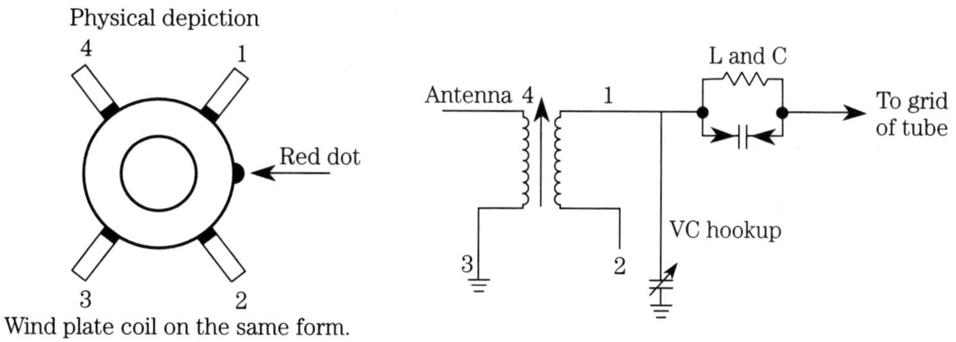

6-13 P-70A antenna coil schematic: physical and electrical.

For a hand-wound coil on a 1½" form, use about 80 turns of #24 to #28 magnet wire for L1 and 30 turns each of #24 to #28 magnet wire for L2 and L3. Coils are close wound. For longwave, use 150 turns of #28 for L1, and 50 turns each for the other coils. Again, these are close-wound coils.

For shortwave reception, use #24 wire. L1 contains about 14 turns, spaced about 1⁄16" apart. L2 has about 10 turns, also spaced, and L3 contains about 8 to 10 close-wound turns. The windings should be separated by ⅛ to ¼" between L1 and L2, and by about ½ to 1" between L2 and L3. The schematic for the set is shown in Fig. 6-14.

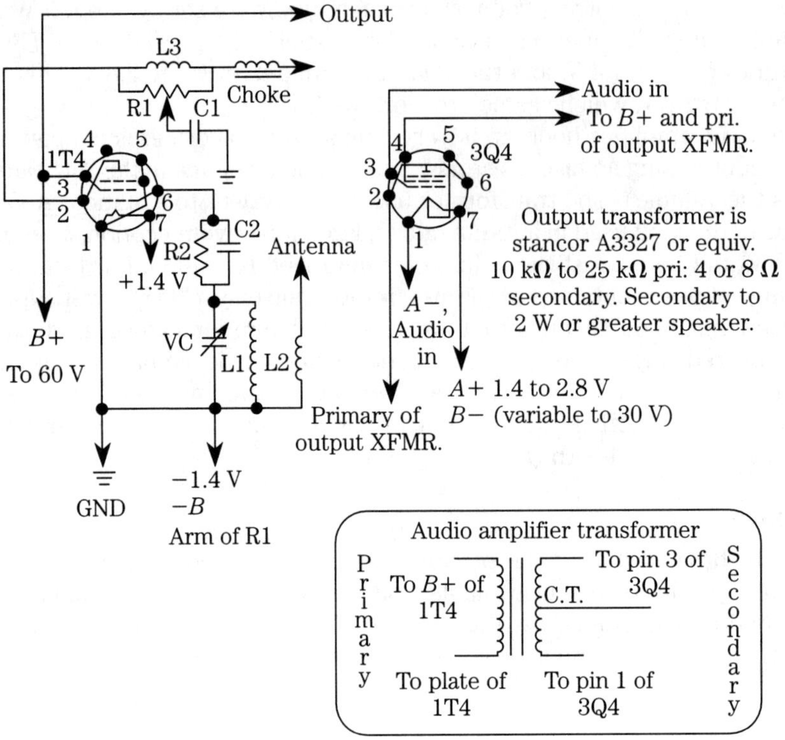

6-14 Schematic of the two-tube harbor master.

The B+ power supply

Three 18-VCT 300-mA transformers were used in the prototype set. One transformer supplied the power to the audio amplifier stage. Upon rectification, about 30 Vdc was obtained from this stage. An LM 317T regulator circuit was used, varying the voltage from 1.2 to 30 V, and this serves as the volume control.

The other two transformers comprise the B+ voltage for the detector stage. The LM 317T regulator circuit is used on the first stage, and it governs the 30 V available from the first stage of the power supply. However, the detector requires more than 30 V to operate at maximum efficiency, and the LM 317T is rated at a maximum of about 37 V. The circuit to this set regulates the voltage to just over 60 V with the final transformer, in the second stage of the supply.

After the first stage is rectified and regulated, it is variable from 1.2 V to about 31 V. An SPDT switch is used to switch in the negative lead from either the first or the second stage of the power supply, effectively using either only the first stage, which remains adjustable, or it selects the full power of the second stage, which is connected in series with the potentiometer of the adjustable first stage, for an additional 31 V.

The resulting circuit allows either 0 or 30 V, adjustable, or a 31- to 61-V adjustable supply to the detector. The higher voltages are particularly useful when tuning the shortwave bands.

The voltmeter, first from the left in the picture, is important. It is a 0- to 100-Vdc meter for monitoring the B supply. It can be used as a point of reference in adjusting the sensitivity of the set. A 0- to 75-Vdc meter would have an even better resolution. The other meters monitor the filament voltages of the tubes, but are not really needed once the voltages are set.

On shortwave, ship-to-shore CW and data can be received, as well as WWV, international broadcasters, amateur radio operators, and the Coast Guard. Use the best antenna system you can rig up, along with a good earth ground.

The schematic diagram of the set shows the basic receiver section as a separate unit, with the amplifier also separate. This arrangement allows you to easily adapt the basic units to a specific building design. The parts list is in Table 6-7.

Table 6-7. Parts List for the Harbor Master

Part	Description
C1	68-pF Dipped Silver Mica Capacitor
C2	100-pF Dipped Silver Mica Capacitor
VC	Tuning Capacitor; 20 to 420 pF
R1	100-kΩ, ½-W Potentiometer
R2	3.3-MΩ, ½-W Resistor
Choke	2.5 mH
Magnet Wire	#24 and #28, for Coil Winding
TC	12- to 100-pF Trimmer Capacitor (See Text)
VT1	1T4 Tube
VT2	3Q4 Tube
AT	A53C, or Equivalent, Audio Amp Transformer
OT	A3327, or Equivalent, Audio Output Transformer
PS	Power Supplies: 1.4 Vdc, 2.2 Vdc A, 0-30 Vdc, 0-60 Vdc B
Meters	0 to 75 or 0 to 100 Vdc, and Optional: Two 0 to 3 Vdc

In the receiver's antenna lead, and not shown in the diagram, a 12- to 100-pF trimmer capacitor should be installed in series with the antenna. After this capacitor is adjusted for best reception, an interesting fine-tuning control can be panel mounted. For fine-tuning, a capacitor stepping circuit, switch selectable in value, can use low-capacitance mica capacitors of 1, 5, 7, 10, and 15 pF, for example, or a variable 5- to 35-pF or 15- to 75-pF trimmer, wired in parallel with the antenna trimmer. This can really help when tuning SSB and CW signals. Along with the variable B voltage control, the fine tuning allows this set to perform well on all bands.

The basic receiver is hooked up according to the following steps:

1. The 1.4 Vdc A– lead, along with one end of a 100-pF capacitor, is connected to pin #1 of the 1T4.

2. The other end of the 100-pF capacitor goes to the arm of a 100-kΩ potentiometer.

3. One lead of the plate circuit coil, and one leg of the potentiometer, go to pin #2 of the 1T4.
4. The B+ power lead connects to pin #3 of the 1T4, which is one audio lead.
5. Pin 6 of the 1T4 connects to both a 68-pF capacitor and a 3.3-MΩ resistor. The other ends of these parts are connected together, and go to the top of coil L1, and the gang of the tuning capacitor.
6. A+ connects to pin 7 of the 1T4.
7. B–, the ground, the bottom of coils L1 and L2, and the frame of the tuner all go to pin 1 of 1T4.
8. The top of coil L2 connects to the antenna lead.
9. The final coil hook-up is the one lead remaining from the coil in the plate circuit, L3. It goes to one end of a 2.5-mH choke, and to leg 3 of the potentiometer. The other end of the choke is the other audio out lead.

The 3Q4 audio stage is hooked up as follows: The audio out from the detector connects to the primary of the amplifier transformer. Pin 7 of the 3Q4 is connected to the A+ and B– power leads of a separate power supply. Pin 4 goes to the B+ and to the primary of an output transformer. Pin 3 is an audio input pin, and is hooked to a secondary lead of the amp transformer. Pin 2 is the plate lead, and it connects to the primary of the output transformer. Pin 1 goes to the A– supply (1.4 to 2.8 Vdc) and to one secondary of the amp transformer, and the secondary of the output transformer connects to the speaker.

The geo-monitor

The geo-monitor receiver is the harbor master circuit with a few modifications and some additional circuitry to make the receiver more efficient and versatile, for seismic wave monitoring, as well as for radio receiving.

The grid capacitor is changed from a 68-pF value to one of from 110 to 220 pF. A solid-state, fixed-voltage B+ supply of 90 V is used for the detector. A 365-pF variable tuning capacitor is used in series with the antenna lead, or fixed-value capacitors with the following values can be used for the antenna trimmer: 1 pF, 3.3 pF, 10 pF, 30 pF, 68 pF, 100 pF, 220 pF, or 330 pF.

In the audio section: A bias supply is used to the 3Q4 tube, which is variable from 0 to 4.5 V. Another audio stage is used, either another 3Q4 tube in series, or in push-pull (see Fig. 6-15), or a solid-state amplifier can be used. The completed geo-monitor shown in Fig. 6-16 has one 3Q4 audio section, plus a second stage of audio based on the LM 380N IC circuit. This provides very satisfactory results in all applications.

With the tuning capacitor installed in the antenna lead, both this capacitor and the tuner have to be adjusted to receive stations, like tuning a two-combination lock. A good antenna and ground will help the radio to perform much better.

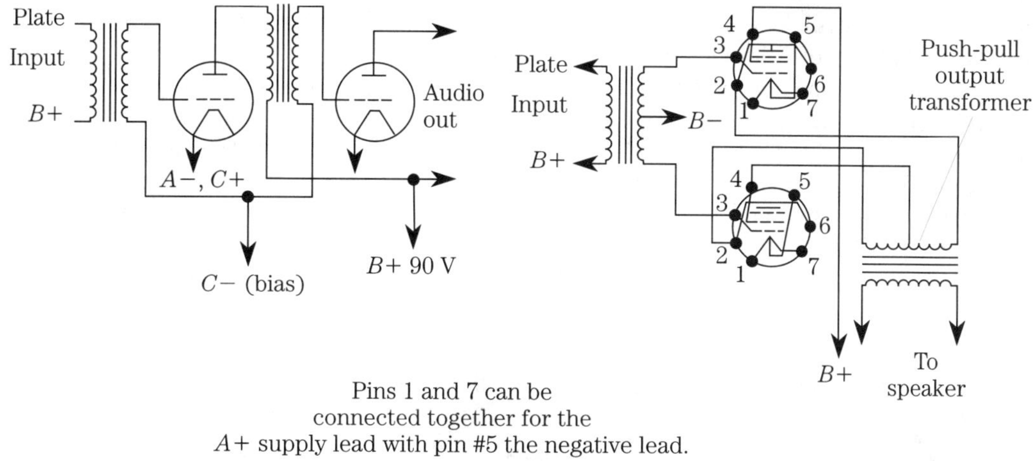

6-15 Audio amplifier configurations.

6-16 The geo-monitor.

The bias supply was obtained by rectifying a 2.5-Vac 30-mA power supply, obtaining about 4.4 Vdc. The bias is connected with the negative lead of the supply connected to the signal grid of the 3Q4, pin 3. The positive lead of the bias connects to the filament (A–) negative of the tube. In the set as shown, the bias' negative lead was connected to one end of a 0.01-µF/400-V capacitor, with the other end of the capacitor going to leg 1 of a 10-kΩ potentiometer. The arm of the potentiometer was connected to the grid of the tube for a variable supply, which was very useful when adjusting the set for seismic monitoring, as well as for improving audio quality on radio receiving.

The B+ to the audio section can be any voltage from 20 to 90 V, but the harbor master's 30-V variable supply is excellent as the volume control, as well as being easy to build and control. The lower supply also seems to work better for seismic monitoring, and plenty of volume can easily be obtained.

The set has no receiving coils, again for the purpose of making things easier when it is used as a seismic monitor. Instead, leads from the various coil connections are routed via shielded cable to the back of the set (Fig. 6-17) along with the antenna and ground connections. Any three-coil circuit can be used for frequencies from VLF to shortwave, and two-coil sensors can be used to receive seismic impulses by audio disturbance. Chapter 9 describes some solid-state circuits for this purpose, and chapter 13 details the building of a sensor, which is briefly described in the following paragraphs.

6-17
Back of the geo-monitor, with a sample sensor transformer being tested.

For receiving seismic waves on the geo-monitor, a two-coil electromagnetic sensor is required, but do not make any connections to the antenna and ground posts. The "water bottle" sensor of chapter 13 is excellent for use with this set, but, in this case, one secondary must be used. A power transformer with a secondary of from 6 to 28 VCT, 1 to 5 A, with a center tap, should be used. Usually, the center tap of the secondary is the preferred hook-up to the receiver. In adapting this configuration to the water bottle sensor, simply route one of the secondary leads out of the bottle in addition to the transformer's primary (115-V input) leads. The transformer primary connects to the grid coil connections of the radio, and the secondary leads go to the plate circuit, used as the "tickler" coil. Here, as with all other regenerative circuits, if no regeneration can be heard, reverse the leads. Because the center tap of the secondary is nearly always the preferred hook-up, simply cut off the unused secondary lead.

No connection should be made to the detector coil leads when the set is used for seismic wave monitoring. For maximum compatibility to various transformers and coils in the grid and detector circuits, the inputs are compensated by using 1-kΩ ½-W potentiometers before they are connected to the circuit board's terminal posts. Be sure to begin all adjustments with each of these two potentiometers at their points of minimum resistance. Then, adjust them only if necessary. These also can be used with radio coils to fine-tune the set; they can really improve the audio quality as well as provide oscillation control.

To adjust for seismic wave reception, proceed as follows:

With the sensor hooked up, rotating the tuning control should produce tones, quiet points, and points of hissing. When used with the two-coil water-bottle sensor, the ideal positions for tuning this set are with the regeneration and tuning set at a point where the receiver just quiets. The bias control can "fine-tune" this position. When the sensor is shaken, hissing or squealing sounds will be produced in the speaker, but the best adjustment is one where the movement of the sensor causes a series of clicks (very rapid in succession), like those of a Geiger counter detecting intense radiation. The bias control can then adjust the final sensitivity of the set, which is stable in operation.

An important point to remember when receiving seismic waves with the geo-monitor: the detector tube's filament supply is critical for maximum sensitivity, which is the reason for the panel mounted meter and the rheostat. It should be set at 1.4 V, not below, and in no case ever above this value. Although a lower voltage will not appreciably affect radio reception, it *will* affect seismic monitoring. A higher than 1.4 V setting will only serve to burn out the tube's filaments.

A signal meter is connected in series with the plate output of the audio tube. Depending on the circuit power used, a value from 200 μA to 2 mA can be used. This connection is also used for digital reading of signals, using a multimeter's 0- to 20-kHz frequency counter range selection (see chapter 13). The schematic diagram for the set is given in Fig. 6-18.

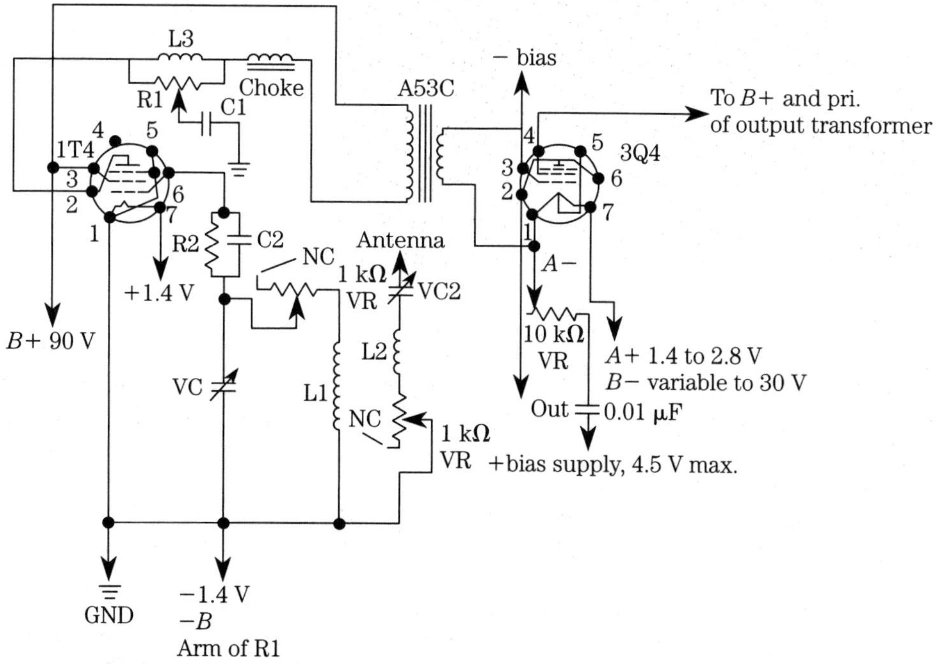

6-18 Geo-monitor schematic with one-tube audio stage.

Rectifier supplies

A couple of representative tube-based rectifier circuits are shown in Figs. 6-19 and 6-20. The 6X5 tube circuit can be used to deliver up to about 325 V at up to 70 mA, in a full-wave rectification circuit. It is shown in a working circuit in the three-tube amplifier schematic. The 117Z6 circuit is used to rectify the B voltage in the dual-detector radio. This circuit is relatively simple; pins 2 and 7 of the tube are connected directly to the ac (115 V) line, and pins 3 and 5 are the tube's plate connections, which are hooked to the B voltage transformer. If the transformer is center-tapped, the tap is connected to ground. Only two large electrolytic capacitors are used to filter the pulsating current; again, be sure to use the proper polarity. With electrolytics, the overwhelming majority of the hook-ups call for the capacitor's – lead to be connected to ground, and the + lead to the circuit. Remember this polarity rule when working with electrolytics—especially in high-voltage circuits.

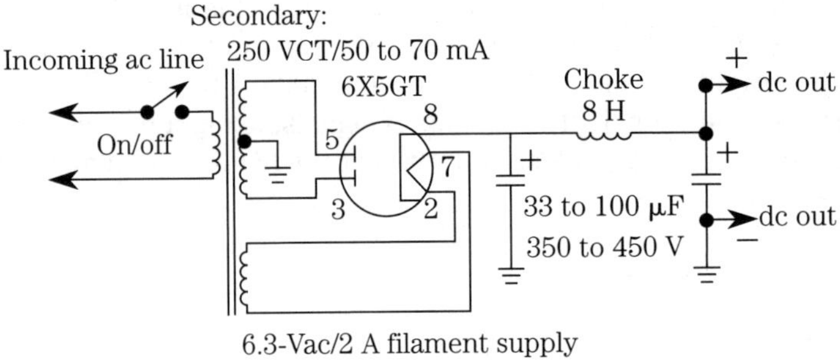

6-19 6X5 tube-based rectifier circuit schematic.

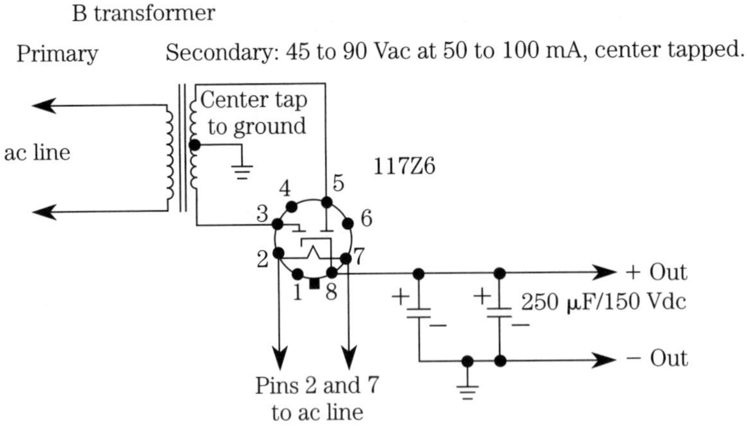

6-20 117Z6 tube-based rectifier circuit schematic.

The navigator

The navigator is a tuned radio frequency (TRF) circuit that is a decided improvement over the primitive regenerative. This radio has no plate circuit regeneration so that a loop antenna can be used with the set, which is not possible with the regeneratives.

An adjustable 36" whip antenna and a water pipe ground provided excellent reception. The schematic of the set is shown in Fig. 6-21, and the parts list is Table 6-8.

Two identical two-winding, solenoid-style coils are used for this set. Separate tuning capacitors are used, one for the RF stage and another for the detector. A smaller tuner is used in the antenna circuit.

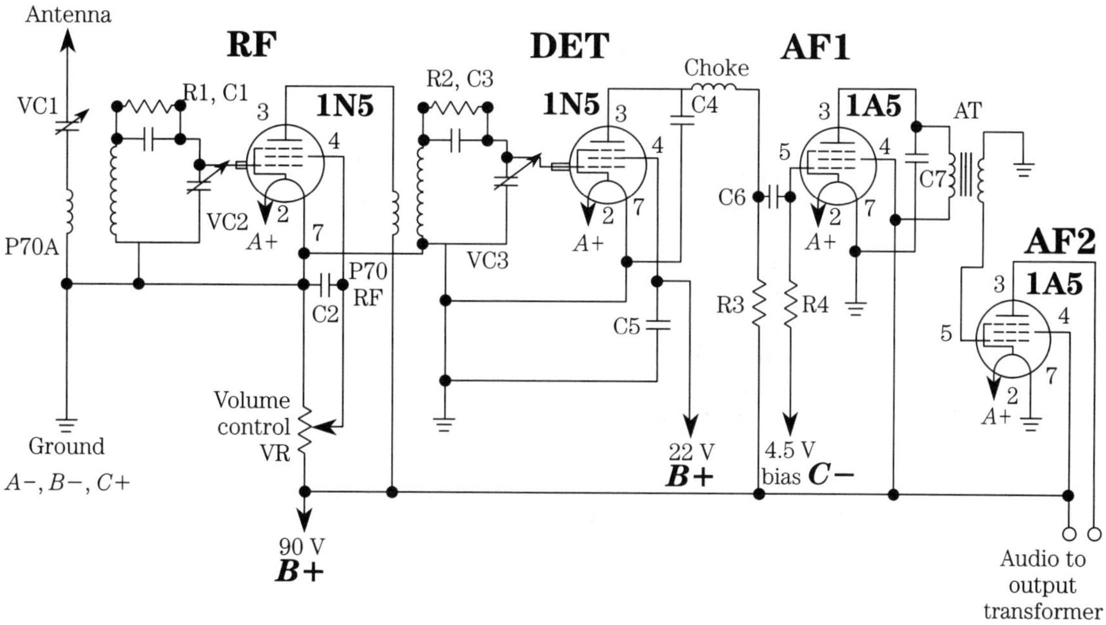

6-21 The navigator.

The commercial P-70A and P-70RF coils can be used for building this set. With a 365-pF tuner in the RF stage and a 450-pF tuner in the detector, these coils will give good reception from about 435 to 850 kHz, good for ship-to-shore, beacon, and NAVTEX data stations. NAVTEX is a Sitor-mode telex network operating on 518 kHz. NAVTEX transmissions originate from coastal areas and contain weather, notices to mariners, and other related information, but it requires a terminal unit to process information. The navigator is also excellent for thunderstorm monitoring.

If massive (48", 200+ turns) loop antennas are used as receiving coils for the set, ELF and ULF frequencies can be received for solar flare and seismic wave studies. A fixed, 0.22-µF low-voltage mica capacitor or a 365-pF tuner can be used to tune the

Table 6-8. Parts List for the Navigator

Part	Description
AT	A53C Audio Amplifier Transformer
C1	0.01-µF, 200-V Capacitor
C2, C5	0.1-µF, 400-V Capacitors
C3	220-pF Dipped Silver Mica Capacitor
C4	500-pF Dipped Silver Mica Capacitor
C6	0.047-µF, 400-V Capacitor
C7	0.033-µF, 400-V Capacitor
Choke	2.5-mH RF Choke
P-70A	Antenna Coil (See Text)
P-70RF	RF Coil (See Text)
R1	Resistor; 1.5 kΩ, ½ W
R2	Resistor; 2.2 MΩ, ½ W
R3	Resistor; 250 kΩ, ½ W
R4	Resistor; 1 MΩ, ½ W
VC1	Variable Capacitor; 270 pF
VC2	Variable Capacitor; 365 pF
VC3	Variable Capacitor; 365 to 500 pF
VR	Potentiometer; 50 kΩ to 250 kΩ, with SPST Switch, for On/Off/Volume
Tubes	Two 1N5, Two 1A5
Output Xfmr	10-kΩ to 25-kΩ Primary : 4- to 8-Ω Secondary
Power Supplies	A: 1.4 Vdc, B: 22.5 and 90 Vdc, C: 4.5 Vdc
Speaker	8 Ω at 2 W
Sockets	Four Octal Sockets for the Tubes
Grid Caps	Two Grid Caps for the 1N5 Tubes

antennas. The loops have to be rotated to null out 60-cycle power line hum, and, because of their size, be mounted outdoors.

When the builder winds the coils, other frequency ranges are possible. Remember, TRF sets usually have better sensitivity over the lower end of the frequency range the coils are designed for.

In the circuit, the plate choke cuts distortion and isolates RF from the first audio stage. The potentiometer in the screen lead of the RF tube is the volume control, and it can be from 50 kΩ to 250 kΩ in value. The power supply switch can be mounted on this control, for an on/off switch.

The RF stage amplifies the incoming signal, but, for ELF reception, an antenna amplifier can really help.

Also, see chapter 15 for an RF stage addition to this set, which provides dramatically improved performance, that was added just before the Northridge earthquake of January 17, 1994.

7
CHAPTER

Basic solid-state power supplies

The variable power supply described in chapter 2, for use in the seismic and flare receivers, is adequate for low-current transistor applications where the current draw is less than 50 mA. For higher current applications, and for the tube circuits, heavier-duty supplies are needed.

Bridge rectifier

Figure 7-1 shows the schematic for a bridge rectifier circuit, providing full-wave rectification even though the transformer is not center tapped. The output voltage is 1.4 times that of the transformer's secondary voltage. The silicon diode rectifiers used in the circuit are rated at 1 A at 1000 PIV, and the filter capacitor is a 2500-μF electrolytic with a voltage rating that is at least double that of the transformer secondary.

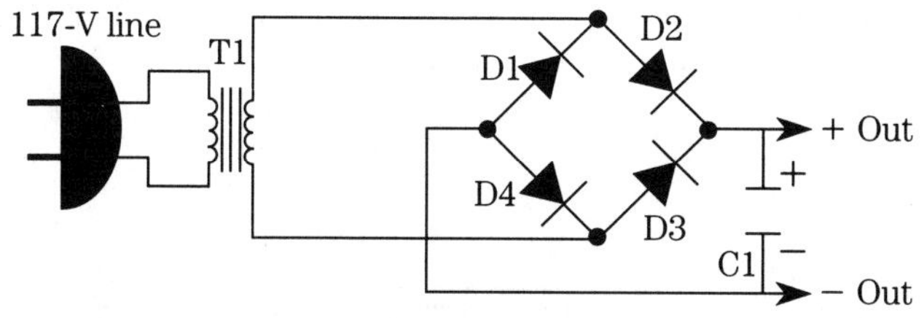

7-1
Bridge rectifier schematic.

Transistor regulator

Figure 7-2 shows a transistor regulator circuit. The circuit uses a 40-W pnp power transistor, and is capable of regulating up to 35 V at 500 mA maximum current.

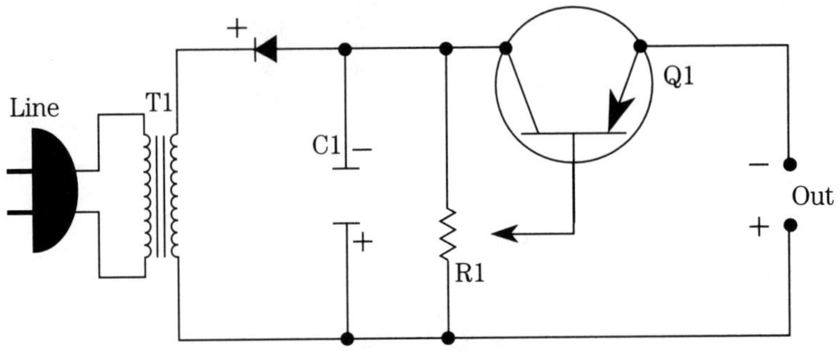

7-2 Transistor regulator schematic.

The power transistor, Q1, must be mounted on a heatsink to dissipate the heat generated in operation. The case of the transistor is also its collector connection, in a TO-3 case. Therefore, if a metal cabinet is used, the transistor must be insulated from the cabinet. The capacitor is rated for 250 µF and at least 50 V. Again, because the capacitor is an electrolytic, be sure to wire it according to the correct polarity.

An IC dc supply

The LM 317T is an adjustable regulator IC for dc outputs of from 1.2 to 37 Vdc. I use this circuit as the filament supply in all of the tube radios that require a dc voltage. The circuit provides a very stable, reliable power supply. Any input voltage from 2.5 Vac to 25 Vac at 1 to 2 A can be rectified and input to this IC, which is mounted on a heatsink. As a voltage regulator, it has a built-in overload protection and thermal shutdown. A schematic for its application is shown in Fig. 7-3, and Fig. 7-4 illustrates the pinouts of the LM 317T and the BR 805D, a typical bridge rectifier.

As used in the dual-detector, the LM 317T is driven by a Stancor P-2945 filament transformer that is rated for 6.3 Vac at 2 A. The voltage available after the circuit is constructed is variable from 1.2 to 5.0 V, at about 1 A.

R1 is a 1-kΩ ½-W potentiometer, the center terminal connected to the – terminal of the BR 805D, and to the – terminal of the electrolytic capacitor, C1. Leg 1 of the potentiometer is connected to the integrated circuit's output.

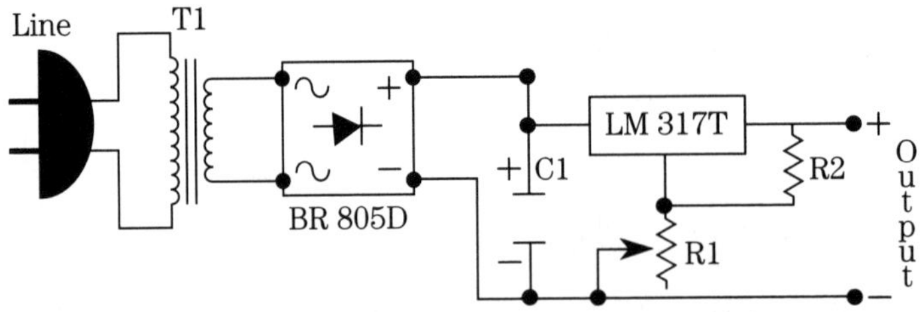

7-3 LM 317T regulator circuit.

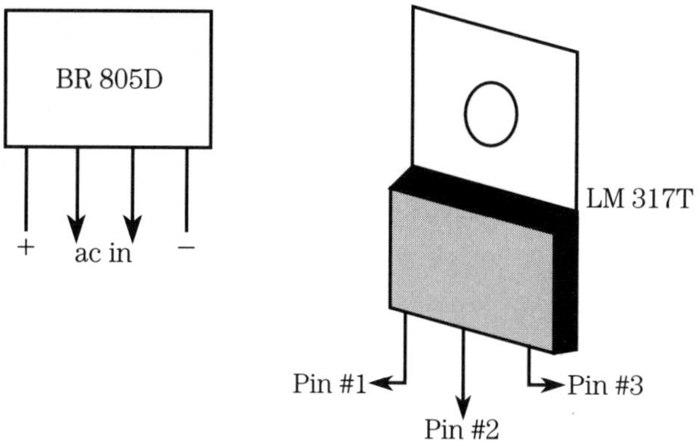

7-4 Pinouts for the BR 805D and the LM 317T.

C1 is a 4700-μF, 25-V electrolytic capacitor. The voltage rating of this capacitor would have to be raised to 50 WVdc if a transformer delivering more than 15 V to the circuit is used.

R2 is a 220-Ω resistor, with at least a 0.5-W rating. BR 805D is a 2-A bridge rectifier rated at 50 V.

Solid-state high-voltage supply

The circuit shown in the schematic of Fig. 7-5 can provide B+ dc voltage for any of the radios described in the text. It is simple to construct and dependable to operate. This supply can be built to provide a dc output at voltages of up to 500 V, with current ratings of up to 1 A, depending on the components used. The circuit shown is for a single-output supply.

The parts needed are: a B+ transformer, with voltage and current ratings as desired, an electrolytic capacitor (or capacitors) with a voltage rating at least 1.5 times the desired output, and a rectifier diode that is rated at least double the dc output voltage with at least 1 A of current.

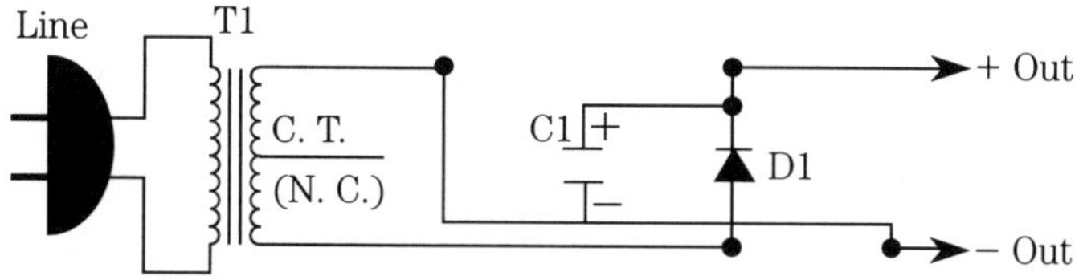

7-5 High-voltage supply.

8
CHAPTER

Antennas and coils

The antenna for VLF radio can be a very difficult article of construction in any system. Antennas and coils at very low wavelengths are closely related, and sometimes the antenna configuration can actually serve as one of the coils of the receiving circuit.

Figure 8-1 details the schematic of a two-coil loop antenna that can be used for VLF reception. Some of the radio circuits can use this kind of antenna as part of the coil assembly, and some of the seismic sensors can use this coil as a receptor.

Loop antennas are capable of deep null reception characteristics. They are highly directional as a result, and are usually low-gain devices.

Figure 8-2 is a schematic of an antenna amplifier that can be used with any antenna and radio. With the use of a wave-trap (rejector circuit) and a suitable antenna, the amplifier should be of help in receiving the low frequencies.

As you can see, switching is used on every critical component. Admittedly, this is overkill. However, every antenna and every receiver is different. Also, the quality of reception differs from one location to another.

The circuit, as described, gives the maximum versatility possible, and can be used with long wires, loops, dipoles, etc. Six-position switches are used in the prototype, and Table 8-1 gives the suggested values for the components in the switch circuits. The capacitors are dipped-silver mica, and resistors are ¼ W, 5% tolerance.

Figure 8-3 shows the circuit schematic for a wave-trap rejector circuit. In the case of broadcast band interference on LF or VLF frequencies, this circuit can be placed in series with the antenna lead, and it can work wonders. It consists of a two-coil circuit wound on a 1½- to 2½-inch-diameter coil form. This circuit is used in the dual-detector set.

Coil winding

Most of the coils for use in these radios are wound on forms between 1½ and 2 inches in diameter. Paper towel tubes, shipping tubes, and PVC or plastic tubes work fine.

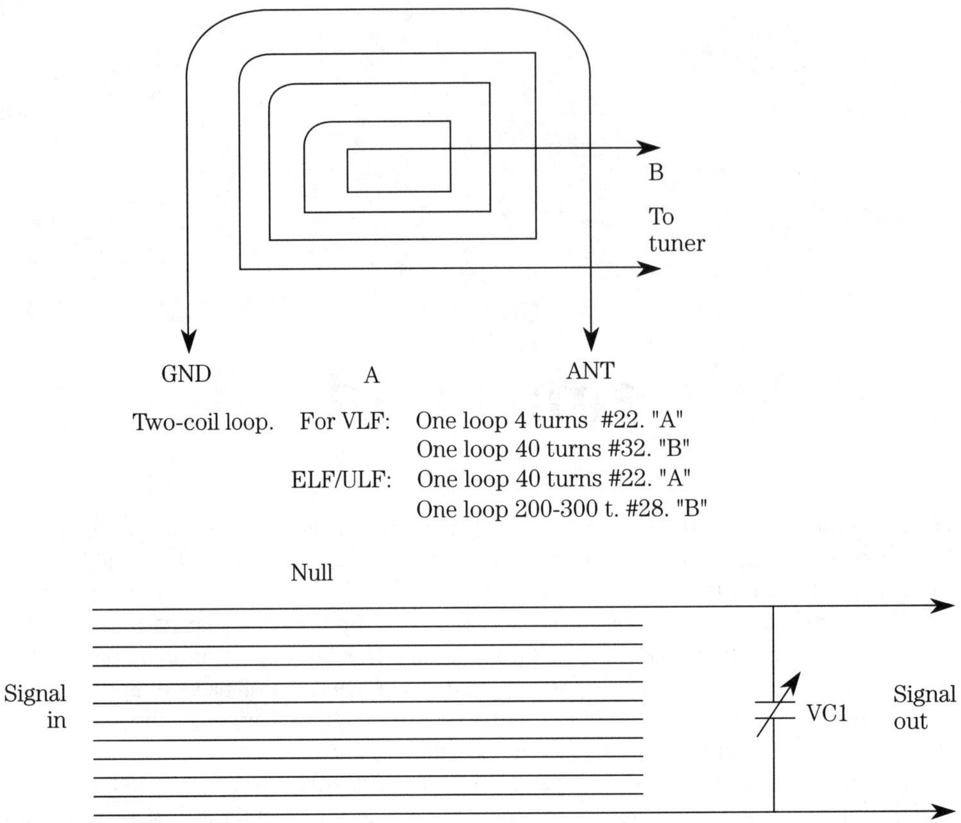

8-1 Loop antenna schematics.

When one coil is wound on top of another, as in the three-coil circuits, be sure to insulate one coil from another. Strong paper can be used, as well as a layer of electrical tape, to insulate between the coils.

Always wind the coils in the same direction, and make note of the starting and finishing ends. Windings are usually close-wound (side-by-side), and enamel-insulated wire is the easiest to use. Always use a good adhesive to cement down the windings.

Table 8-1. Antenna Amplifier Switch Parts

Pos.	L1 (μH)	C1 (μF)	R1 (Ω)	C2 (μF)	C4 (μF)	R3 (Ω)
1	Direct	Direct	1kΩ	Direct	Direct	47
2	0.22	0.0001	10kΩ	0.0001	0.0001	69
3	0.68	0.0005	50kΩ	0.0005	0.0005	100
4	10.0	0.001	100kΩ	0.001	0.001	270
5	100.0	0.005	150kΩ	0.005	0.005	470
6	1000.0	0.01	200kΩ	0.01	0.01	680

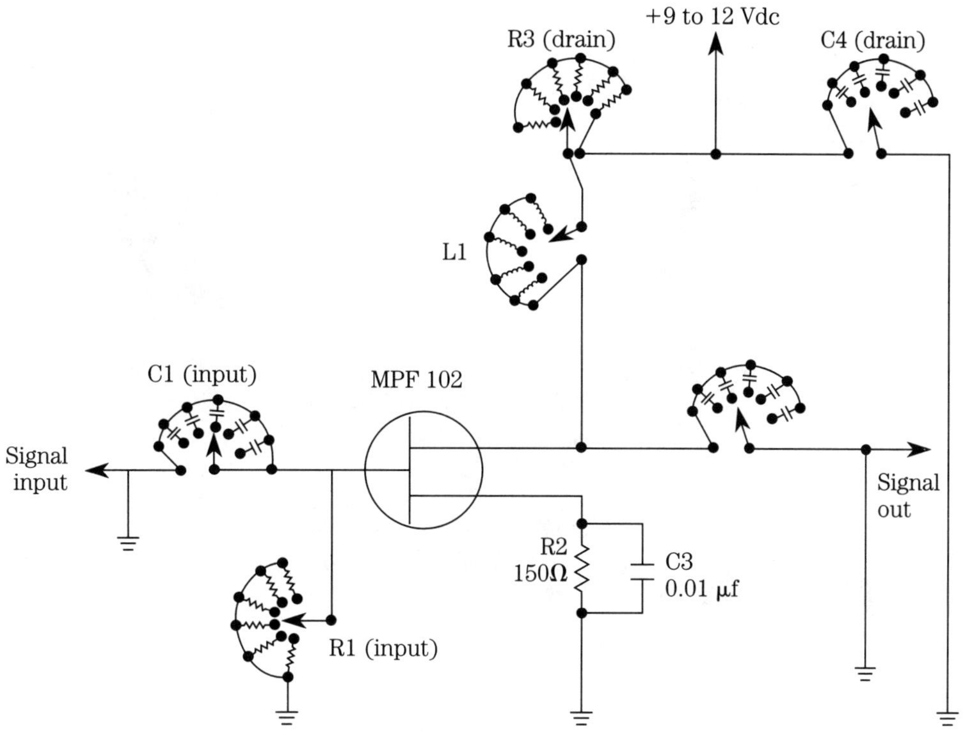

8-2 Antenna amplifier.

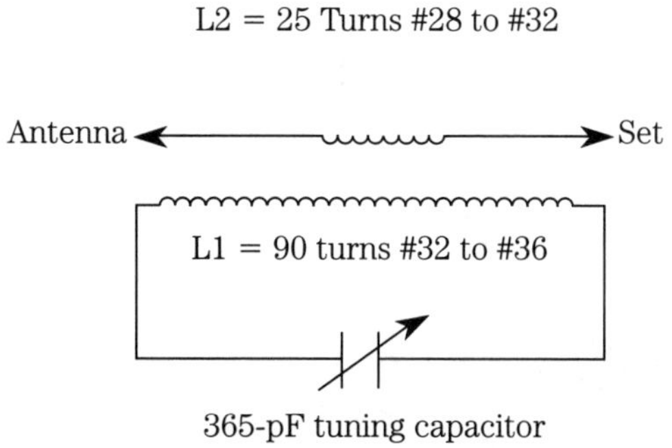

8-3
Wave-trap rejector circuit.

Tapped coils are easy to make. Simply count the number of windings until the tap number is reached, then make a loop in the wire, raise it slightly, and continue to wind until the desired number is reached. Strip the insulation from part of the loop and solder the lead to serve as the tap point, which then can be used in a switching arrangement.

Figure 8-4 shows an assortment of hand-wound coils. Refer to Table 8-2 for the turns ratio for various magnet wire gauges, for coils wound on a two-inch form. Fig-

8-4 Hand-wound coils.

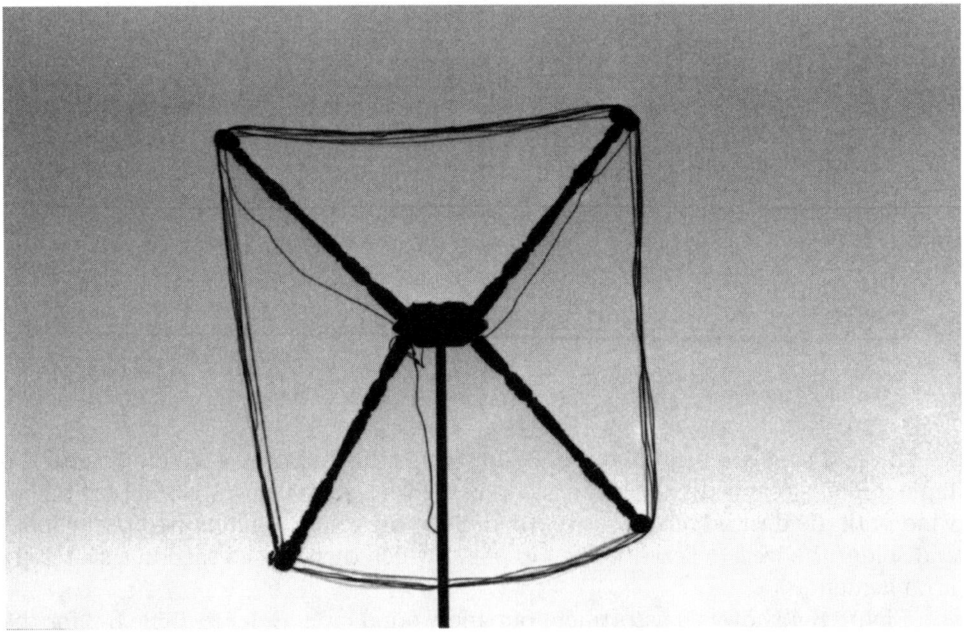

8-5 Loop antenna.

Table 8-2. Magnet Wire Table

Gauge	Diameter	Turns Per Inch
16	0.0508	18.9
17	0.0453	21.2
18	0.0403	23.6
19	0.0359	26.4
20	0.0320	29.4
21	0.0285	33.1
22	0.0253	37.0
23	0.0226	41.3
24	0.0201	46.3
25	0.0179	51.7
26	0.0159	58.0
27	0.0142	64.9
28	0.0126	72.7
29	0.0113	81.6
30	0.0100	90.5
31	0.0089	101.0
32	0.0080	113.0
33	0.0071	127.0
34	0.0063	143.0
35	0.0056	158.0
36	0.0050	175.0
37	0.0045	198.0
38	0.0040	224.0
39	0.0035	248.0
40	0.0031	282.0

ure 8-5 shows a homemade loop antenna of the two-coil variety that can also be used as a VLF inductor coil. The "spokes" are wooden table spindles 21 inches in length, and the overall diagonal measurement of the loop is 50 inches. #22 wire is used on the secondary and #24 wire is used on the primary.

Be very careful about the circuit to which a loop is connected. Some sets are "chassis hot." Often, especially in old radios made in the late 1930s and 1940s (as well as some homebrew radios and all regenerative sets), the sets have wiring with one side of the power supply connected to the chassis, and the chassis is connected to the ground connection. Do not connect a single-element loop between the antenna and ground terminals of such a set because a dangerous voltage potential exists. A two-coil loop could be used in such a set, with the loops connected as follows: the larger loop would be connected in series with the antenna terminal; the smaller loop would be connected in series with the ground. The external antenna and ground would then be connected to the other end of each loop, and tuned with variable capacitors.

Also, with the ground connection, *never* connect to a gas line or a hot-water pipe. Connect the ground to either a cold-water line or to a ground rod driven deep into the ground. The earth around the ground rod should always be kept moist. The most efficient ground scheme accomplishes a "0 Ω" potential to earth ground. An efficient ground system is very important when receiving low-frequency signals, so pay particular attention to this part of the hook-up.

2
PART
Advanced projects, earthquakes, and seismicity

9

CHAPTER

Advanced projects

The advanced projects discussed in this section involve some new concepts, along with adaptations to other, sometimes unrelated circuitry. In this chapter, transistors and integrated circuits are the foundation of the detector and display units, and specialized sensors are often used.

BFO signal processors

The BFO signal processors use transistor circuitry to provide an audible indication of seismic activity. In the circuit diagram of Fig. 9-1, a two-transistor circuit is shown. One transistor, along with its associated components, comprises an oscillator. Coil L1 is used as a tuned inductor. The other transistor, along with its associated components, is also an oscillator, but coil L2 is tuned by a panel-mounted broadcast-type tuning capacitor. When the coils in the sensor are judiciously positioned and L2 is correctly tuned, ground movement will correspondingly change the oscillation frequency and a tone, or a change in tone, will be present at the audio output. A regulated, single-stage power supply is used in the set, which is shown as a finished unit in Fig. 9-2. Table 9-1 is the parts list.

The coils for this set must be wound "solenoid-style," that is, the turns must be side-by-side, and not "jumble wound."

L1 can be wound on any size form from 1½ inches to 6 inches in diameter, and L2 should be wound on a 1½- to 2½-inch form. Magnet wire, #20 to #24, is recommended. The number of turns required on each coil is quite variable, anywhere from 50 to 100 turns on each coil works well. If a large form is used for L1, wrap at least 80 turns on the form.

In this configuration, the sensor requires a large, nonmetal, hermetically sealed container so that coils can be placed from 12 to 18 inches apart. Any kind of metal, including a magnet, can be used as the sensor's plumb. When small-diameter coils are used, the plumb should be adjusted by height so that it hangs within 1 inch of L1. When a large-diameter coil is used, the plumb is positioned to move within the cen-

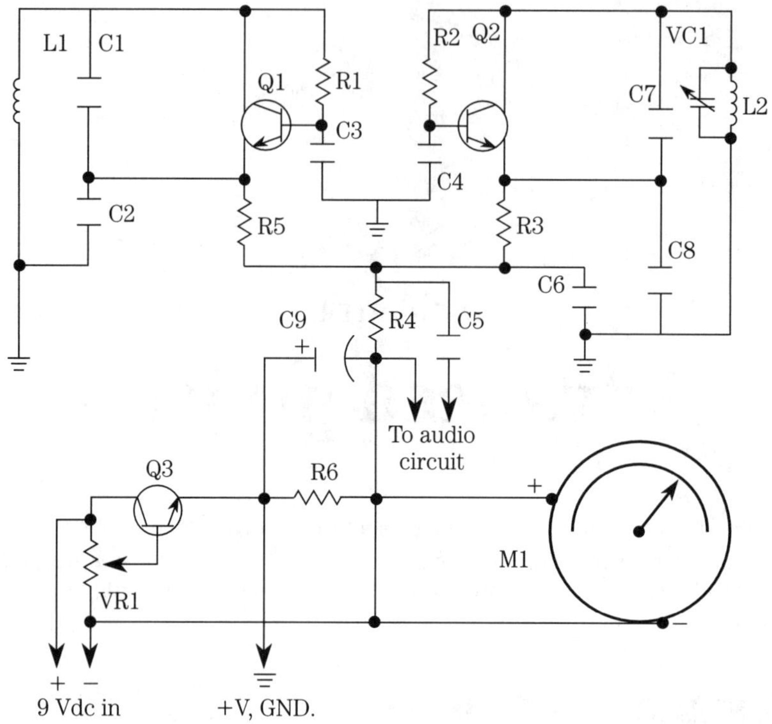

9-1 Two-transistor BFO signal processor.

ter of the form, in the field of the inductor. Only one plumb line is used, and it is placed over or within L1. L2 is tuned via the panel-mounted capacitor.

In tuning, the plumb-line relationship to L1 has to be adjusted (positioned) so that the circuit is in quiescent status. That is, the quiet portion of the plumb is over the area of the coil that balances the induction of L2.

9-2 Completed BFO processor.

Table 9-1.
Parts List for the BFO Seismic Signal Processor

Part	Description
C1, C2	0.0022-μF, 100-V Mylar Capacitors
C3–C6	0.01-μF, 50-V Ceramic Disc Capacitors
C7, C8	560-pF, 100-V Dipped Silver Mica Capacitors
C9	47-μF, 35-V Electrolytic Capacitor
M1	0- to 10-Vdc Meter
L1, L2	Coils
Q1, Q2	2N222A Transistors
Q3	2N3904 Transistor
R1, R2	200-kΩ, ¼-W Resistors
R3–R5	1.1-kΩ, ¼-W Resistors
R6	6.8-kΩ, ¼-W Resistor
VC1	365 pF Tuning Capacitor
VR1	5-kΩ or 10-kΩ ½-W Potentiometer
PS	Power Supply; 9 Vdc at least 100 mA
Misc.	Line Cord, SPST Power Switch, Enclosure, Perfboard, Knobs, Hi-Z Phones or Audio Amp Stage with Speaker, and Terminal Posts

For example; at the bottom of the container, glue and stabilize L2 at one end. Turn on the power to the unit and adjust the voltage to 7.5 V. Then tune the variable capacitor. Several areas of tuning will produce "beat tones." Tune for the loudest and clearest of these tones. Then adjust the sensor's plumb to the most sensitive area of L1, to a position where movement in either direction causes a change in the tone at the audio output. When a large-diameter coil is used for L1, the plumb adjustment is via its height inside the inductor's field. When the correct positioning is found, cement the adjustment, hook the sensor to the signal processor with shielded cables, and peak the signal with fine tuning of the capacitor and the power supply. The correct power setting can be anywhere from about 6.0 to 8.5 V. The unit is now ready for operation. Sensor diagrams are shown in Fig. 9-3.

Single-oscillator processor

The single-oscillator processor is a variation of the BFO circuit that uses just one transistor as an oscillator. It serves as a receptor of seismic impulses, as well as being able to transmit the signals a short distance to any longwave or AM radio.

Only one coil is used in the sensor for this unit, and the sensor can be placed some distance away from the electronics, and hooked via coaxial cable. The receiver for the audio is any receiver that is capable of receiving the harmonics of the oscillator.

The sensor uses any kind of metal for the plumb. The coil of the sensor serves as the inductor, and its value changes as the metal plumb moves within its field. The two 0.001-μF capacitors work with the inductor to determine the frequency output.

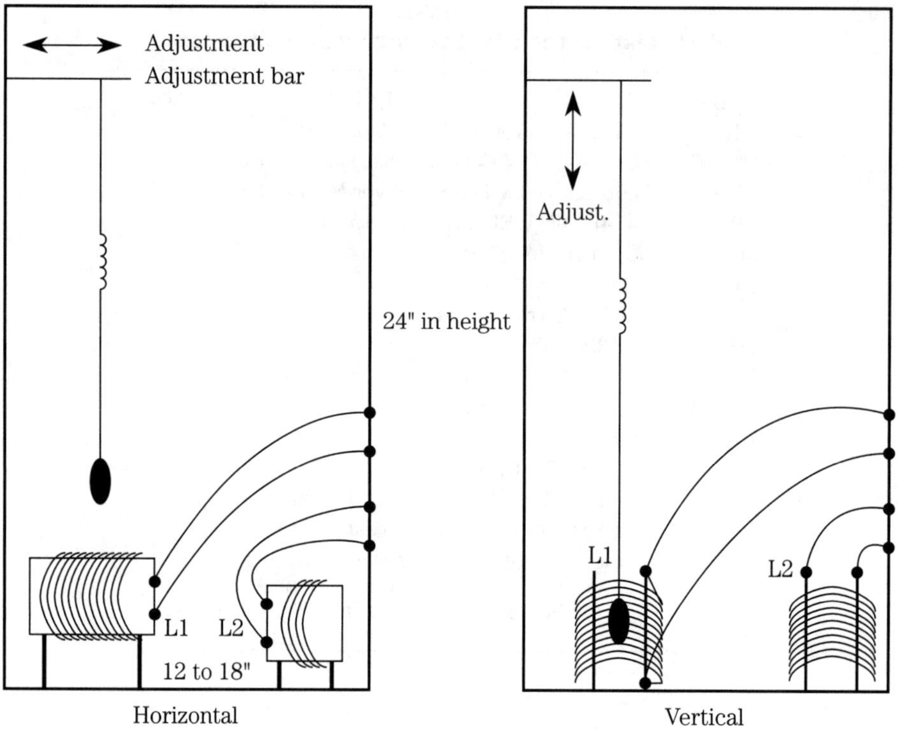

9-3 Specialized sensor examples for the BFO processors.

The coil for this set requires a relatively small amount of wire. 25 to 50 feet of #28 magnet wire wrapped around a 6-inch form works very well. The more wire used, the higher the number of turns, and the lower will be the corresponding output frequency of the oscillator. A high number of turns will put the output in the longwave region, and fewer turns will place it in the broadcast band where a regular AM radio can be used for monitoring.

The 47-kΩ resistor going to the base of the transistor provides bias voltage. The transistor used is a pnp type. This circuit, along with a transistor radio, makes a nice portable unit, good for field use. Table 9-2 is the parts list and Fig. 9-4 is the schematic.

Magnetic transducer detector

The magnetic transducer detector is a very simple detection unit that uses the LM 311 integrated circuit. The LM 111/LM 211/LM 311 ICs are voltage comparators that have input currents nearly a thousand times lower than other ICs, such as the LM 106 or LM 710. They are designed to operate from the standard + or – 15 V op-amp supply, down to the single 5-V supply used in logic circuits.

Both the inputs and the outputs can be isolated from the system ground, and the output can drive leads referred to the ground, the positive, or the negative supply.

Table 9-2. Parts List for the Single Transistor BFO Processor

Part	Description
C1, C2	0.001-µF, 50-V Capacitors
C3, C4	0.01-µF, 50 V-Capacitors
Q1	2N3906 PNP Transistor
PS	9-Vdc Power Supply
R1	100-Ω, ¼-W Resistor
R2	47-kΩ, ¼-W Resistor
Coil	25 to 50 Feet of #28 Magnet Wire
Form	Form for Inductor (Sensor)
Misc.	Hook-Up Wire, Coaxial Cable, Perfboard
Radio	Transistor, or Tube Type, for Receiving Signals from the Oscillator

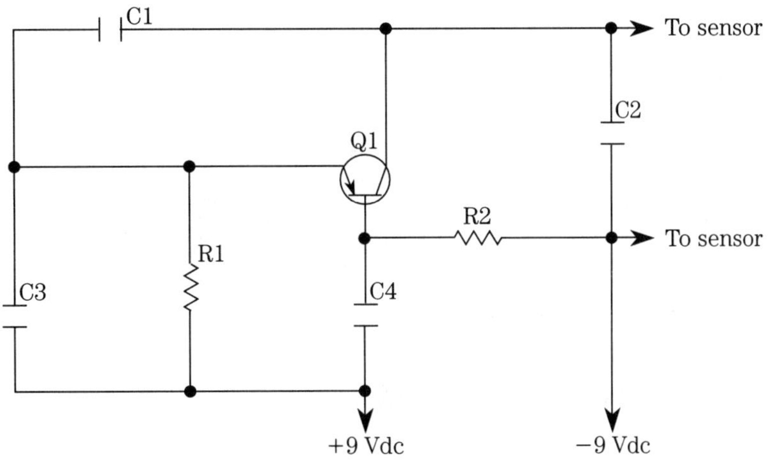

9-4 One-transistor BFO signal processor.

The LM 211 is identical to the LM 111, except that its performance is specified in a temperature range of –25°C to +85°C, instead of from –55°C to +125°C. The LM 311 operates in a temperature range of 0 to 70°C. Pinouts are shown in Fig. 9-5.

In building the magnetic transducer detector, just a few parts are used, and the construction is very straightforward. Any sensor that uses a magnet in its construction can be used with this detector. Sensors that use longer lengths of wire will generally provide the best results.

In operation, adjust the potentiometer to read about mid-scale on a 200-µA meter with the sensor in quiescent status. Test the detector by slightly jiggling the sensor. Movement should be observed via the meter. The sensor can then be returned to its observation status. The schematic for this sensitive detector is shown in Fig. 9-6,

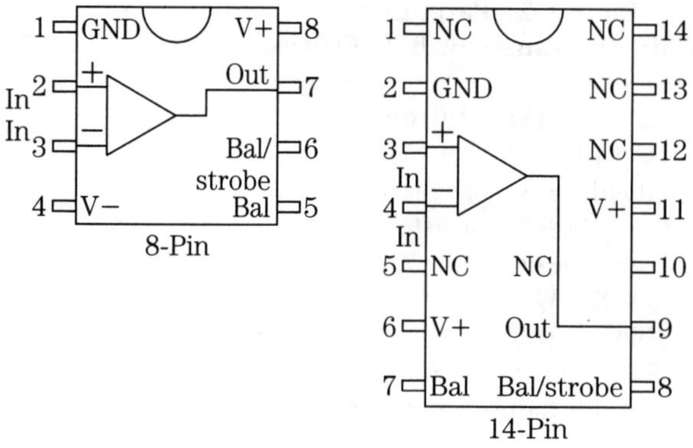

9-5 Pinouts of the LM 311 IC.

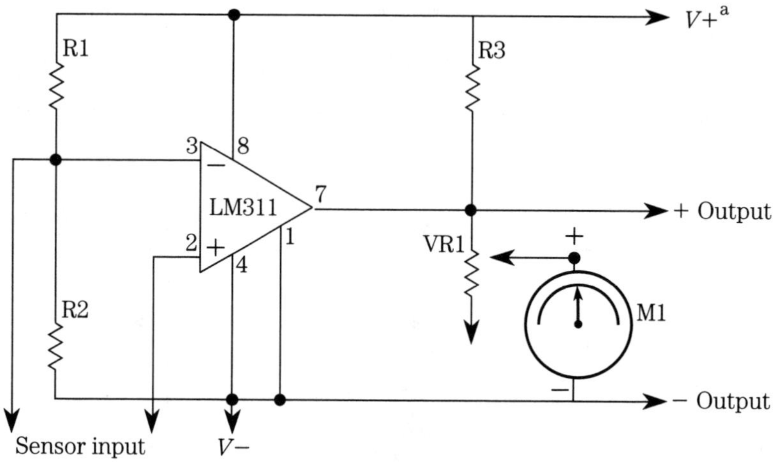

9-6 Schematic of the magnetic transducer detector.

the parts list is given in Table 9-3, and the inside and outside of a finished set are shown in Figs. 9-7 and 9-8, respectively.

Seismometer-receiver circuits based on the LM 324 IC

The LM 324 is a quad op-amp integrated circuit. It is very well suited to the building of high-gain, low-cost circuits for measuring magnetic field disturbances, receiving RF (radio) signals, and for the amplification and display of seismic signals. Pin connections for the 14-pin IC are shown in Fig. 9-9.

Table 9-3. Parts List for the Magnetic Transducer Detector

Part	Description
IC1	LM 311 Integrated Circuit
M1	0- to 200-μA dc Meter
PS	Power Supply (See Schematic)
R1	3.9-kΩ, ¼-W Resistor
R2	1.0-kΩ, ¼-W Resistor
R3	2.0-kΩ, ¼-W Resistor
VR1	500-kΩ, ½-W Potentiometer
Misc.	Socket for the IC, SPST Power Switch, Line Cord, Enclosure, Four Terminal Posts (Two for Input, Two for Output), Knob, Hook-Up Wire, and Perfboard

9-7 Interior of the magnetic transducer detector.

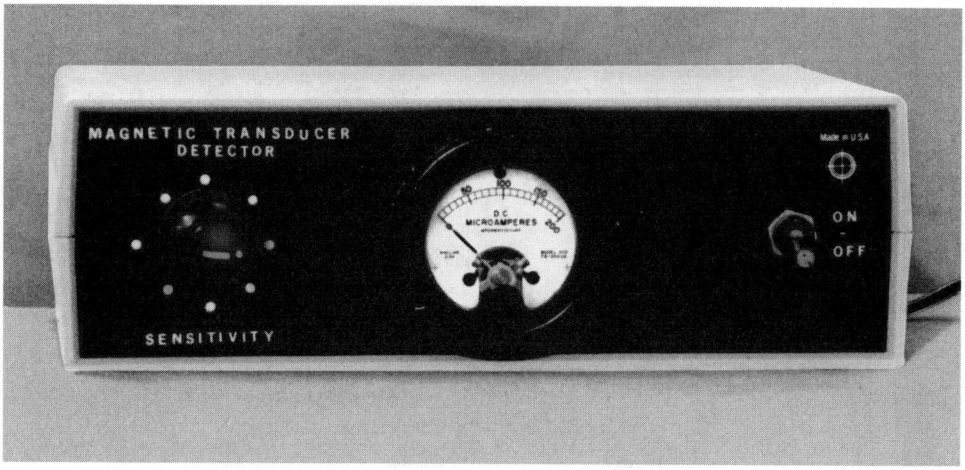

9-8 Exterior of the magnetic transducer detector.

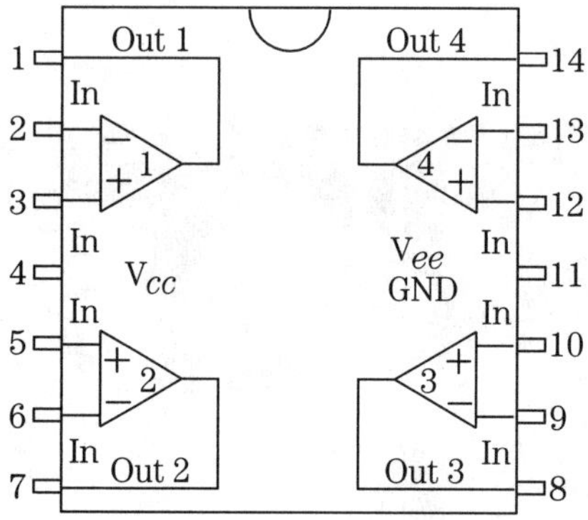

9-9 Pin connections for the LM 324 IC.

The magnetic field disturbance detector

The first circuit described based on the LM 324 is the magnetic field disturbance detector. The schematic diagram and parts list are in Fig. 9-10 and Table 9-4, respectively, and a finished unit is shown in Fig. 9-11.

Any disturbance within the inductive field of the sensor (small when a ferrous object and much larger when a magnet is used in sensor construction) is detected and boosted by three stages of amplification from the IC.

Disturbances are fed into stage 1 of the op amp. Here the signal is boosted, and its output connected to the potentiometer, functioning as the circuit's gain control. Stage 2 of the op amp further amplifies the signal, which is then fed to stage three for the final stage of amplification.

The 0.083-µF capacitors act to neutralize oscillation that would otherwise occur because of the high gain of the circuit. Therefore, in spite of its high gain factor, the circuit is very stable in operation.

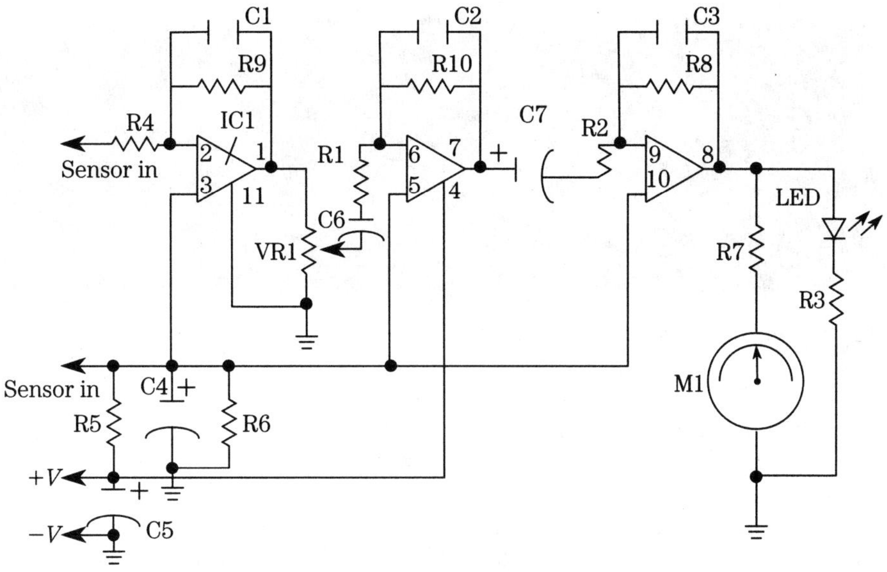

9-10 Schematic diagram of the magnetic field disturbance detector.

Table 9-4. Parts List for the Magnetic Field Disturbance Detector

Part	Description
C1, C2, C3	0.083-μF, 100-V Mylar Capacitors
C4, C5	220-μF, 25-V Electrolytic Capacitors
C6, C7	10-μF, 25-V Electrolytic Capacitors
IC1	LM 324 Integrated Circuit
LED1	General-Purpose LED
M1	0- to 1-mA FS dc Meter
PS1	6-Vdc, 200-mA Power Supply
R1, R2, R3	1-kΩ, ¼-W Resistors
R4	2.2-kΩ, ¼-W Resistor
R5, R6	4.7-kΩ, ¼-W Resistors
R7	10-kΩ, ¼-W Resistor
R8	100-kΩ, ¼-W Resistor
R9, R10	470-kΩ, ¼-W Resistors
VR1	100-kΩ, ½-W Potentiometer, with SPST Switch
Misc.	Line Cord, Coaxial Cable, Perfboard, IC Socket, Knob, Enclosure

When movement is sensed, it is detected as a low-frequency pulse. The meter should read about one-half of full-scale during quiescent periods, but the meter readings can rise and fall over the full scale during active periods. The LED will

9-11 Magnetic field disturbance detector.

flicker and/or change intensity in response to the motion. The resistor value between the LED and the ground connection might have to be changed, depending on the type of LED and the input power supply used.

Any electromagnetic sensor variety can be used; generic or commercial. The sensor should be buried several feet deep and connected to the apparatus with coaxial cable. The circuit is very sensitive, and any ferrous object can be used as the plumb weight. However, using a magnet as the weight will result in most sensitive detection.

An LM 324-based ULF/VLF radio receiver

A completely solid-state, coilless receiver for the ULF/VLF region can be built as shown in the schematic diagram of Fig. 9-12. The parts list is given in Table 9-5.

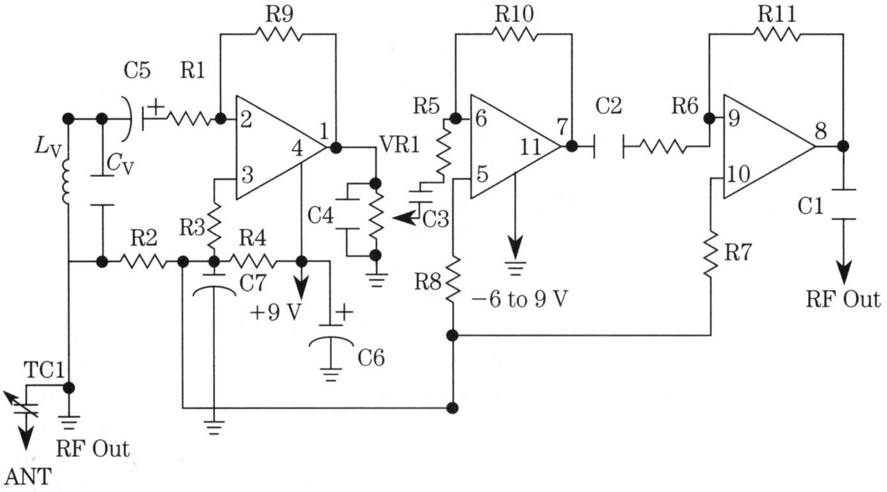

9-12 LM 324-based radio receiver.

Table 9-5. Parts List for the LM324-Based Receiver

Part	Description
C1, C2, C3	0.1-µF, 100-V Mylar Capacitors
C4	0.0047-µF, 50-V Mylar Capacitor
C5	10-µF, 25-V Electrolytic Capacitor
C6, C7	100-µF, 25-V Electrolytic Capacitors
C_v	Frequency-Determining Capacitor
TC1	12- to 100-pF Trimmer Capacitor
R1–R8	4.7-kΩ, ¼-W Resistors
R9	47-kΩ, ¼-W Resistor
R10	220-kΩ, ¼-W Resistor
R11	100-kΩ, ¼-W Resistor
VR1	1-kΩ to 5-kΩ, ½-W Pot, with SPST Power Switch
Misc.	IC Socket, Phones (or Audio Amp), Line Cord, Enclosure, 6- to 9-V Power Switch, Knob, etc.

Various inductors or coils, L_v, serve as the tuning inductors. C_v tunes the inductor for the frequency selection. All active circuits in the receiver are served by three stages of a single LM 324 IC.

The output of the tuned circuit is coupled to the input of section 1 of the IC. Output from this stage is connected to a 1-kΩ to 5-kΩ potentiometer, which is the gain control. Section 2 of the IC further amplifies the signal, and the output is coupled to the third section of the IC. After further amplification here, audio is provided at pin #8 and signals are ready to be analyzed.

The schematic shows the circuit as used for headphone reception. The output impedance of the phones should be about 32 Ω, but from 8 to 40 Ω can be used. An audio amplifier can be connected to the output if it is isolated from the basic circuit. A meter can be used to monitor the output as well, if it is isolated from the rectifier circuit. For this, the fourth section of the IC can be used. Input is to pin 12, and the output is from pin 14, and a jumper wire needs to be installed from pin 13 to pin 14. This requires diode isolation between pin 8 and ground, and in the signal input to pin 12.

Miniature epoxy-coated chokes can be used with this receiver, and effective reception with various values can be obtained down to below 25 kHz. With wire coils, ULF (down to below 1 kHz) can be received.

The use of a good earth ground or a large longwire antenna is recommended. Hook-up is made to the ground terminal of the receiver. A 12-pF to 100-pF trimmer capacitor should be installed in series with the antenna connection.

Because L_v and C_v are the frequency-determining components, a single-channel receiver is shown in the diagram. A stepping switch; single-pole, eight-position, can be installed at the connection points of these components for multi-channel reception. A variable capacitor will effect continuous tuning over a band when it is used. Leaving one of the inductor switch points with a spare position can allow the user great flexibility because a seismic sensor can be input via outside terminal posts; in this way, the VLF band can be heard with the sensor used as the receiving inductor.

Table 9-6. Sample C_v and L_v Values for the LM 324-Based Receiver

Inductors (μH)	Switch Position[a]	Capacitors
4.7	1	15 to 500 pF C_v[b]
8.2	2	2000 pF
100	3	3300 pF
180	4	4700 pF
1000	5	0.1 μF
2000	6	0.47 μF
3000	7	1.0 μF
External	8	2.2 μF

a Switch position uses an 8-position, single-pole rotary switch.

b Suggested value when using a tuning capacitor as one of the C_v capacitor selections.

Sample L_v and C_v component values are given in Table 9-6. An inside view of a completed receiver is shown in Fig. 9-13, with the outside view in Fig. 9-14.

With appropriate use of capacitors and coils, this set can operate from the ULF frequencies to short waves. It is ideally suited for experimentation with seismic wave and solar disturbance studies.

Interfacing the receiver to a seismometer

The basic LF 356 seismometer circuit can be interfaced into the LM 324 receiver quite easily and effectively. Figure 9-15 shows just such a circuit adaptation. The seismometer should be constructed with its signal meter being of a 0- to 2- or 0- to

9-13 Interior view of the LM 324 receiver.

9-14 Exterior view of the LM 324 receiver.

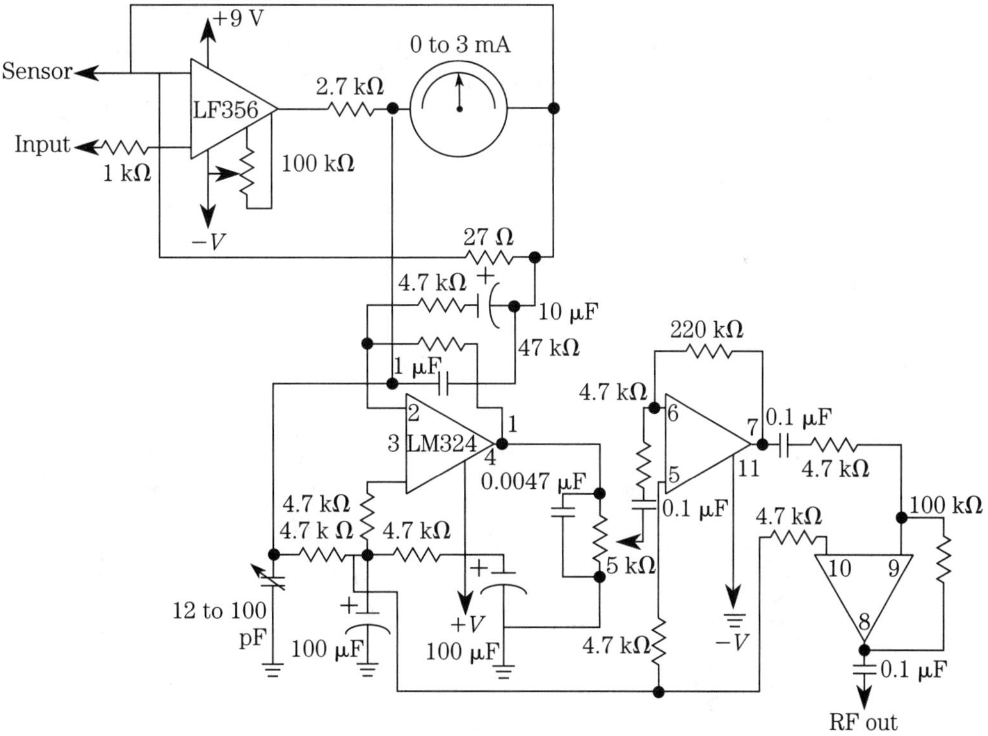

9-15 Seismometer interfaced to the receiver.

3-mA full-scale type, and set to about 1.25 mA in operation. The receiver's tuned inductor is not used, output B of the seismometer is not necessary, and the attenuation circuit is either omitted, or replaced with a single low-value resistor. Any sensor can be used in this adaptation, and the resulting audio indications of seismic impulses are quite clear and distinct.

Induction/oscillation disturbance detector

The induction/oscillation disturbance detector uses a circuit similar to the regenerative radio receivers of the early 1920s. In this detection circuit, the input of an amplifier is coupled to its own output, which allows oscillation. More accurately, this is a feedback oscillator circuit. Figure 9-16 is the schematic for the set and Fig. 9-17 shows a finished unit.

Sections 1 and 2 of the LM 324 IC are connected in a two-stage amplifier circuit. Section three rectifies the signal and allows for the meter readings. The circuit uses two coils, each of which is adjustable in balance. Feedback is controlled by the value of the neutralizing capacitor, also adjustable, which is installed between pins 8 and 9 of the IC.

The oscillation frequency is in the VLF frequency range, and its exact properties are determined by the composition and tuning of the coils. When movement occurs within the field of L1, the inductive field is disturbed, and oscillation of the circuit occurs.

This circuit allows for experimentation because a wide variety of coil sizes and shapes can be used in the sensors, as well as the use of either metal or magnetic plumb weights. The use of capacitor stepping in the tuning of L1, L2, and in the value of the neutralizing capacitor allows for the balancing of various inputs, and for the fine-tuning of the sensitivity of the set. Suggested values for the components of these stepping circuits are given in Table 9-7.

As for the specialized sensor for this circuit, coil movement is one method used effectively. Figure 9-18 shows just such a sensor. L1, containing the plumb weight

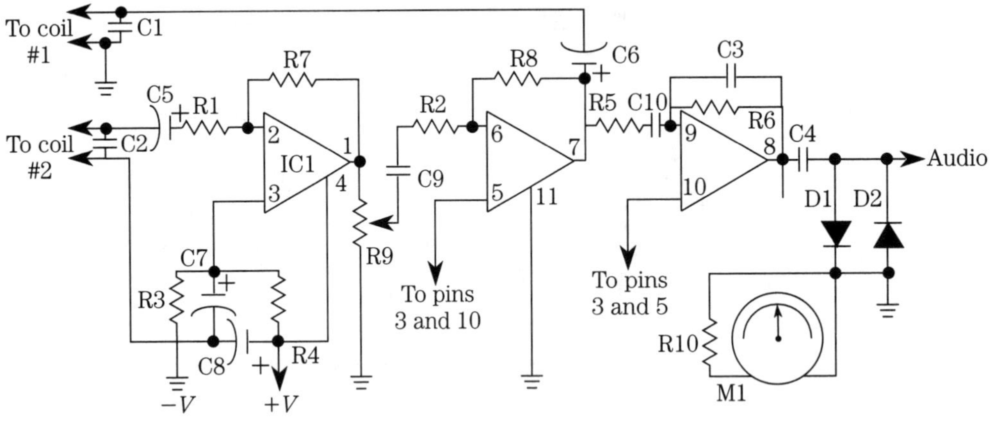

9-16 Schematic of the induction/oscillation disturbance detector.

9-17 Completed induction/oscillation disturbance detector.

Table 9-7. Parts List for the Capacitor Stepping Values of the Induction/Oscillation Disturbance Detector

L1 Capacitor (C1)	L2 Capacitor (C2)	Feedback (C3)[a]
0.1 μF	0.1 μF	470 pF
0.2 μF	0.2 μF	560 pF
0.3 μF	0.3 μF	680 pF
0.47 μF	0.47 μF	820 pF
0.56 μF	0.47 μF	910 pF
0.68 μF	0.56 μF	1000 pF

a Capacitors here should be 100-V Dipped-Silver Mica.

within it, is suspended within the field of L2 and its movement is reflected on the meter, which responds to the fluctuations of oscillation produced by movement.

In the height adjustment, adjust for the slight dip in meter reading by *slowly* adjusting the height in relation to the large, stationary coil. This will be the most sensitive setting for the quiescent status of the sensor, and even very slight movement of the coil will show up as fluctuations on the meter. Figure 9-19 shows the set with all of the fine-tuning controls in the capacitor sections of the set. The parts list for the basic unit is given in Table 9-8.

Inductive energy field monitor

The inductive energy field monitor, also based on the LM 324 IC, uses a sensor with a winding that comprises an inductive bridge circuit. Two versions are presented. The first is very tricky in both construction and in operation. It is the only circuit in this book that has not undergone testing under actual field conditions. Therefore, it is strictly an experimental, speculative project.

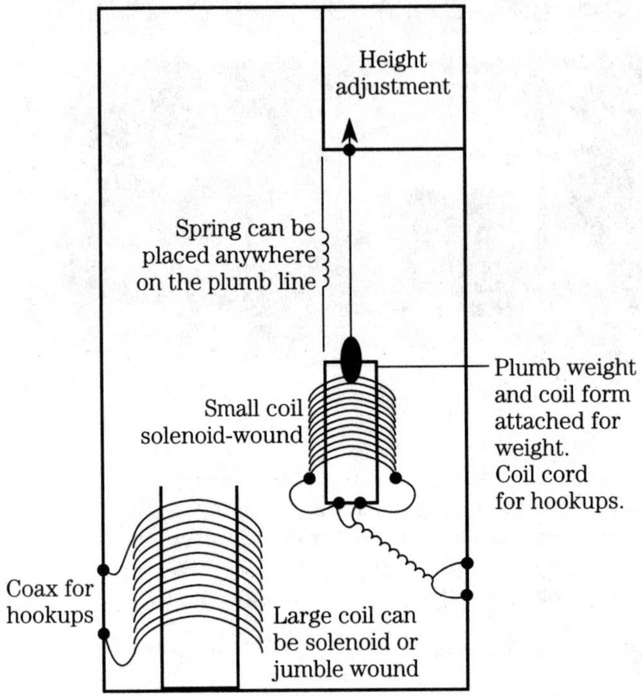

9-18 Example of a specialized sensor for the induction/oscillation disturbance detector.

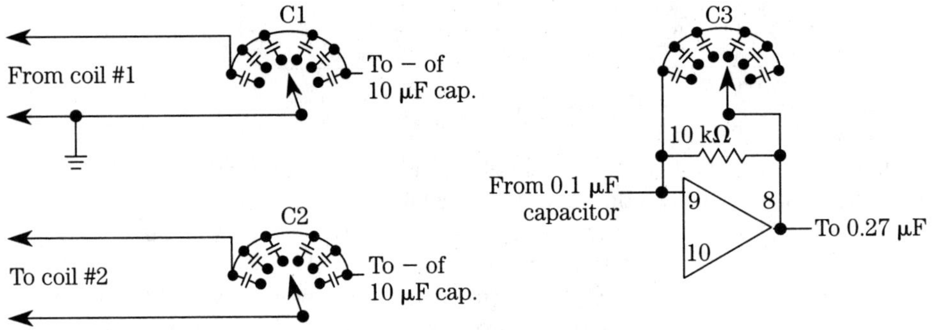

9-19 Induction/oscillation disturbance detector: fine-tuning controls.

In this version, the bridge circuit is electronically balanced to a null condition. The design is such that any movement upsets the balance of the circuit, with the disruption then amplified to drive meter and audio circuits.

Figure 9-20 shows the schematic for the circuit. The unit is essentially three modular units, and it uses two separate 9-Vdc power supplies. One supply powers the receptor circuit and the other supply powers the signal amplifier and audio circuits.

Table 9-8. Induction/Oscillation Disturbance Detector Parts List

Part	Description
C1, C2	0.47-µF, 100-V Mylar Capacitors
C3	680-pF, 100-V Mylar Capacitor
C4	0.27-µF, 100-V Mylar Capacitor
C5, C6	10-µF, 25-V Electrolytic Capacitors
C7, C8	220-µF, 25-V Electrolytic Capacitors
C9, C10	0.1-µF, 100-V Mylar Capacitors
D1, D2	1N4001 Diodes
IC1	LM 324 IC
M1	0- to 1-mA dc Meter
PS	Power Supply; 6 to 9 Vdc, 50 to 500 mA
R1, R2	1-kΩ, ¼-W Resistors
R3, R4	4.7-kΩ, ¼-W Resistors
R5, R6	10-kΩ, ¼-W Resistors
R7, R8	120-kΩ, ¼-W Resistors
R9	5-kΩ, ½-W Potentiometer
R10	2.7-kΩ, ¼-W Resistor—Installed in Series with the "+" Meter Lead
S1	SPST Power Switch
Misc.	IC Socket, Line Cord, Enclosure, Perfboard, Terminal Posts, Coils, Knob, etc.

The receptor is an oscillator. The specially wound coils are connected as the inductors of the unit. Both coils must provide the same level of signal so that the circuit can be balanced for optimum gain.

Both coils are wound in the same direction to achieve this balance. They are also wound with the same wire size, around the same form, and contain the same number of turns. The size of the coils, wound on nonconductive forms, can be anywhere from 4½ to 12 inches in diameter. Coils are wound "solenoid-style," not "jumble-wound." #24 to #28 magnet wire is recommended, and the first and last windings of the coils must be terminated in binding posts. The "finish" lead of the first coil is connected to the "start" lead of the second coil, providing a three-terminal hook-up. Figure 9-21 shows a sensor example usable with this version of the inductive energy field monitor.

Tapped coils can be constructed, if you desire, for observing different response periods. If tapped coils are used, they should be switch-selectable, using an SPDT switch, as the circuit balances the inductive fields electronically. Use about 150 to 175 feet of wire on each coil, making sure both coils are wound in exactly the same fashion. Make sure taps are on the same turn number, as well.

When constructing an untapped coil, 100 feet of wire around a 9-inch-diameter form is sufficient. On the larger, tapped coil, make the tap on turn #50, then continue to the end of the length of wire.

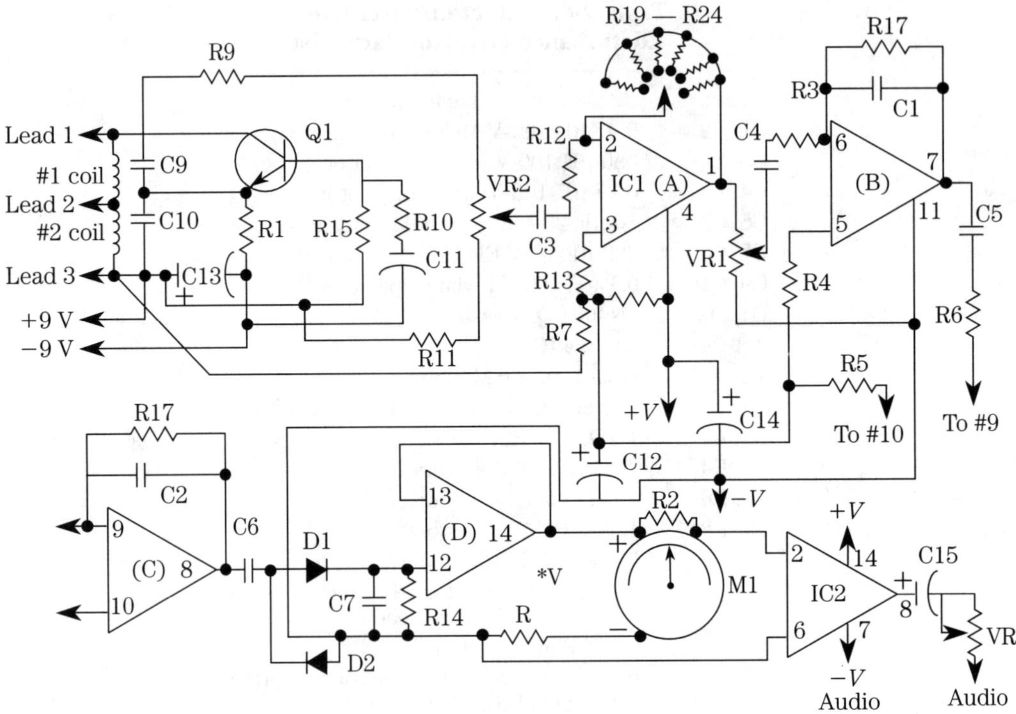

9-20 Inductive energy field monitor schematic.

VR1 is the balancing control and VR2 is the gain control. With the gain of the amplifier circuit set by a six-position resistor stepping net connected to section A of the IC, VR3 is used as the volume control.

The resistor net is set to select the 100-kΩ resistance setting. VR2 is tuned to about one-half scale, and VR1 is tuned for the deepest null, where any adjustment of VR2 in either direction causes a sharp rise in the meter reading. The resistor selection governs the overall gain of the circuit; the higher values represent higher gain factors. Tune VR2 as high as possible while adjusting VR1 for null conditions, then the final overall setting for the circuit's gain can be set.

It will take some time and patience to adjust this circuit, as each control has to be peaked and readjusted for best performance. At least in theory, very high gain and sensitivity can be achieved with this circuit.

The inductive energy field monitor's other version is a much easier to use model that is closely related to the induction/oscillation disturbance detector. Here, the gain of the circuit is determined by the six-position resistor selection switch that is contained within section 1 of the IC.

A single 9-Vdc power supply powers the unit, and its feedback capacity is governed by an SPDT two-position switch in section 3 of the IC, between pins 8 and 9, and no audio circuit is used. Coils for the sensor used in this version are similar in construction to those used in the induction/oscillation disturbance detector.

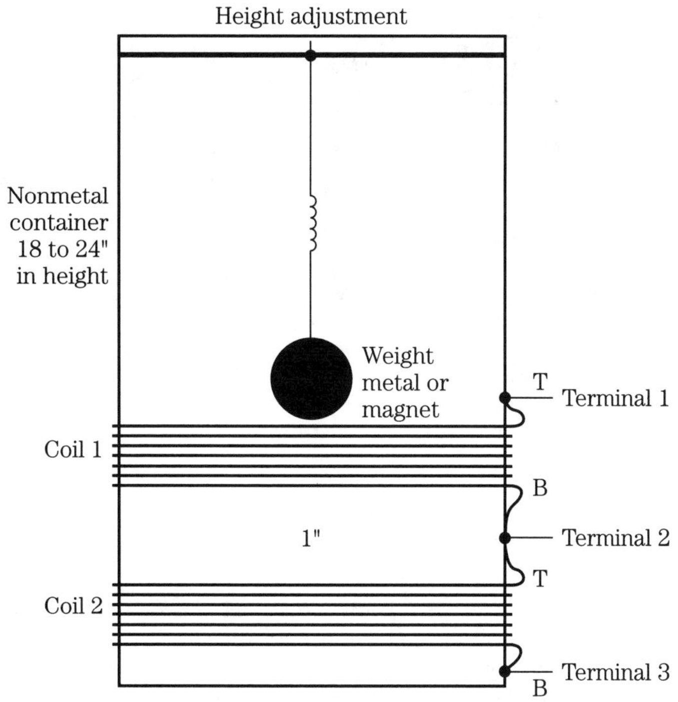

9-21 Sensor example with a five-gallon container serving as both the sensor frame and the coil form.

This particular version offers many advantages over the first circuit presented. It also offers high, selectable gain, but it is much easier to construct and use. It is more economical because fewer parts are used, and less power is required. This version of the inductive energy field monitor has been field tested with excellent results. Figure 9-22 shows the schematic for this version of the set, and Fig. 9-23 is a sensor diagram. Tables 9-9 and 9-10 are the parts lists for the first and second versions, respectively.

Signal processors based on the TL 080 series ICs

The TL 081, TL 082, and TL 084 integrated circuits are JFET input operational amplifiers that work very well as low-cost devices in various detectors.

This family of ICs is available in single, dual, and quad op-amp packages, which are pin compatible with industry standards, such as the LM 324.

Figure 9-24 shows the pinouts for the various amplifiers. Some sample circuits for use as seismic signal processors follow. Two versions of a basic circuit using the TL 080 series integrated circuits are described in circuits A and B.

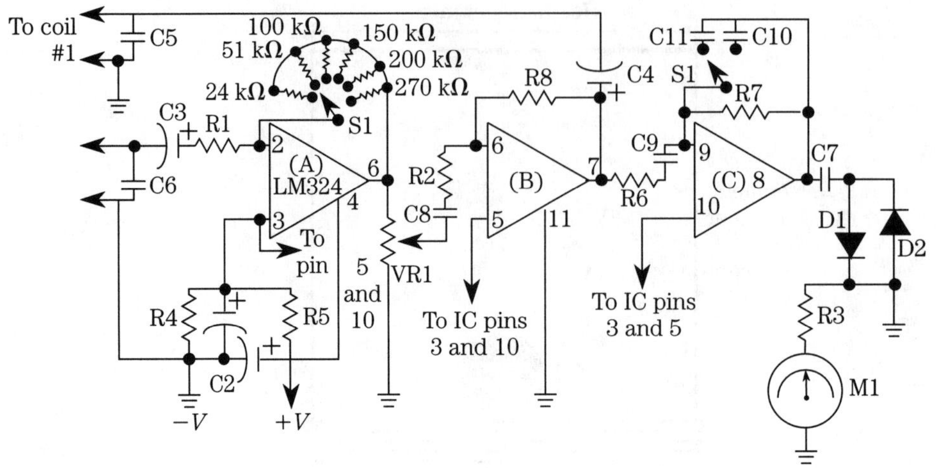

9-22 Inductive energy field monitor: simplified version.

Figures 9-25 and 9-26 show the circuits, with Fig. 9-25 showing circuits A and B using the TL 081, and Fig. 9-26 diagramming their use with the TL 082.

The TL 081 is the only member of this IC group to have the null/offset circuit. Circuit A uses this feature, and is most sensitive to smaller, more rapid vibrations, while B

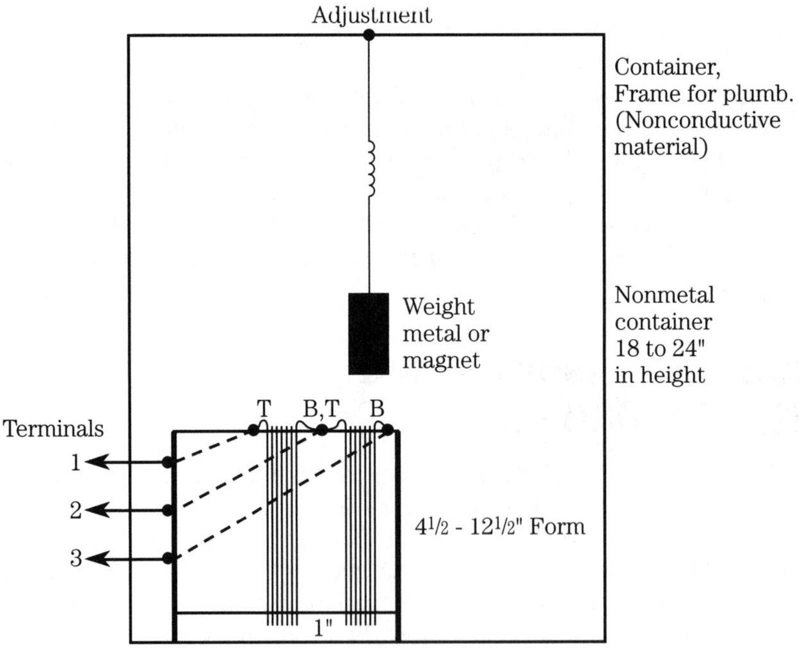

9-23 Specialized sensor for the simplified version.

Table 9-9. Parts List for the Inductive Energy Field Monitor—Experimental Version

Part	Description
C1, C2	33-pF, 50-V Mylar Capacitors
C3–C7	0.1-µF, 100-V Mylar Capacitors
C8, C9	1-µF, 100-V Mylar Capacitors
C10	10-µF, 25-V Electrolytic Capacitor
C11	47-µF, 25-V Electrolytic Capacitor
C12, C13, C14	1000-µF, 25-V Electrolytic Capacitors
D1, D2	1N4001 Diodes
IC1	LM 324 Integrated Circuit
IC2	LM 380N Integrated Circuit
M1	Meter; 0 to 50 or 0 to 100 µA dc, and Appropriate Resistor; ½ W, 91 kΩ for Use with a 0- to 50-µA Meter, or 47 kΩ for Use with a 100-µA Meter
PS1, PS2	9-Vdc Power Supplies
Q1	2N2222A Transistor
R1, R2	1-kΩ, ¼-W Resistors
R3–R6	2.2-kΩ, ¼-W Resistors
R7, R8	4.7-kΩ, ¼-W Resistors
R9–R14	10-kΩ, ¼-W Resistors
R15	100-kΩ, ¼-W Resistor
R16, R17	220-kΩ, ¼-W Resistors
S1	SPST On/Off Switch
S2	Single-Pole, Six-Position Rotary Switch, with One Each of the Following ¼-W Resistors: R18–R23; 24 kΩ, 51 kΩ, 100 kΩ, 150 kΩ, 200 kΩ, 270 kΩ
VR1, VR3	1-kΩ ½-W Potentiometers
VR2	5-kΩ ½-W Potentiometer
Speaker	8 Ω, at least 2 W
Misc.	IC and Transistor Sockets, Line Cord, Enclosure, Knobs, Binding Posts, etc.

is sensitive to longer period vibrations, and is best utilized when used with sensors using pendulum-arc movements. Both circuits can be used with the same IC, by installing an SPST switch between pin 4 of the IC and the center terminal of the potentiometer.

The TL 082 can handle two circuits of the B version, to display readings from two sensors, and the TL 084 can handle up to four circuits, along with some of the configurations used with the LM 324-based circuits: the voltage doubler, audio amplifier, meter display, and receiving circuits.

These ICs can also use a single-stage power supply in the power input circuit, so that many different dc sources can be easily input. Batteries, ac/dc adapters, and solar cells can be used for power regulation for the seismometers. Table 9-11 is a general parts list for these circuits.

Table 9-10. Parts List for the Inductive Energy Field Monitor—Version Two

Part	Description
PS	9-Vdc Power Supply
C1, C2	1000-μF, 25-V Electrolytic Capacitors
C3, C4	10-μF, 25-V Electrolytic Capacitors
C5, C6	0.47-μF, 100-V Mylar Capacitors
C7	0.27-μF, 100-V Mylar Capacitors
C8, C9	0.1-μF, 100-V Mylar Capacitors
C10	1000 pF, 100-V Mylar Capacitor
C11	680 pF, 100-V Mylar Capacitor
D1, D2	1N4001 Diodes
IC1	LM 324 Integrated Circuit
M1	Meter; 0 to 1 mA dc
R1, R2	1-kΩ, ¼-W Resistors
R3	2.7-kΩ, ¼-W Resistor
R4, R5	4.7-kΩ, ¼-W Resistors
R6, R7	10-kΩ, ¼-W Resistors
R8	120-kΩ, ¼-W Resistor
R9–R14	One ¼-W Resistor of Each of the Following Values; 24-kΩ, 51-kΩ, 100-kΩ, 150-kΩ, 200-kΩ, 270-kΩ
S1	SPDT Switch
S2	SPST On/Off Switch
S3	Single-Pole, Six-Position Rotary Switch
VR1	5-kΩ, ½-W Potentiometer
Misc.	IC Socket, Terminal Posts, Coils, Line Cord, Enclosure, Perfboard, Knobs, etc.

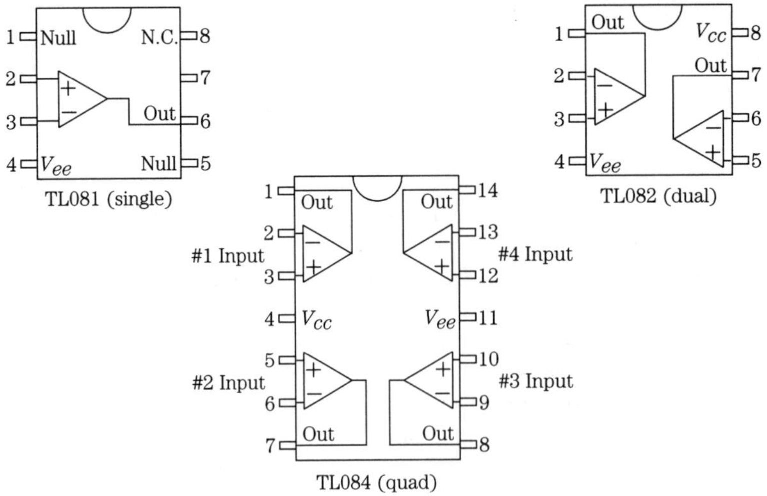

9-24 TL 080 family IC pinouts (top views).

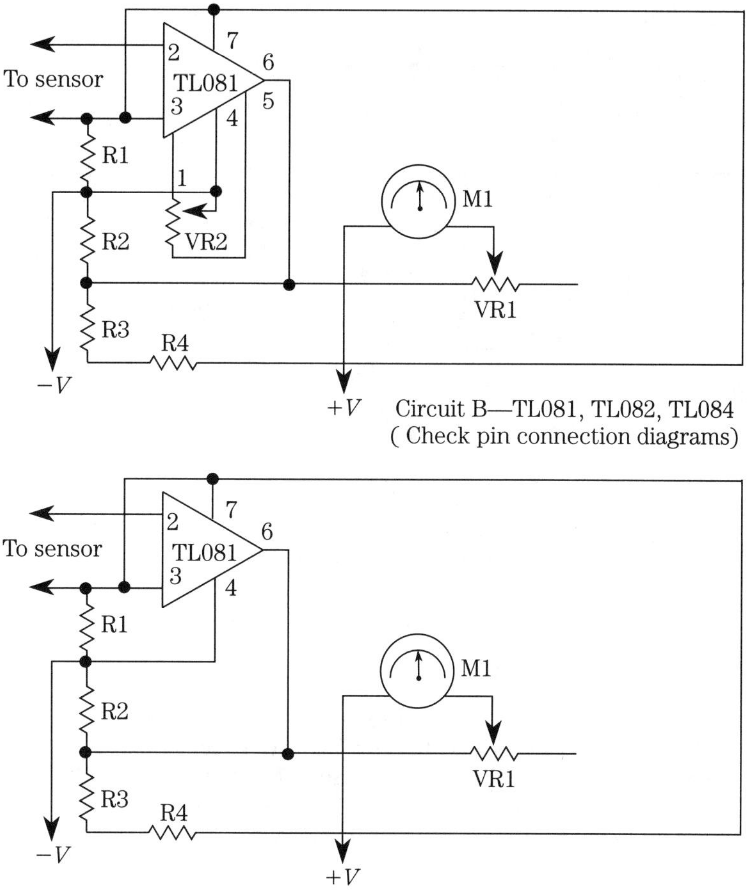

9-25 Single-channel hook-ups for the TL 080 series ICs.

Figure 9-27 shows a completed unit that utilizes both circuits A and B with a single TL 081 IC. In operation, both controls are set for minimum readings to avoid pinning the meter movement. An SPST switch selects between the circuits.

First, hook a sensor to the input terminals. Circuit A uses the null circuit, so balance via VR2 to read about half-scale. When using circuit B, adjust for a meter reading of about 75% of full scale. With a little fine tuning, both circuits can be switched in or out of use with no further adjustments.

Figure 9-28 shows a sample test sensor that is used for this unit, and also to test many of the other seismic receptor circuits described in this book. It has been used under actual field conditions with the LM 311-based magnetic transducer detector, the LM 324-based magnetic field disturbance detector, and with the TL 081-based reactance monitor.

This test sensor is very versatile for other uses, too. It can be used as one of the coils in the BFO seismic signal processors, as well as in the LM 324-based ULF/VLF

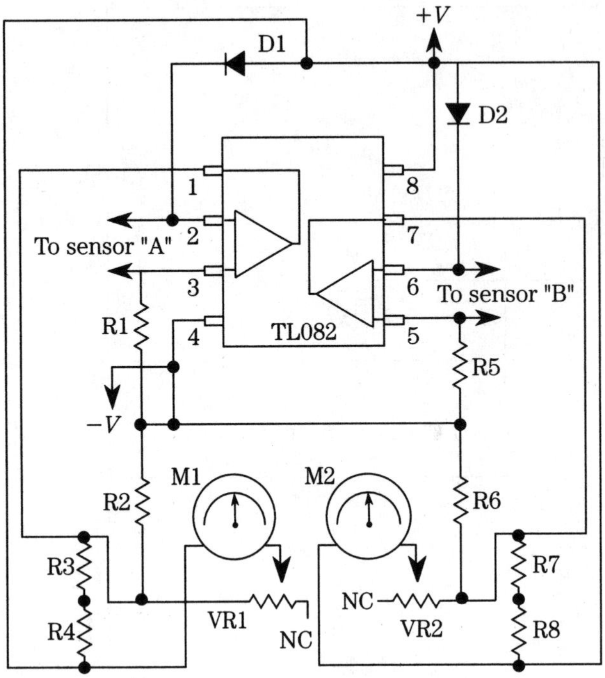

9-26
Sample TL 082-based dual-channel schematic.

Table 9-11.
Parts List for the TL 080 Series Processors: Reactance Monitor

Part	Description
D1, D2	1N4001 Diodes
IC1	TL 081, TL 082, or TL 084 IC
M[a]	Meter; 0 to 1 mA dc
PS[b]	9 Vdc, at Least 100-mA Power Supply
R1, R2[c]	8.2-kΩ, ¼-W Resistors
R3	91-kΩ, ¼-W Resistor
R4	100-kΩ, ¼-W Resistor
VR1	100-kΩ, ½-W Potentiometer
VR2	500-kΩ to 1-MΩ ½-W Potentiometer
S[d]	SPST Switches
Misc.	Socket for the IC, Enclosure, Terminal Posts, Knobs, Etc.

a The number of meters used in a particular application depends on the number of sections of the IC being used.

b The use of a single 9-Vdc supply is possible. When using a multichannel unit, if interchannel interference occurs, use two separate power supplies, and discard common negative connections and the diodes in the circuit.

c When multichannel circuits are constructed, each channel requires its full complement of resistors.

d The number of switches is dependent on the circuitry used.

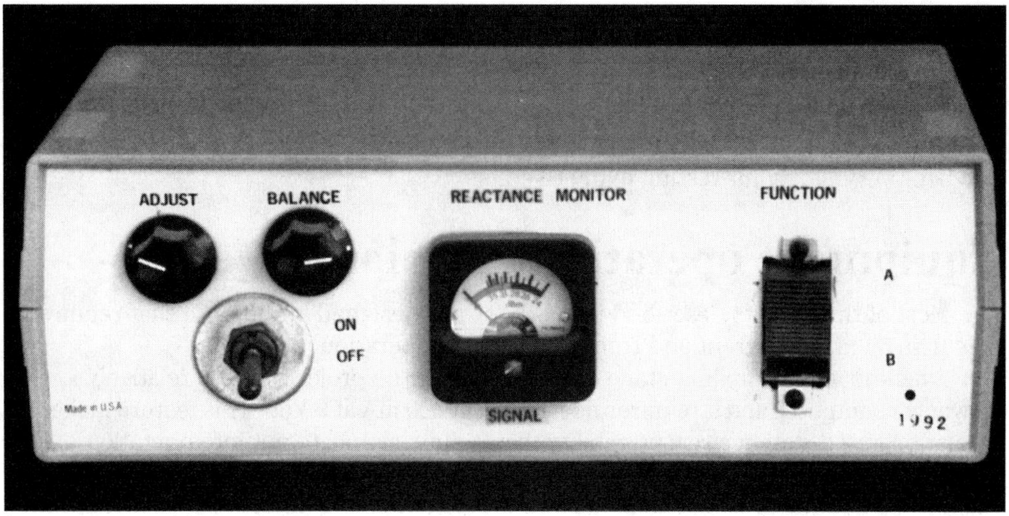

9-27 Completed TL 081-based unit with both A and B circuits.

9-28 Sample test sensor: a TV yoke coil.

receiver. It can also be used as coil L1 in the Pentaflex radio circuit; when it is used along with a 470-μH choke coil as L2, and a 500-pF tuning capacitor, it turns the Pentaflex into a broadband receiver down into the Omega navigation band, and below, to around 10 kHz.

The coil was found ready-made; it is a yoke coil from an old, discarded TV set. It has been very handy for testing many circuits.

Equipment operation on solar power

Most of the circuits described in this book are designed around ICs that require less than 50 mA of current and from 6 to 12 V for operation.

Even the audio amplifier stage used in many of the projects is of a relatively low power consumption, and it requires no more than 200 mA at 9 Vdc. This feature makes a solar-operated station a real possibility that is quite simple to put into operation.

A small-in-size (9" × 11"), single panel can power several projects simultaneously. A commercially available panel of this size, supplying up to 9 Vdc at 1 A or 18 Vdc at 500 mA, provides excellent, dependable performance.

All that remains necessary for a solar-operated set-up is an appropriate voltage rechargeable battery pack, some hook-up wire, and a blocking diode, such as a 1N5400, installed in series between the battery pack and the solar panel so that the battery cannot discharge through the panel when there is no sunlight.

10
CHAPTER

The Southern California earthquakes of 1992

Two strong earthquakes, one of them the strongest in California in 40 years, hit in southern California on June 28, 1992. The first measured 7.6 on the Richter scale, the second measured 6.7. Figure 10-1 is a map of the area, showing the relative locations of the quakes and the major fault lines.

These major earthquakes spawned over 6000 aftershocks in the first month following their occurrence. Most were under magnitude 3.0. Ninety were over 4.0, and 11 measured over 5.0. Major aftershocks continued throughout 1992 and smaller ones were still being recorded in early 1994.

Much is unknown regarding the triggering mechanisms of earthquakes and the behavior of fault lines. The basics are known; continental plates grind against each other, accumulate stress, and the resulting slips and ruptures create earthquakes.

Some questions are still unanswered: "Why do earthquakes happen when and where they do?," "How and why do they stop?," and "Why are some earthquakes bigger than others?"

One huge difficulty stems from the fact that researchers have to study something they cannot see; the faults, fissures, and ruptures exist miles beneath the ground surface. As in the major southern California earthquakes of 1992, many faults remain undiscovered until a quake occurs. The one in the Big Bear Lake area left no visible rupture on the surface, as opposed to the 7.6 Landers quake, which opened an approximately 43-mile-long crack in the desert floor. The shift of movement on either side was as great as 21 feet. Faults not leaving a surface trail have to be charted by plotting the series of aftershocks in the region.

The large Landers earthquake apparently triggered swarms of small temblors hundreds of miles away. In fact, farther away than any other known quake in history.

Within just a few minutes, swarms of small earthquakes were recorded from the Owens Valley to the Mammoth Lakes region, the Long Valley area, and other activity rattling under California's two largest dormant volcanoes, Mt. Lassen and Mt. Shasta.

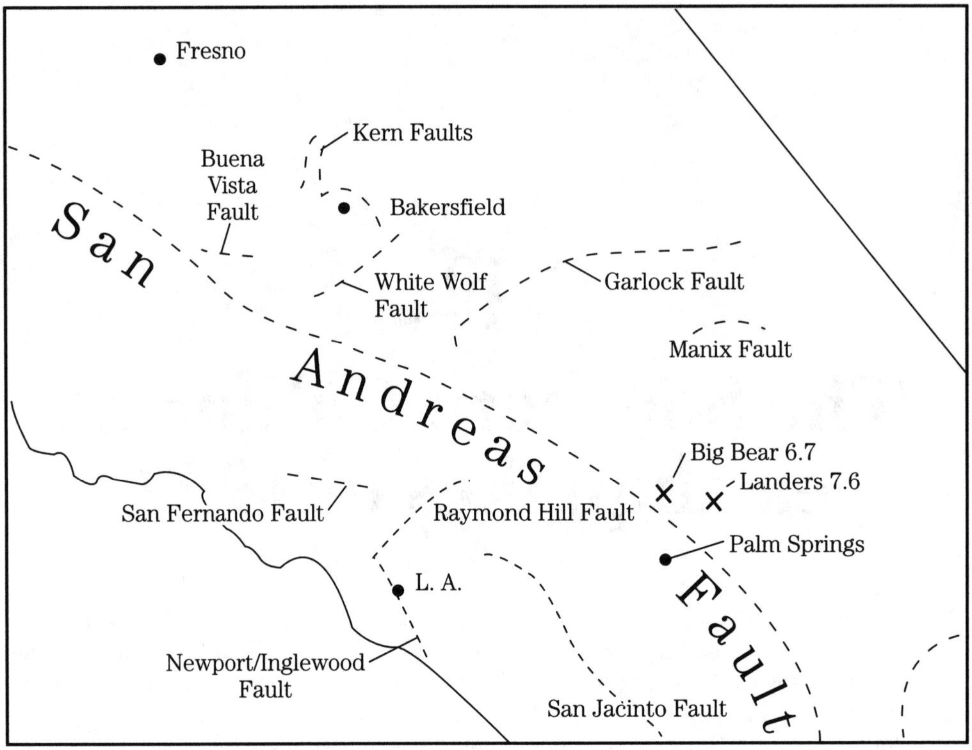

10-1 Southern California fault lines and the June 28, 1992 earthquakes.

Activity was widely distributed throughout the Long Valley Caldera, a large depression located over a chamber of magma that has been restless for the past 10 years, marked by nearly continuous low-level activity. About one dozen small quakes had been registered per day in the region, and the frequency and intensity of these rose abruptly following the Landers quake.

The two volcanoes are more than 500 miles from the epicenter of the Landers quake. Mt. Shasta registered one quake as high as 3.2 on the Richter scale, and Mt. Lassen, which last erupted in 1911, registered a 2.5 shaker.

A separate earthquake, occurring at 11:14 AM PDT on July 12, 1992, registered about 5.2, and was centered 12 miles north-northeast of Mojave (about 90 miles north of Los Angeles). In some way, this earthquake was probably related to the June 28 quakes, although it was not located in their aftershock zones. Although the earlier earthquakes occurred near, but not on the San Andreas Fault, this latter one struck near, but not on the Garlock Fault, which is also one of California's most dangerous faults. Some scientists say that this fault is capable of producing major earthquakes of from 7.5 to 8.5 in magnitude. The "Big One" frequently referred to would have a magnitude of over 8.0.

The Landers earthquake broke the surface in many places, probably because it was an unusually shallow-focus event, occurring from one to two miles beneath the surface.

The two major quakes, just 20 miles apart, are complex phenomena, and are indicating their relationship in ways thus far not understood. Although not on the San Andreas Fault, the proximity is of great concern. The San Andreas stretches over 800 miles from south of the Mexican border to offshore from Cape Mendocino. It is one system, but has about 10 distinct segments, each with peculiar characteristics, and various historical rupture patterns.

There are two very dangerous portions of this fault. One is in the San Francisco Bay area to the Hayward Fault in the East Bay, and the other dangerous area is in the southernmost part of the fault, running from San Bernardino, through Palm Springs, to the Mexican border. The main area of concern is the Coachella Valley segment, starting about 10 miles southwest of the June 28 quakes, and continuing about 60 miles to the border. This section has not had a large earthquake since European and Mexican settlement, but geologic evidence indicates that large quakes occurred in about 1020, 1300, 1450, and 1680.

The two June 28 quakes came after a series of events in the region near this southern end of the San Andreas Fault, starting with a 6.4 shaker in the Imperial Valley in 1979. Major quakes occurred in 1982, 1986, and 1987, and in April 1992 (which measured at 6.1) in Joshua Tree. The June 28 quakes appear to be an extension of the Joshua Tree rupture.

The twin earthquakes were very complicated in nature, and the resulting movements have been very complex.

In the middle of July 1992, a set of small quakes were centered comparatively very deep, from 15 to 20 miles deep, beneath the central valleys. They included 2.8-magnitude quakes near Fresno and Avenal, a 2.5 near Redding (on July 17), and a 4.0 near Red Bluff on July 16.

Using measurements based in part on the Defense Department's Global Positioning System (GPS) satellites, it has been measured that Los Angeles moved about one-half inch to the northwest.

Nearer to the epicenters; San Bernardino moved northwest about two inches, Palm Springs moved about four inches to the north-northeast, and, near the surface rupture caused by the Landers quake, the land shift was at least five feet.

Also, five major earthquakes have struck in a line north of Joshua Tree since 1947. Figure 10-2 shows the positions of these quakes relative to existing documented fault lines, the location of the Mojave quake of July 12, 1992, and the line of the suspected formation of a new, major fault. Geologic features of the area are shown in Fig. 10-3.

The Parkfield activity

On October 19, 1992, a magnitude 4.7 earthquake occurred in Parkfield, California. It prompted a seven-county "Level-A" alert, which caused the California State Office of Emergency Services to warn disaster and law enforcement officials in nearby counties that there was a one-in-three chance of a magnitude 6.0 or greater earthquake in the next 72 hours. Parkfield is located on the San Andreas Fault.

The alert applied to a 30-mile radius around Parkfield, and included parts of Fresno, Kern, Kings, Monterey, San Luis Obispo, San Benito, and Santa Barbara counties.

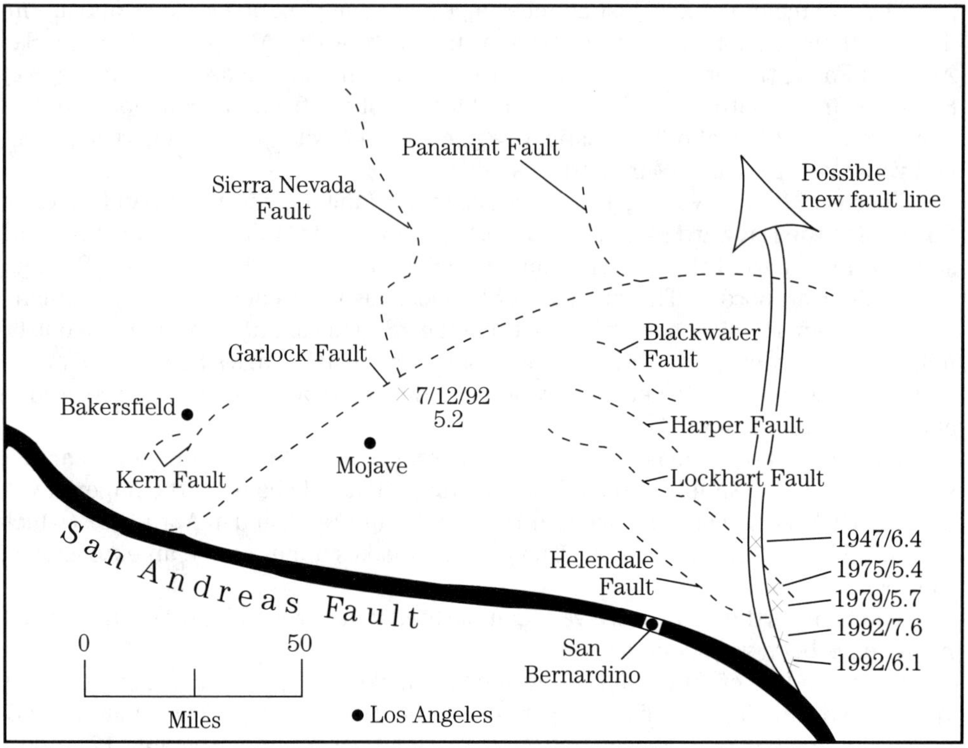

10-2 Possible new fault area in California.

Leading up to the October 19 quake; a magnitude 3.1 hit October 4, two quakes of 2.7 hit on October 5, and smaller shakes occurred on the 7th. So, the 4.7 earthquake was no surprise to scientists participating in the nearly $20 million Parkfield project, considered the world's premier earthquake prediction experiment.

Because of the regular seismic activity near Parkfield, it is the only area where the U.S. Geologic Survey has issued a specific prediction on the approximate time and magnitude of an earthquake. Since 1857, Parkfield has had a major quake on the average of every 22 years. The last 6.0 or greater quake occurred on June 28, 1966.

However, the 72-hour alert expired without incident. Further examination of the quake's record showed that it was a virtual match of a quake recorded on January 5, 1975. Therefore, it seems to be a quake of its own character, with its own series of aftershocks, and most likely not a precursor of a larger, imminent earthquake.

Private laser measurements showed nothing unusual before and after the quake, thus adding to the evidence that the 4.7 earthquake was not a prelude to a larger event in the immediate future. Rather, it seems more likely that the earthquake is associated with the local activity on a larger timescale, on the order of a few months, instead of a few days.

As a result, after a series of very small quakes measuring from less than 1.0 to

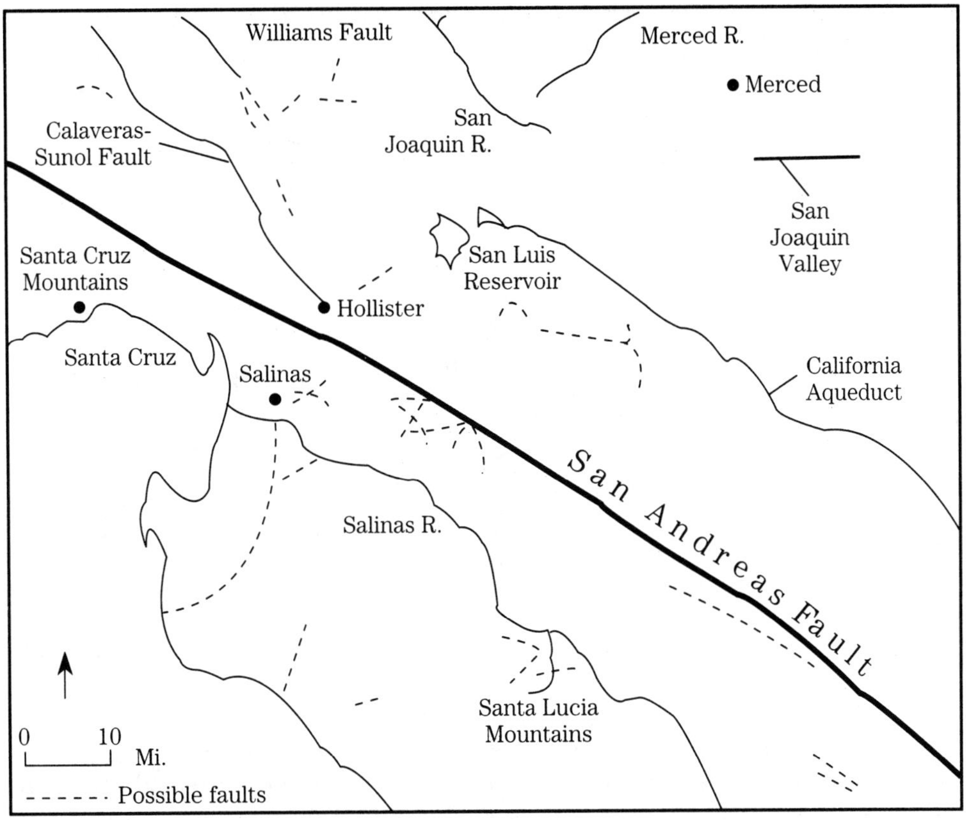

10-3 Geologic features of central California.

nearly 2.0 on October 22, the warning was allowed to expire that evening. A few very small aftershocks continued to jostle the area throughout the remainder of 1992.

Another similar earthquake, 4.8 at 4:28 A.M. local time on November 14, 1993, also prompted the same alert status. Very small magnitude quakes began late in the day on November 12, and, once again, scientists swarmed to Parkfield in anticipation of a larger magnitude earthquake.

The pattern was a carbon copy of the October 1992 episode, and again after 72 hours, the level A alert was allowed to expire. Parkfield had "teased" the scientists again. A map of the general area around Parkfield is shown in Fig. 10-4.

Earthquake prediction

Predicting earthquakes, as you have seen, is far from being an exact science. In fact, it is still in the realm of the unknown. But, the unknown phenomenon is part of the attraction for scientists and researchers. Earthquake prediction has proven very elusive. Whether it is even possible is subject to great debate.

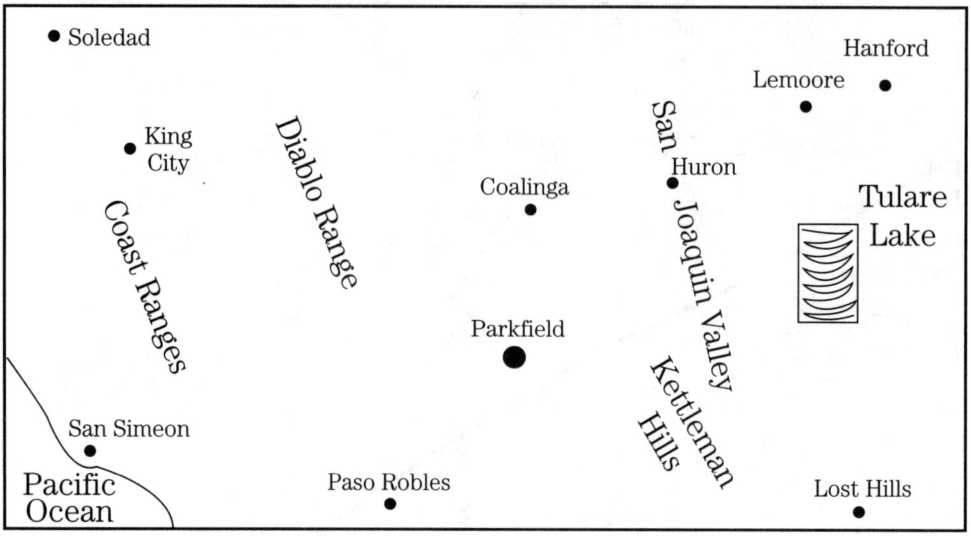

10-4 Location of Parkfield and the surrounding communities.

Some of the circuitry described in this book is "maverick" and speculative with regards to earthquake-prediction efforts. However, all data has been useful in monitoring and measuring, detecting, and displaying related data.

The idea is that with an inexpensive, relatively simple, and easily operated apparatus, an observation network might be set up so that amateurs and serious researchers, along with scientists, can have access to data that might contribute to further experimentation and study. Table 10-1 is an earthquake-preparedness chart, of common-sense information in dealing with an emergency situation. The danger from earthquakes is so great that attempts to predict them must continue.

Networking information

The networking of information involves the acquisition and transfer of data among participating stations. For networking to be effective, monitoring stations should be set up in a grid pattern, with the capability to transmit data to the member stations, quickly during active times.

A computer network would be ideal, but networks can be implemented using fax, telephone, or two-way radio as well. Similar instrumentation at each site is a must: a pyrheliometer, seismic sensor and seismometer, perhaps a Geiger counter at each station. The seismic unit should be "on-line" continuously, and the basic seismometer circuit is probably the best suited for this use. Stations in active earthquake or fault zones should be set up at least at 20- to 30-mile intervals.

During an earthquake, each station can observe the time of an event, calculate the relative intensity by observing local readings, and relay the information to other stations along the network. In a large quake, some stations will receive data before the seismic waves reach their locations. Rapid intensity and epicenter calculations

Table 10-1. Earthquake Preparedness Guide

Before the quake

Purchase supplies:

Battery-operated radio	Crowbar
Batteries	Pipe wrench
First-aid kit	Flashlights
Canned foods and drinks	Bottled water

Locate utilities shutoffs: gas, electricity, water

During the quake

If indoors: Stay there. Get under a desk, table, or a doorway. Stay away from windows.

If outdoors: Get in an open area. Stay away from trees, buildings and power lines.

After the quake

Check for, and treat injuries	Listen to the radio
Shut off utilities	Use outdoor grills to cook
Check for damages	Boil water to disinfect
Check for downed utility lines	Avoid travel until damage is assessed

can then be made. Fault lines could be pinned down. New faults might be discovered, which would provide useful information for future earthquake preparedness.

A ULF radio receiver should also be installed to monitor a possible "seismic electromagnetic resonance frequency" within the earth. When the audio output of a ULF frequency below 1 kHz is input into a computer or chart recorder, electromagnetic anomalies can be monitored for their potential correlation to earthquakes. This is a versatile application scheme because the same display of ULF information can indicate changes in atmospheric and radio propagation characteristics, along with information about seismic and solar disturbances.

Other station possibilities include water well monitoring to detect changes in the water's resistance or pH, monitoring radon emissions, ground conductivity, and inclinometer readings.

Pyrheliometer and Geiger counter readings could be taken once daily for relay among member stations. The frequency of observations could be changed during turbulent conditions, and could aid in forecasting.

Keys to the successful implementation and use of such a network include: reliability of the information, well-constructed, similar parameter equipment, and cooperation among member stations.

11
CHAPTER
United States seismicity: map section

The map section presented in this chapter is a collection of hand-drawn maps containing approximate fault line patterns and/or locations of earthquake epicenters in historically active areas. This information was garnered from locally reported observations, studies of geologic structures, and data from reconnaissance satellites. The scales are approximate, as are fault line traces of regional activity.

Earthquake activity in the United States is widely distributed. Figure 11-1 shows the geographical distribution of the areas most prone to seismic activity.

Figures 11-2 through 11-8 show California's most active areas, and the offshore fault lines. California, along with Alaska (Fig. 11-9) and Hawaii (Fig. 11-10), are the most active states. However, there are several other very dangerous areas.

Dangerous geologic zones exist along the coasts of Oregon and Washington. The areas around Memphis, Tennessee, and Little Rock, Arkansas, are also areas of possible heavy damage from earthquakes.

The earth's underlying formations off the Pacific Northwest and the southern Alaska coasts consist of subduction zones (see chapter 1) similar to that off of Mexico's western coast, which produced a devastating earthquake in September of 1985.

In fact, along the western coast of North America, from offshore northern California to Vancouver Island, Canada, the Cascadia subduction zone is a very dangerous region; the only question is whether part of the zone, or all of it will buckle at one time, which could produce a great earthquake with a magnitude over 9.0. If smaller pieces buckle, earthquakes would be more regional in nature. A very large series could also signal the increase of volcanic activity in the mountains near the coast.

The subduction zone, where two sections of the earth's crust are pushing against each other, is active, causing changes in land surveys inland and influencing changing tidal patterns in the coastal areas.

The smaller Juan de Fuca plate is undercrusting the larger North American plate along the Oregon and Washington coasts as a result of volcanism associated with the

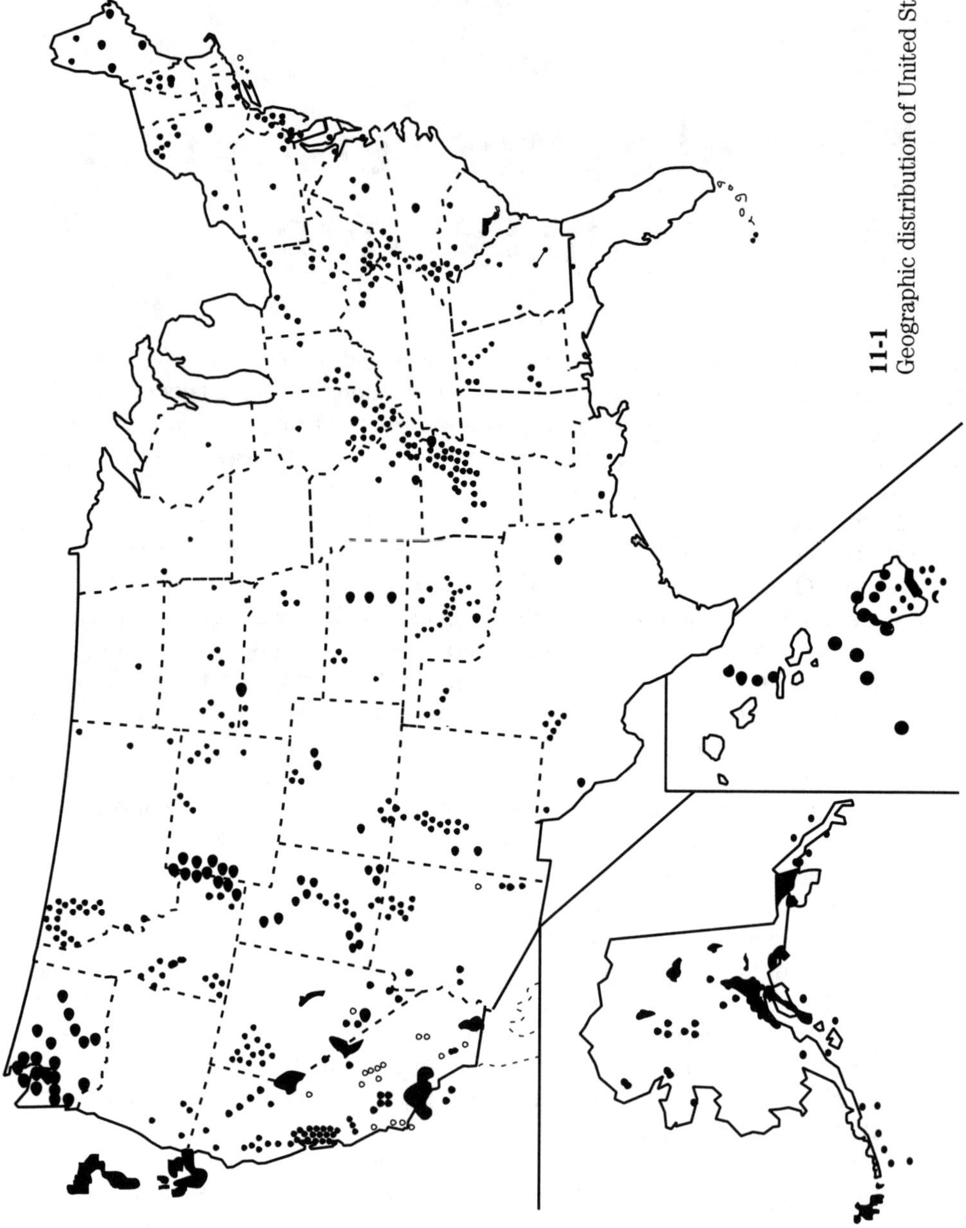

11-1
Geographic distribution of United States Seismicity.

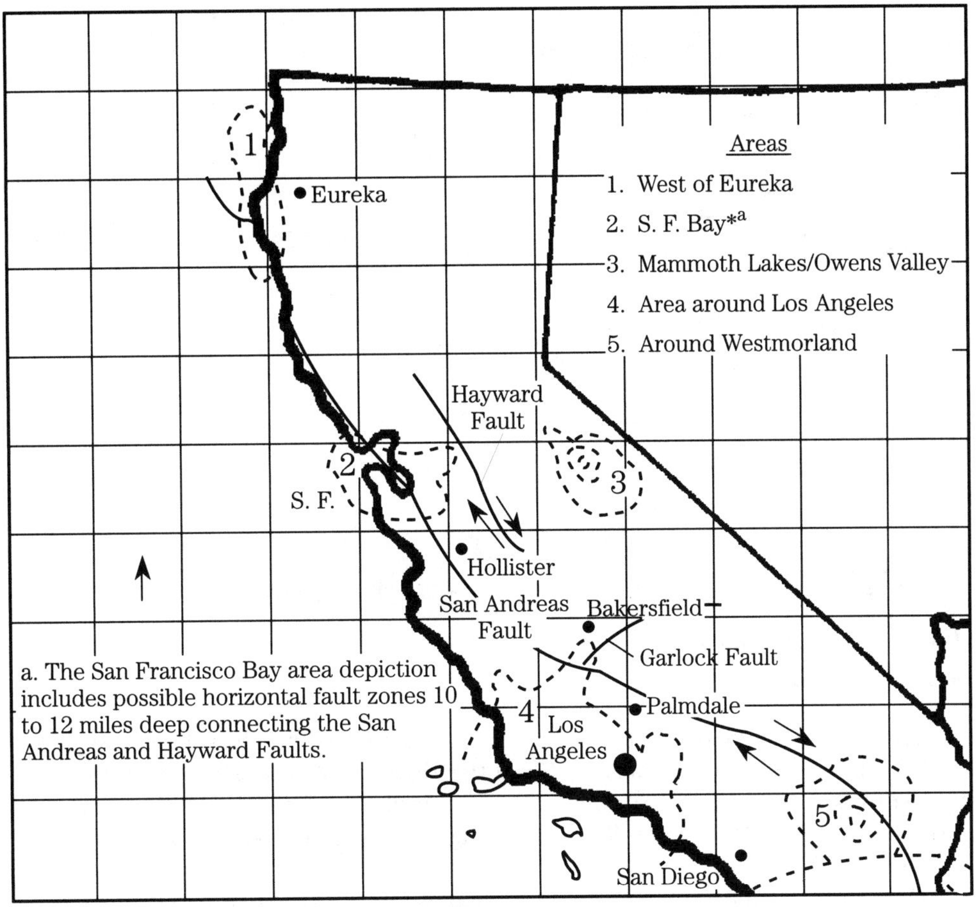

11-2 Dangerous, earthquake-prone areas of California.

Cascade mountains. In contrast, California's San Andreas Fault is a strike-slip fault, where subterranean plates shift sideways, and trigger earthquakes when the stresses are released.

The area bounded by Memphis, Tennessee; Little Rock, Arkansas; Paducah, Kentucky; Poplar Bluff, Missouri; Evansville, Indiana; and Carbondale, Illinois is perhaps the most dangerous area outside California. The worst earthquake series in the continental United States occurred near New Madrid, Missouri in late 1811 and early 1812, when three earthquakes, measuring an estimated Richter scale magnitude of 8.6, 8.4, and 8.7, hit the central United States.

One quake was so powerful that some people claim the Mississippi River ran backwards, and that shock waves from the earthquake rang church bells in Boston, Massachusetts. Figures 11-11 and 11-12 depict these areas.

In California, five earthquakes measuring 7.0 to 8.3 estimated magnitudes have ruptured the San Bernardino Mountains segment of the San Andreas Fault, near Wrightwood, north of San Bernardino, in the last 500 years.

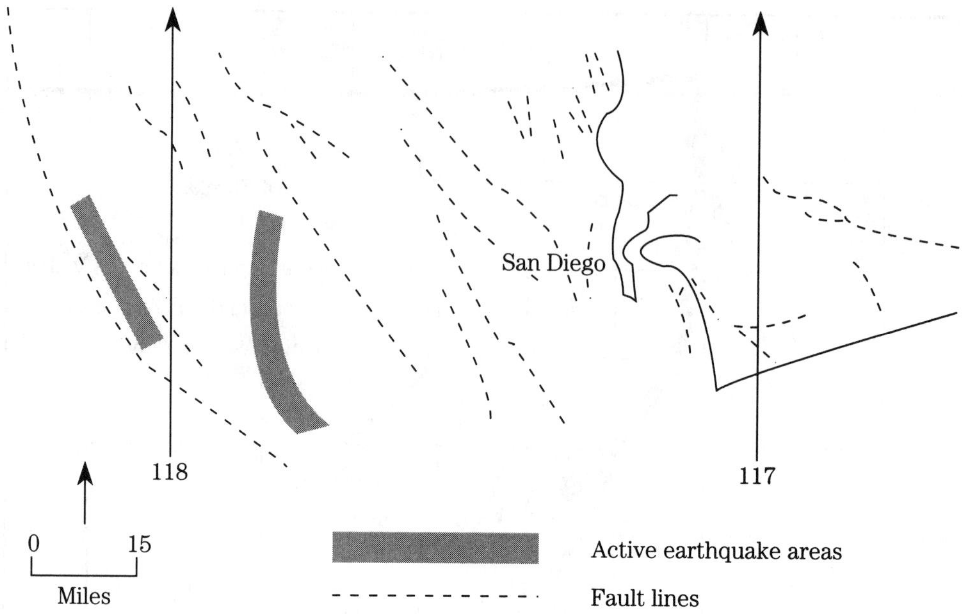

11-3 San Diego area and offshore faults.

These quakes happened in 1857 and 1812, and approximately in 1700, 1610, and 1480. Three of these quakes also ruptured the Mojave segment of the fault, the part nearest to Los Angeles. One of the quakes also broke the Coachella Valley segment of the fault, near Palm Springs. The most dangerous areas of California are shown in Fig. 11-2.

Elsewhere in California, the Mammoth Lakes area is a high-activity zone, where movement of a subterranean pool of magma has been detected and is being monitored. This activity led to the issuing of a low-level volcanic activity alert for a time in the 1980s.

The Imperial Valley segment of the San Andreas is also a high-risk area, and is one of the most complex regions in the world. The activity around Westmoreland in this zone often involves at least six distinct fault planes.

Intense activity is also noted in the offshore waters of northern California. Though most is well offshore, vibrations are often felt across northern and central California, and in southern Oregon. This area is shown in Fig. 11-8.

Central California is an area of few large earthquakes. However, the underlying bedrock, part of the Sierra Nevada Batholith, is an excellent transmitting medium for seismic waves, and moderate-to-large earthquakes from nearly anywhere in the state can be felt throughout the region.

Hawaiian earthquakes, shown in Fig. 11-10, are mostly concentrated around Kilauea volcano on the island of Hawaii. Intrusions of magma are responsible for most of the seismic activity here, producing volcanic "swarms" of earthquakes. Kilauea also produces a steady output of "tectonic" activity on its flanks.

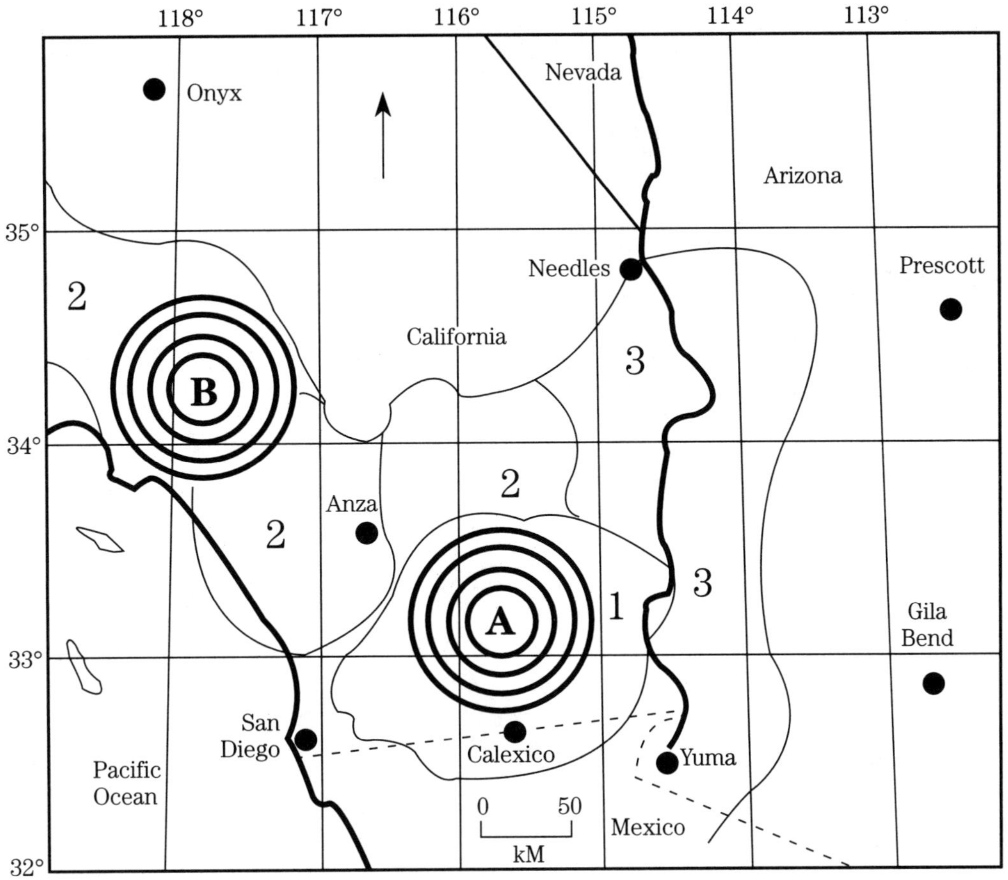

A = Potential Great Quake Epicenter
B = Potential Severe Quake Epicenter
1 = Boundary of Moderate to Severe Damage Potential
2 = Boundary of Moderate Damage Potential
3 = Area of Moderate of Lesser Damage Potential

11-4 Dangerous areas in extreme southern California.

Mauna Loa also spawns occasional shallow-focus earthquakes. Like those at Kilauea, most are caused by the subterranean movement of magma between eruptions.

Earthquakes are also widely scattered around the island of Hawaii, principally south of the island including the areas around the active submarine volcano Loihi. Another cause of seismicity around Hawaii is caused by the deformation of the land under the weight of the volcanic debris produced by the active growth of the island.

Figure 11-13 depicts fault lines in Kansas and Nebraska. Earthquakes in this region are small and sparse. Some fault lines extend into Iowa and Missouri, as shown

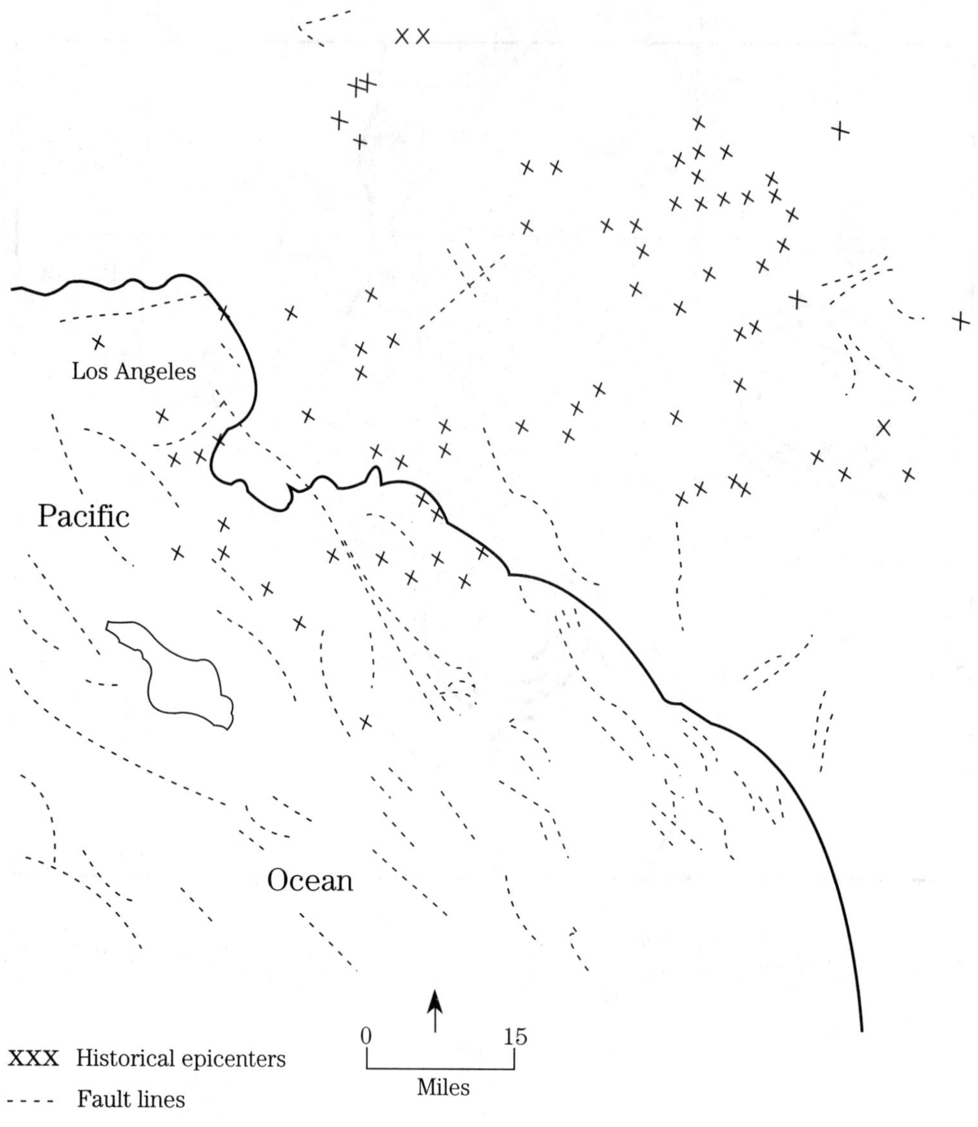

11-5 Southern California historical epicenters and offshore fault lines.

in Fig. 11-11. The larger earthquakes are usually limited in magnitude from about 1.5 to about 3.0 Richter.

Figure 11-14 depicts areas of seismicity in Nevada and in eastern California. Most of Nevada's activity is just east of the border with California, from the Long Valley area eastward. Active areas also extend from the central part of the state through a broad zone southwestward throughout Nevada.

Figures 11-15 and 11-16 show the New England and New York areas of seismic activity. Thrust-faulting exists on a broad NE/SW trending plane in Long Island

11-6 Hollister, California zone of seismic activity and faults offshore to the west.

Sound. On October 21, 1981, the first instrumentally detected area earthquake was detected and recorded. It measured nearly 4.0 on the Richter scale.

Prominent activity in New York is noted around Attica. Massena recorded a 5.9 in 1944, and several small shocks occur in the area intermittently. Southwest of Oswego, earthquakes, usually too small to be felt, occur regularly with instrument detection.

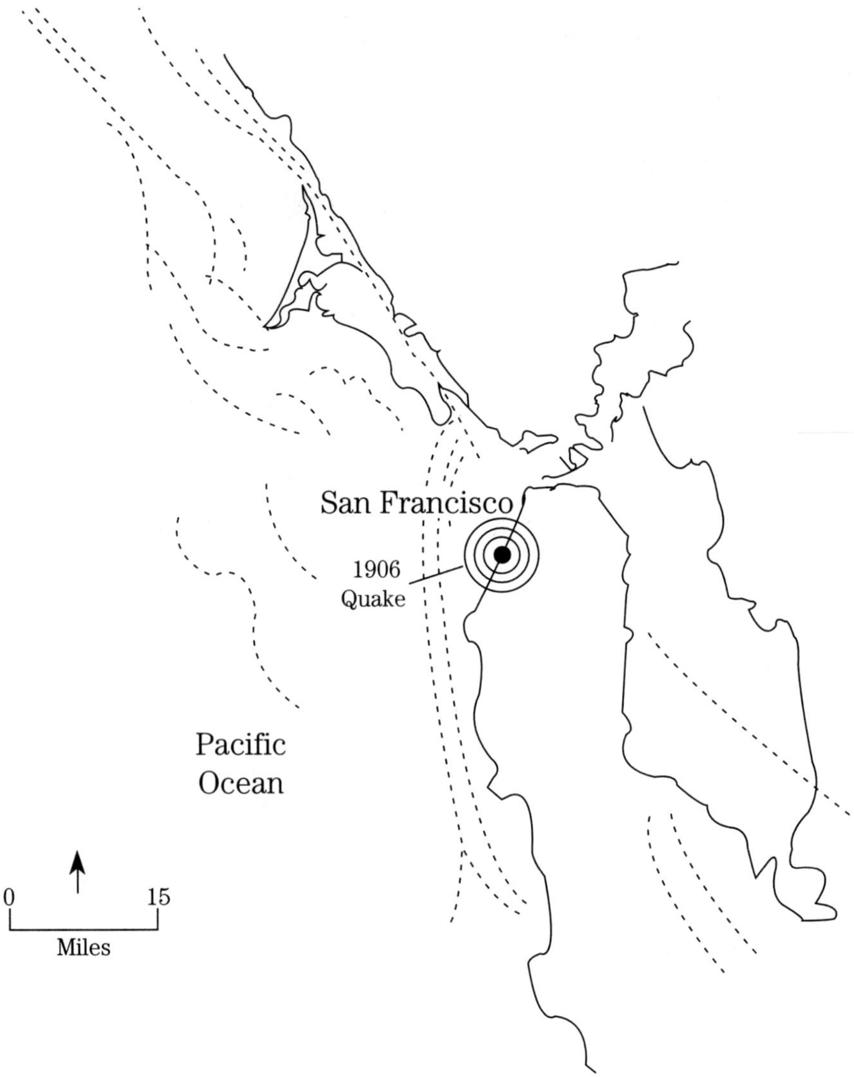

11-7 San Francisco Bay area and offshore faults.

An area of historically active seismicity exists near Moodus, Connecticut. This area is characterized by numerous small earthquakes. These events have been observed for hundreds of years. Other historically active areas of New England exist around Laconia, New Hampshire, and the Dover-Foxcroft areas of Maine.

Oklahoma fault lines are shown in Fig. 11-17. Here, again, earthquakes are usually too small to be felt, and must be instrumentally recorded. An earthquake zone extends northeastward in the north-central part of the state. A second area is located between Norman and Pauls Valley, paralleling the McClain County fault zone.

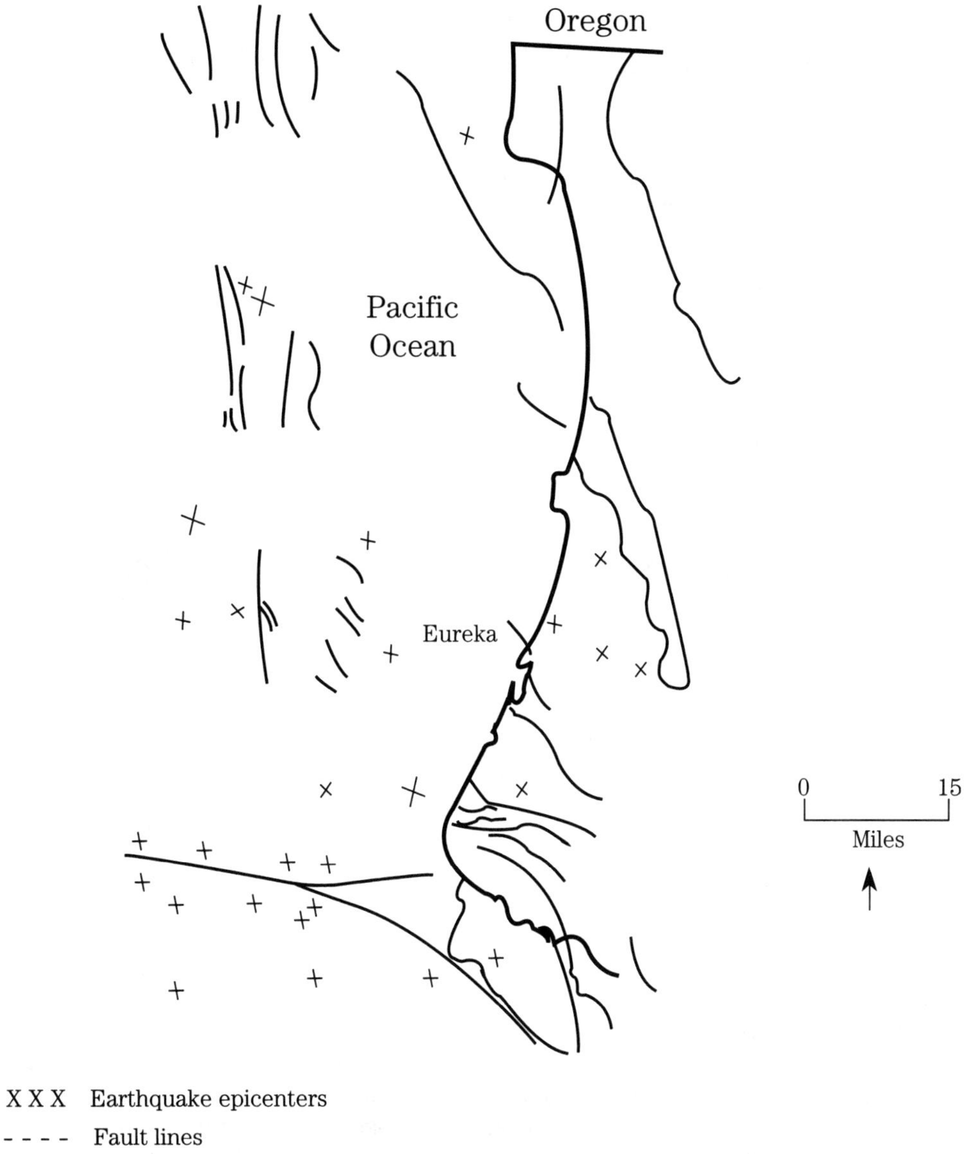

11-8 Northern California and offshore fault lines.

Still another active area is located just north of the Ouachita Mountains, and there is also activity in the extreme south-central part of the state.

In the southern and eastern United States, shown in Figs. 11-18 and 11-19, earthquakes are common near the North Carolina/South Carolina border areas, and in the southern Appalachians. Especially vulnerable to seismic activity is southeastern Tennessee, southwestern North Carolina, and northern Georgia. Other areas exist in northeastern Alabama, and, recently discovered, in central Virginia.

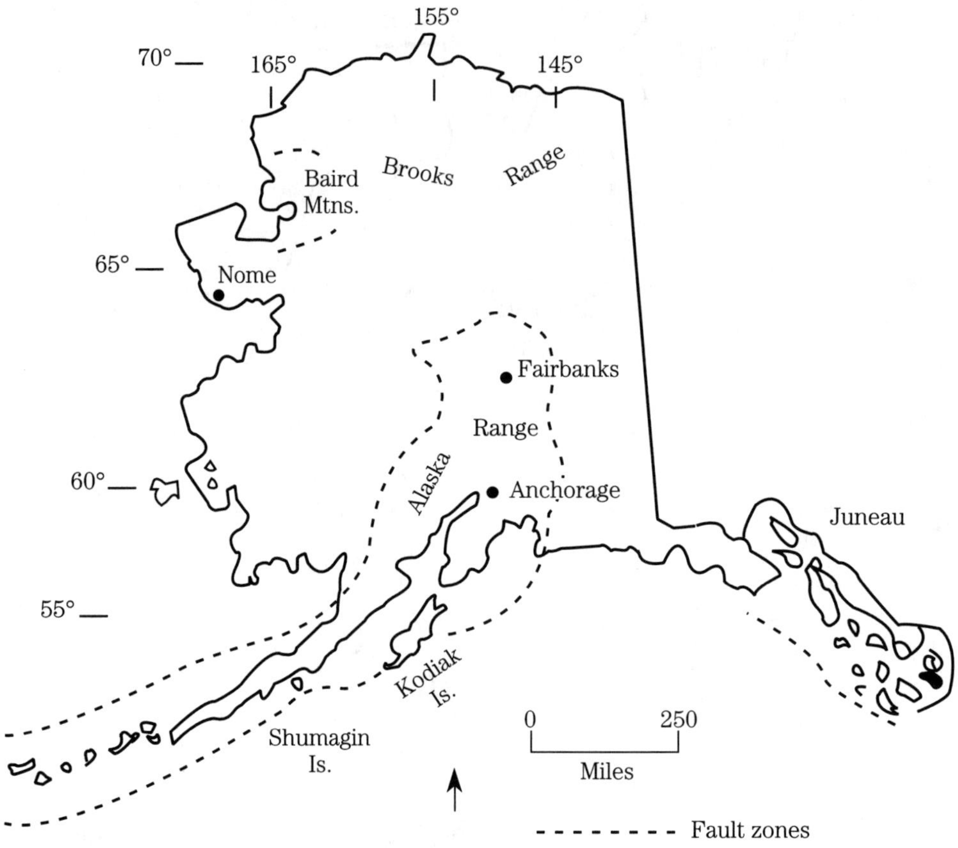

11-9 Alaska seismicity zones and geologic features.

Figure 11-20 shows the state of Utah. Part of the intermountain west, Utah has quite a bit of seismic activity.

A very active region lies along the border with Idaho, influenced by the geology beneath the Pocatello Valley. Other active areas exist beneath the Bear River Range, in a north/south line just east of Logan, occasional activity near Provo, and throughout south-central and south-western Utah, in a broad NE/SW trending area, cutting across several fault zones. Earthquakes also occur just south of Cedar City.

In Washington, Fig. 11-21, earthquakes are very closely monitored around Mt. St. Helens. Most quakes above magnitude 3.0 occur in the Cascade Province. Activity is usually associated with either volcanic activity at Mt. St. Helens, or to tectonic activity within the Cascades.

Historically, large magnitude earthquakes in the Puget Sound area are the most dominant seismic feature of the area. Eastern Washington earthquakes are usually smaller and shallower-focus than those in the western part of the state.

The map index for this section is listed in Table 11-1. Refer to the individual maps for scale and information keys.

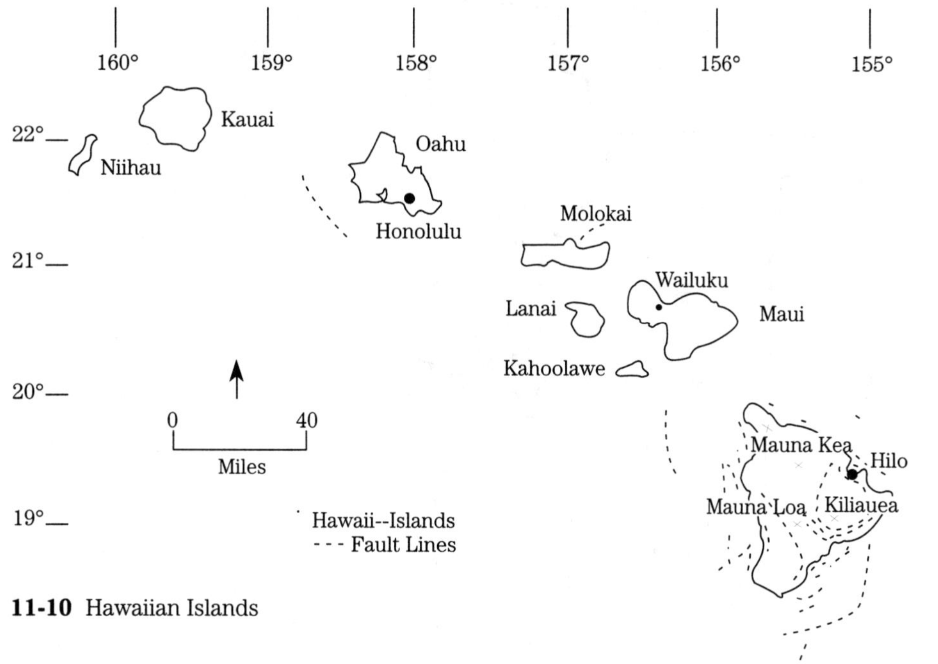

11-10 Hawaiian Islands

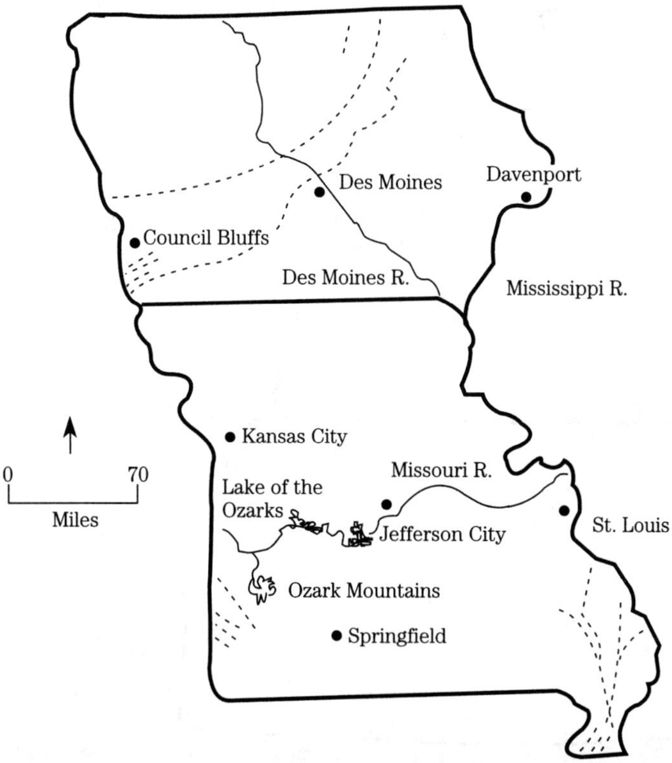

11-11
Fault lines in Iowa and Missouri.

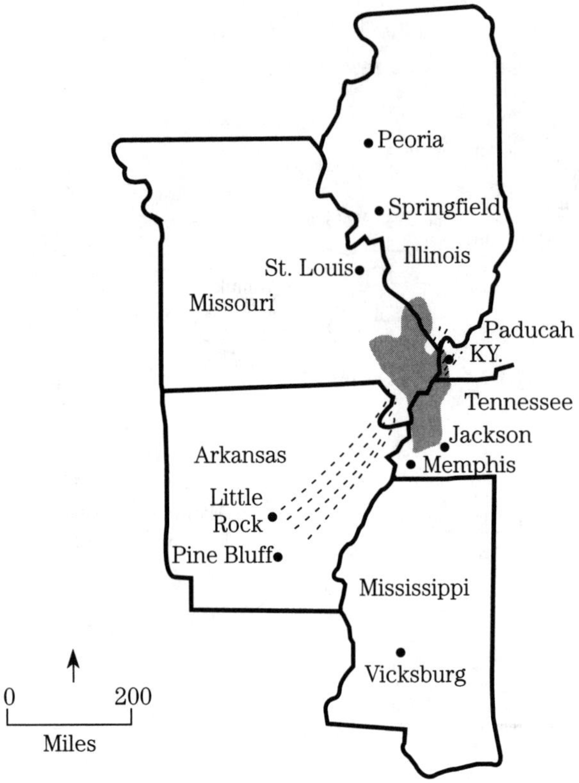

11-12
Mississippi Valley seismic areas.

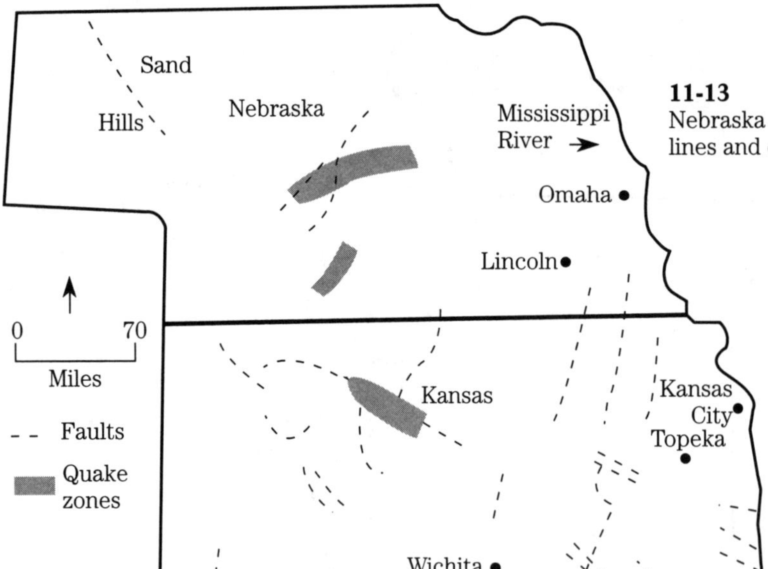

11-13
Nebraska and Kansas fault lines and earthquake zones.

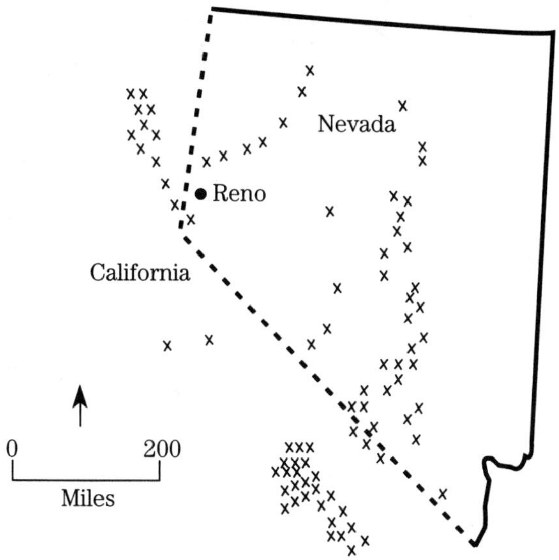

11-14
Nevada and eastern California earthquake epicenters (historical).

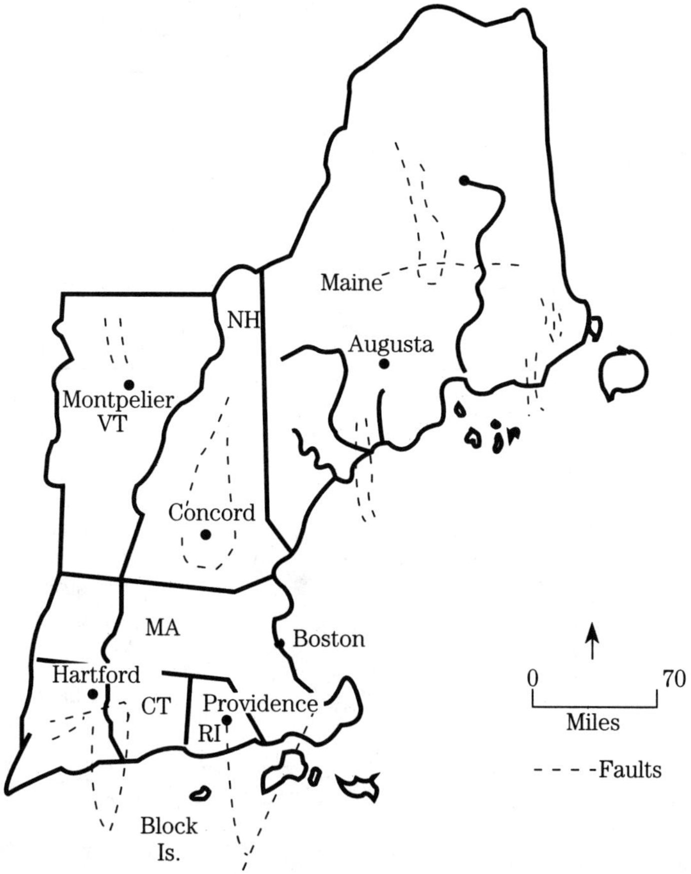

11-15
New England fault lines.

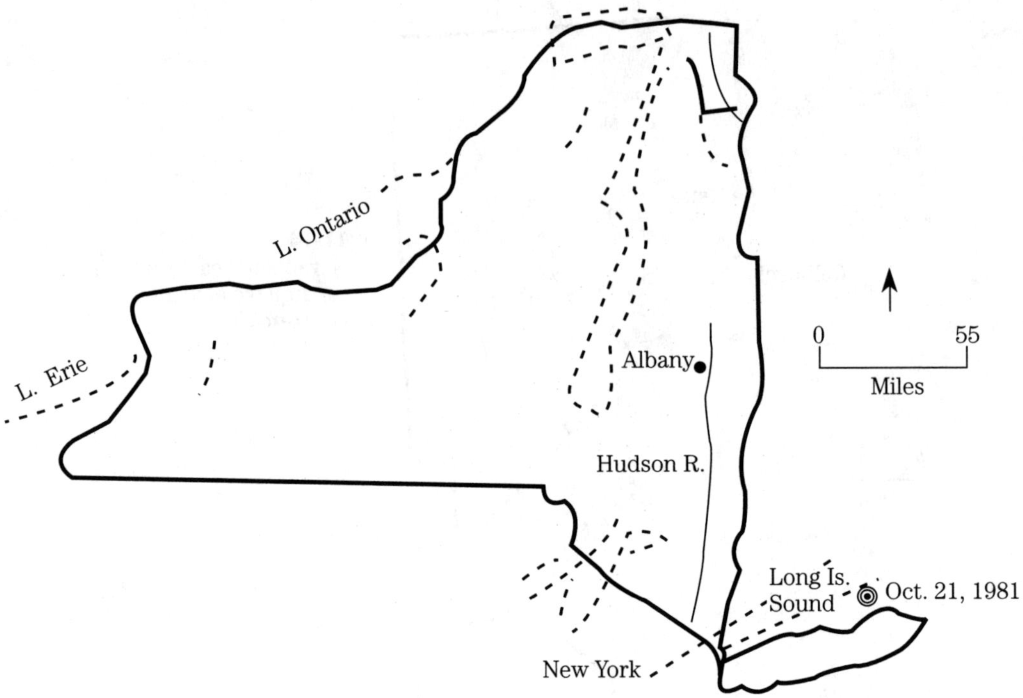

11-16 New York fault lines.

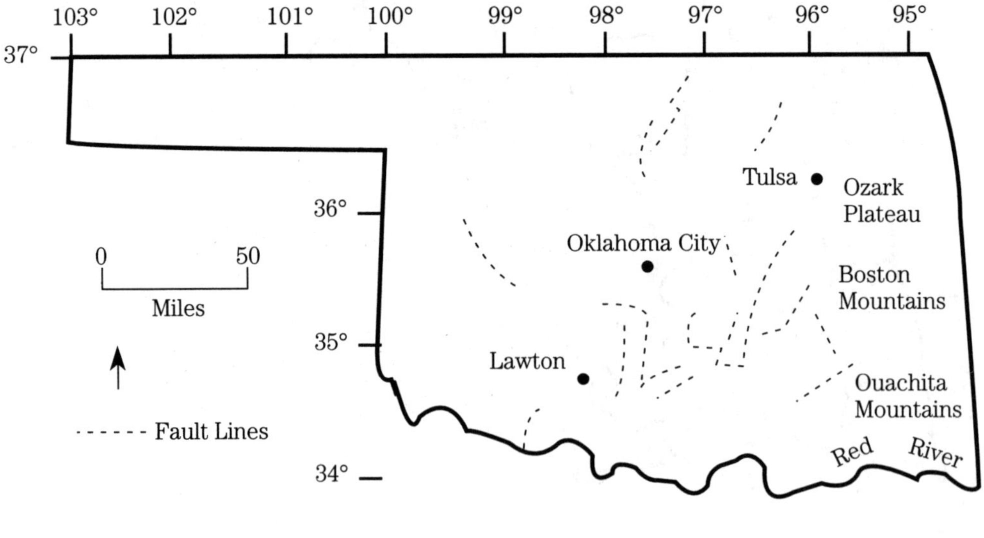

11-17 Oklahoma fault zones.

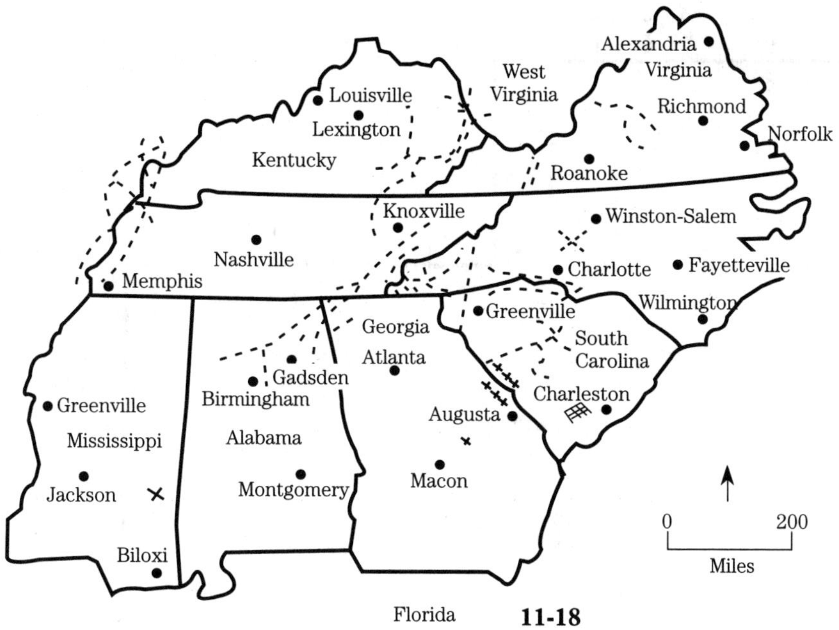

11-18
Fault lines in the Southeast United States.

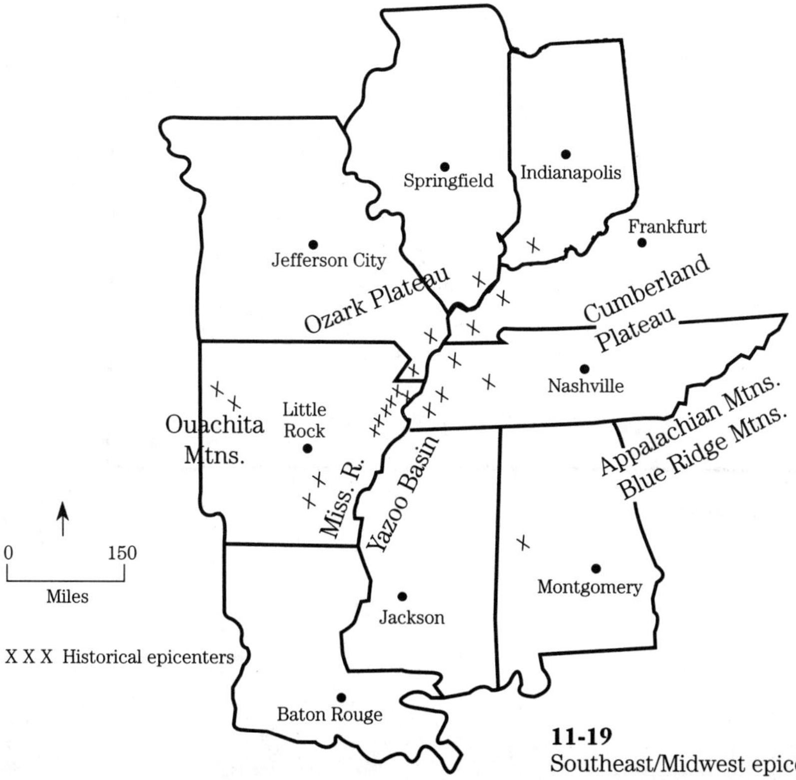

11-19
Southeast/Midwest epicenter zones.

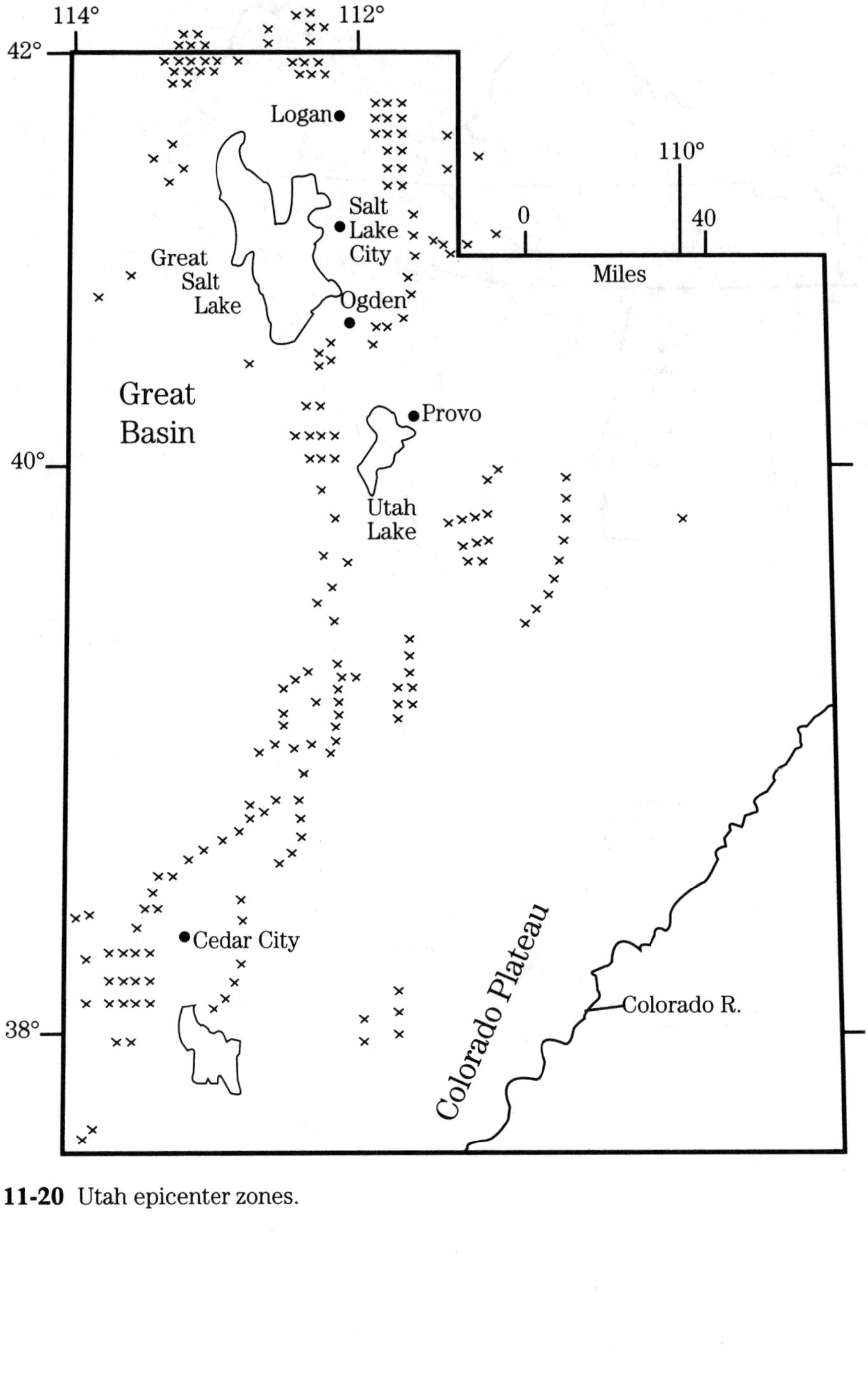

11-20 Utah epicenter zones.

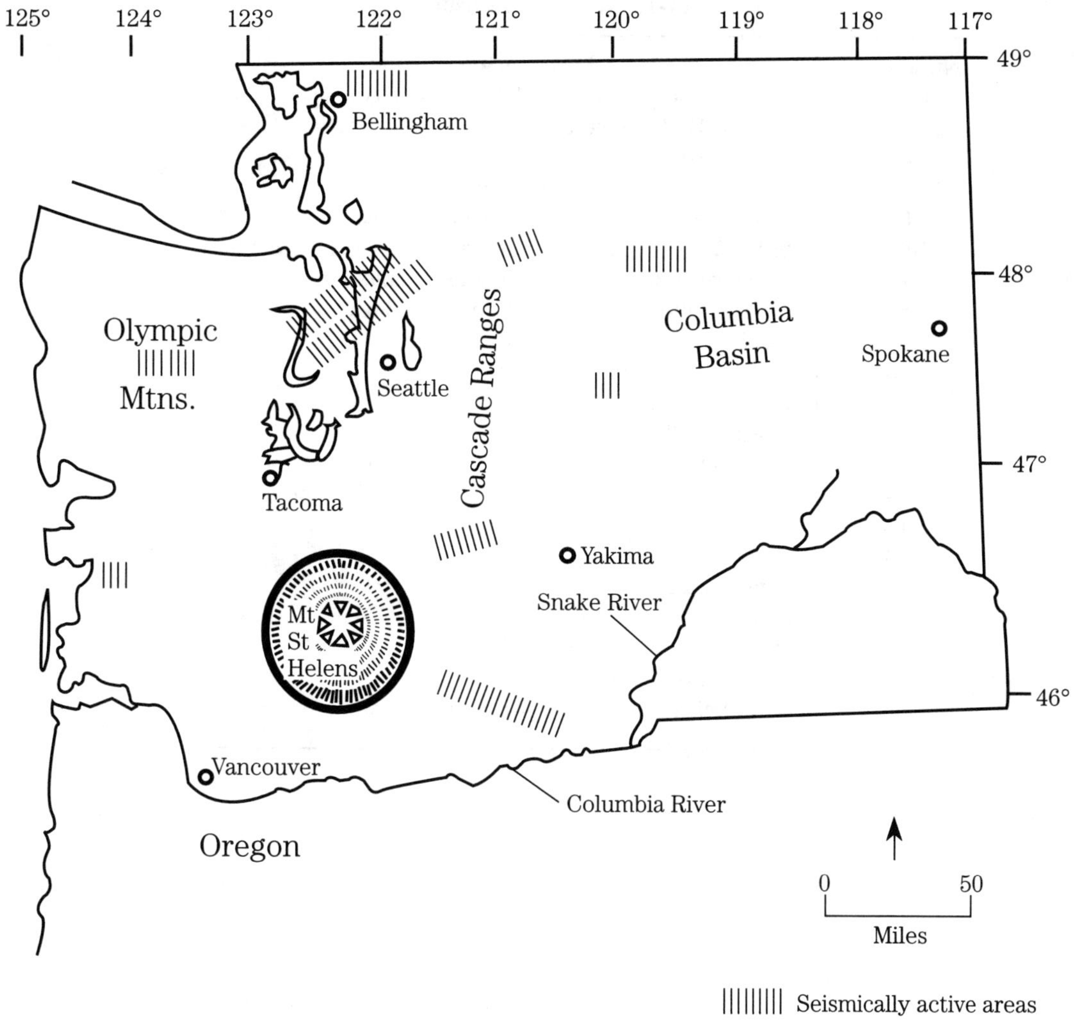

11-21 Areas of seismic activity in Washington State.

Table 11-1. Map Figure Index

Number	Description
11-1	Geographical Distribution of U. S. Seismicity
11-2	Dangerous Earthquake-Prone Areas of California
11-3	San Diego Area, and Offshore Faults
11-4	Dangerous Areas in Extreme Southern California
11-5	Southern California and Offshore Fault Lines
11-6	Hollister Zone of Earthquake Activity, and Offshore Faults
11-7	San Francisco Bay Area, and Offshore Faults
11-8	Northern California, and Offshore Fault Lines
11-9	Alaska Seismicity and Geologic Features
11-10	Hawaii: Islands and Fault Lines
11-11	Iowa and Missouri Fault Lines
11-12	Mississippi Valley
11-13	Nebraska and Kansas Fault Lines and Earthquake Zones
11-14	Nevada and Eastern California Earthquake Epicenters
11-15	New England Fault Zones
11-16	New York Fault Lines
11-17	Oklahoma Fault Zones
11-18	Southeast United States Fault Lines
11-19	Southeast/Midwest Epicenter Zones
11-20	Utah Epicenter Zones
11-21	Areas of Seismic Activity in Washington State

3
PART
Maverick observations, theories, and tools

12
CHAPTER
Maverick observations

Southern California's 1992 earthquakes provided scientists with added information about the effects of the increased underground pressure that quakes can create.

The twin earthquakes in June triggered flows of crude oil from mountains up to 150 miles from the nearest epicenters.

Natural seeps exist in the Santa Susana mountains, 35 miles northwest of Los Angeles. They have been observed for decades, but were aggravated by the earthquakes. The increased flow was first reported 10 days after the two quakes occurred, the oil flow having increased from seeps to a measured flow of about 2300 gallons per week.

Previous temblors have also provided insight on the effects of earthquakes on subterranean systems:

Springs (of water) in the San Francisco Bay area showed a marked increase in flow after the 7.1 magnitude Loma Prieta earthquake in 1989. The flow of geysers and streams in Yellowstone National Park were affected by a major earthquake in the region in 1959.

It has been established that large earthquakes affect water pressure, and, possibly, water levels. Because underground pools of both oil and water are closely related (oil being the lighter material), it is reasonable that when water moves around in response to a quake, the oil moves also.

On March 27, 1964, the greatest earthquake ever recorded on the North American continent occurred in Alaska. It measured about 8.4 on the Richter scale, and lifted over 25,000 square miles of the earth's surface three to eight feet, and pushed one end of Montague Island as much as 33 feet upward. Also, a 35,000 square mile area around Kodiak and Anchorage sank. Some areas sank as much as six feet. Buildings in Anchorage fell 30 feet in just a few seconds.

The quake's effects were felt around the world. In Iran, the ground surface oscillated one-third of an inch. In Houston, Texas, the streets rose almost five inches. In Georgia, well water levels oscillated 10 to 20 feet. In Vicksburg, Mississippi, a boat was torn from its moorings. In the six months following the earthquake, the number of aftershocks amounted to over 10,000.

In February, 1969, the National Academy of Sciences published the first of eight volumes in a study of the great Alaskan earthquake. Much of the study dealt with hydrology. Part of the study showed that such a great quake can permanently alter water levels thousands of miles away. This has led some experts to believe the observation of well and lake water levels might serve as both an index of an area's vulnerability to earthquakes, and as a warning of an impending quake.

The following is one of the related theories. A type of seismic wave slows down, then speeds up to normal just before an earthquake. As stress on the rocks in a fault builds to a critical level, cracks and fissures develop. Ordinarily, ground water flows into these cracks as soon as they form, which permits the wave to travel at its normal speed. But, in the case of fault-zone pressures, the new cracks open faster than the water can fill them, and the wave slows down. Then, when the water finally fills the cracks, the wave speed returns to normal. However, an earthquake can occur because of the weakening of the rocks by the new cracks.

In central California, an 83-foot-deep water well has been used to test these theories, and to observe earthquakes for over 15 years. Because of heavy pumping of the aquifer, the water level has decreased gradually from about 56 feet to its current 61-foot depth, and the water temperature is about 69.6 degrees (F) in average.

Slight oscillations of water level, variations in water temperature, and many instances of increased water pressure have been observed during times of seismic activity. Turbidity and the water's mineral content have also varied. Relative to its influence on seismicity, water can prove to be either a quake stimulator or an inhibitor.

Measurements of the friction scars on rocks of past quakes along the San Andreas Fault have indicated that earthquakes along the fault tend to release only about 20% of the energy that would be expected, according to calculations.

As a quake inhibitor; a very hot, pressurized level of water exists about seven miles in depth along the rocks at the base of the fault. This water causes the rocks to be more pliable, and thereby causes these segments of the fault to stretch, rather than to crack. Water seems to be the mechanism that reduces friction.

Scientists have experimented simulating these deep conditions, with a piece of granite that was sliced, simulating a fault, placing water into the fracture, and subjecting the rock to various influences of temperature and pressure.

Under high pressures and temperatures over 500°F, the granite assumed clay-like characteristics, stretching and bending easily. When the heat and pressures were reduced to conditions like those near the surface, the lubricating action of the water was observed.

Shallow-focus earthquakes can be stimulated when water seeps into the rocks near the surface, where the pressure is lower and the water is cool. Water acts more like a mechanical "spring" forcing rocks to shatter apart.

This feature is exhibited at the surface, where mechanical weathering has been observed for years. Frost wedging is one way, thermal change is another.

Laboratory tests consistently show that rocks have high compressive strength, but weak tensile stress tolerance. Rock is a rather poor conductor of heat, and expansion and distortion can set up thermal stresses. These stresses act at right angles to the surface. These stresses are relieved with the least expenditure of energy, by upward or outward movements, which rupture the rock parallel to the surface and pull away from the expanded rock.

A few observation tools and circuits are covered in the following section.

Observation tools

The following tools are readily obtainable and very simple to maintain and use. They serve many functions other than for seismic observation, and the average homeowner might already have one or more of them.

Hydrometer

A *hydrometer* is used to find the specific gravity of liquids. The specific gravity is a determination of the density of the liquid as compared to water. The weight of water is expressed as 1.000. Soluble solids will increase the weight.

The hydrometer is usually a long, slender glass tube with a weighted bottom. It is placed in a container of the liquid to be measured, and the level it sinks to is read on a scale affixed on the side of the tube. Some hydrometers are electronically operated, but inexpensive models that operate by gravity are just as accurate, are easy to use, and require no maintenance.

One example of a hydrometer is the *acidimeter*, which is used to test automobile batteries. This instrument measures the concentration of acid, which increases the density of the liquid in the battery.

pH meter

The *pH* is the number of hydroxyl ions present in the soil or water. It is a chemical term for measuring amounts of acid or alkali in a solution. *pH* means "Potential of electricity for positive hydrogen ions." The measurement is for the quantity of free hydrogen ions, the foundation for all acids present in an unknown solution.

On the pH scale, pure water is 7.0. Substances that read below 7 are acids and those above 7 are alkalies.

There are various forms of this meter, as well. There are electronic meters, various chemical indicators (such as litmus paper), and reactive chemicals that are measured by amount into the solution being tested, referencing the amount of chemical needed to achieve neutral readings.

A very simple direct-reading meter is shown in Fig. 12-1. This portable meter measures the pH on the meter and requires no power supply. A hydrometer is also shown on the photo.

Thermometer

The thermometer is (in this case) used to measure water temperature. Special sensors are used for this, and some water temperature thermometers are used by fishermen, farmers, weather observers, swimmers, and others.

Some very accurate sensors are made for use with electronic multimeters, and many hobby and home-based indoor-outdoor thermometers contain sensors that can be used to monitor water temperature.

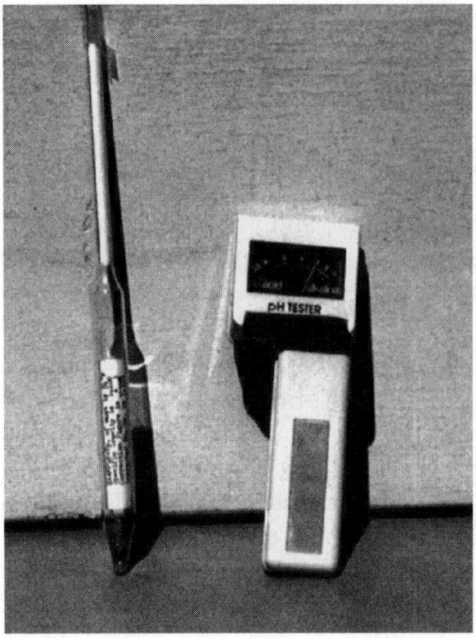

12-1
pH meter on the right, hydrometer on the left.

For the seismic study application, the measurement of water temperature can be done via a cable long enough to be immersed into the water table, or the temperature can be measured at the discharge of a pumping plant.

Accuracy is important, but the resolution is most critical. The thermometer should read to at least a resolution of one-tenth of one degree.

Circuits

The following electrical circuits are somewhat exotic in their applications, but are very simple to build and operate. The circuits are based on easy-to-obtain integrated circuits and are excellent low-cost research tools.

Water-resistance monitor

Water has a resistance, which varies with the amounts and compositions of materials that are dissolved in the water.

The simple circuit shown in Fig. 12-2 gives an audible indication of resistance. As shown, with a sample containing a low mineral content, like distilled water, the resistance is very high and no tone will be heard. The frequency output will lower as the concentration of dissolved solids increases, as this will lower the resistance of the water.

The values of the resistors in and out of pin 3 of the IC can be changed to give an audible indication of the high resistance of distilled water, and comparisons can be made from this value, or the circuit can be used as shown, where a change in the frequency of the tone would be an indication of the changing mineral content.

Output can be observed on a frequency meter, on a multimeter, an oscilloscope, or just by listening. The 2N2222A transistor is the audio amplifier for the circuit.

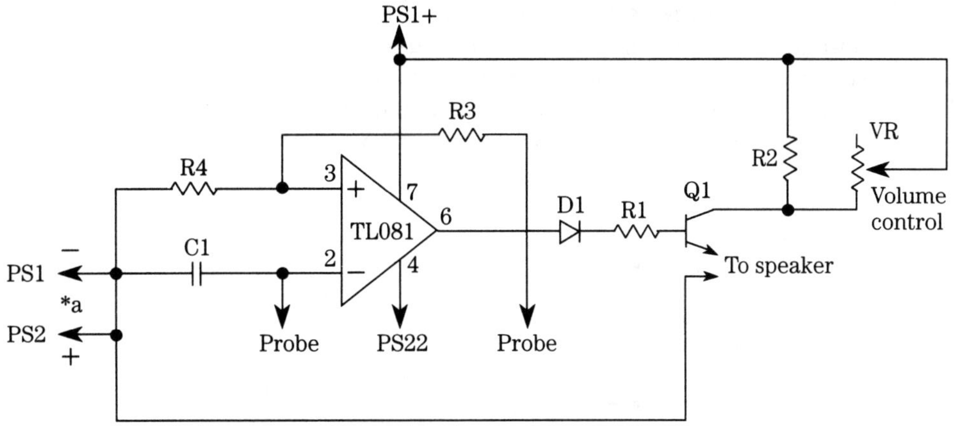

a. The SPST power switch is installed on the line where the "−" of PS1 and the "+" of PS2 join, for the on/off switch.

12-2 Water resistance monitor schematic.

Probes in the sensor should be kept at a constant distance apart, adjusted to about one inch in separation. The sensor diagram is shown in Fig. 12-8.

The 741 op-amp IC can be used. However, the circuit is shown with the TL 081, which is recommended. Do *not* use potentiometers to adjust the resistance values; use fixed resistors of at least a 2% tolerance. The parts list for the unit is given in Table 12-1.

In use, the circuit operates as an audio tone generator. As detailed, the response of the circuit will be from about 3 kΩ to about 15 MΩ, with the varying tones. When used with a digital multimeter, ranges above and below those of human hearing range can be displayed, allowing even more sensitivity. Try various resistance values between the input probes to get an idea about the varying responses the circuit is capable of producing.

Table 12-1. Parts List for the Water Resistance Monitor

Part	Description
IC1	TL 081 Integrated Circuit
SW1	SPST Power Switch
D1	1N4001 Diode
PS1, PS2	9-Vdc/100-mA Power Supplies or Batteries
VR	5-kΩ ½-W Potentiometer
Q1	2N222a NPN Transistor
C1	0.01-µF/50-V Disc Capacitor
R1	510-Ω, ¼-W Resistor
R2	620-Ω, ¼-W Resistor
R3, R4	91-kΩ, ¼-W Resistors
Misc.	IC Socket, Speaker, Cabinet, Hook-up Wire, Binding Posts, Wires for Probes

Conductivity and resistance change monitor

The circuit of Fig. 12-3 is sensitive to changes in conductance and resistance of a body, such as soil. In operation, adjust the potentiometer to a setting where both LEDs are off. If the setting is not sensitive enough, replace the potentiometer with one of a smaller value; 10 kΩ to 50 kΩ. Parts as shown in the diagram.

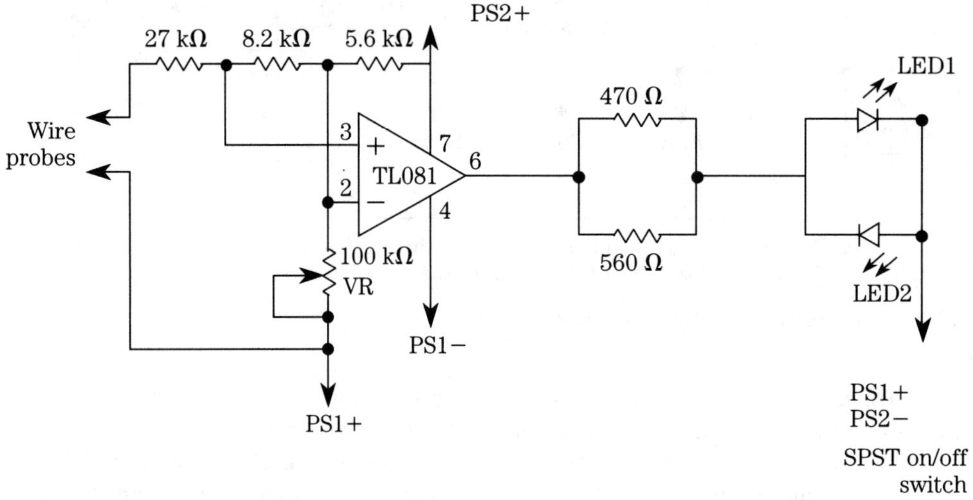

12-3 Schematic for the conductivity and resistance change monitor.

Circuits based on the 555 timer IC

The 555 timer circuit is a stable controller that is capable of producing accurate time delays or oscillations. Terminals are provided for triggering or resetting, if desired.

The 556 IC is a dual 555. The two timers operate independently of each other, sharing only the ground and the V_{cc} pins.

The supply voltages can be from 4.5 to 18 Vdc, and 9 Vdc battery operation is common. The temperature stability is about 0.005% per degree C, which makes these components ideal for field use.

The circuits here use the 555 timer as a "clock"; when activated, the circuits produce a series of pulses. The frequency of the pulses depends on the resistance and capacitance values used, and that is the basis for these monitoring circuits. Figure 12-4 is a graphic representation of what these pulses look like on an oscilloscope. Pinouts are shown in Fig. 12-5.

Water seepage detector

Based on the 555, the detector circuit of Fig. 12-6 can detect and display, via the speaker, the incursion of water into formerly dry areas. It is useful for monitoring spring flow patterns and areas of soil saturation.

In use, as moisture begins to permeate strata, the circuit begins to deliver a series of pulses, which become more rapid as the percentage of water in the test area

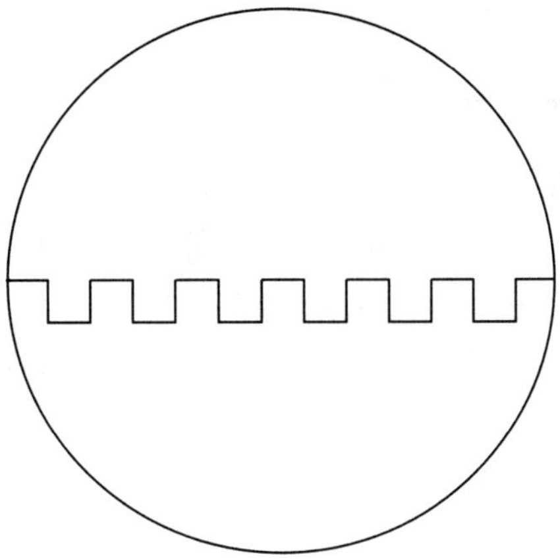

12-4 555 timer IC pulses on an oscilloscope would look like this.

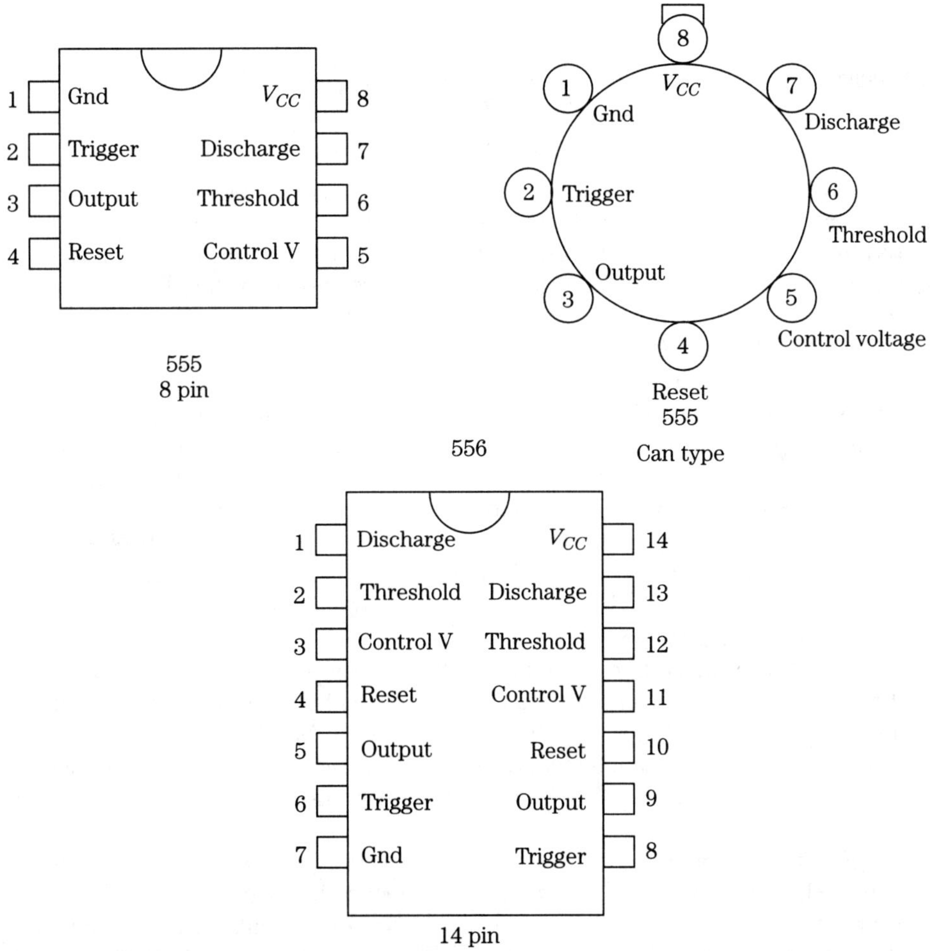

12-5 Pinouts for the 555 and 556 timer ICs.

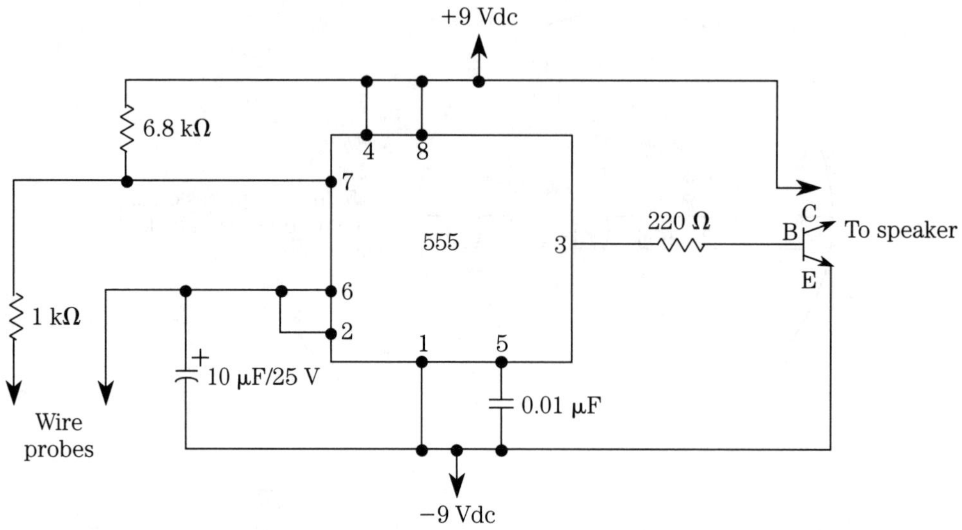

12-6 Schematic of the water seepage detector.

increases. The transistor amplifies the pulses; as shown, the circuit can drive a large speaker with good volume. An LED can be used as a visual indicator instead of the audio speaker. In this case, the transistor is omitted. The – lead of the LED is connected to the lead of the 220-Ω resistor formerly connected to the base of the transistor, and the + lead is connected to +9 V. Parts are shown on the diagram.

Water level indicator

Figure 12-7 is the schematic for a simple continuity tester that uses the 555 IC's features as a "clock." The circuit can be used as test equipment, but it also makes an excellent water detector. For this function, its probes are positioned about 0.5 to 1 inch apart at the end of a suspension-type sensor, and placed within 0.5 inch of the ambient water level, or even closer (depending on sensitivity requirements).

A rising water level needs only to make continuity between the probes to activate the circuit and produce a tone. If the probes are placed just within the body of water to be monitored, a falling water level will stop the tone. Again, the parts are shown in the diagram.

Figure 12-8 shows construction hints for a sample sensor used to monitor water level characteristics. The sensor uses a metal box as both the weight and the positioning vehicle. To monitor the water level, its height is adjusted so that it is suspended at a measured distance just above or just below the water level. Any change in the level makes or breaks continuity. The observations are made at the display unit.

To monitor seepage and saturation of strata, the probes are lowered into the ground in close proximity to known areas of perforation. Changes in the moisture content of the surrounding soil are indicated as a series of pulses, or a change in resistance, depending on your choice of circuits. A completed display unit that uses

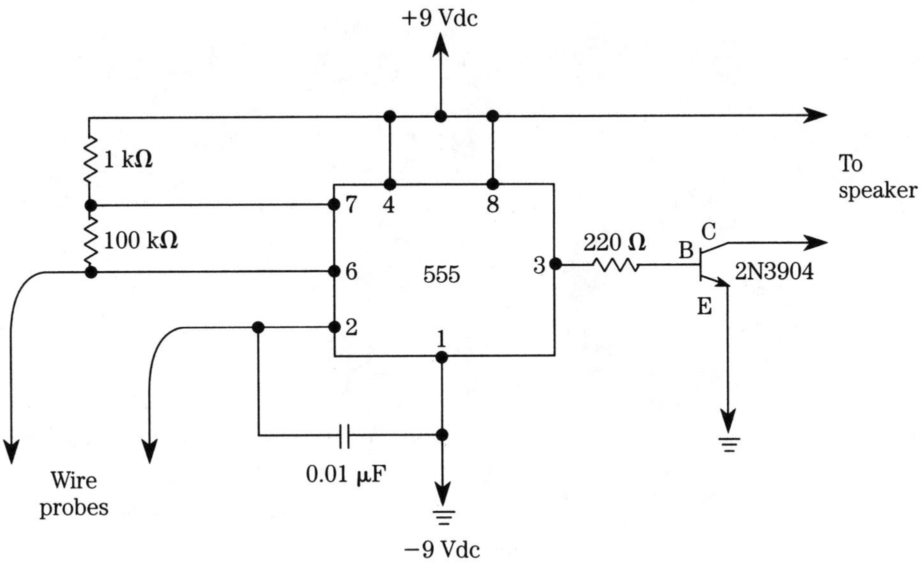

12-7 Water-level indicator schematic.

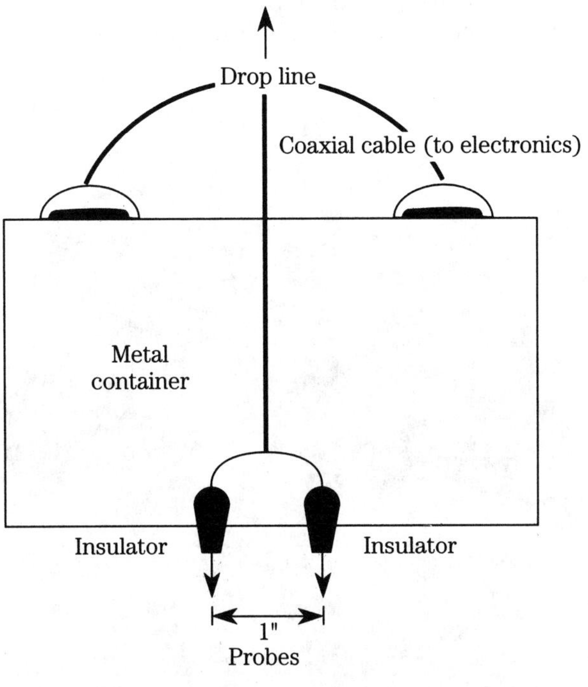

12-8 Sample water observation sensor.

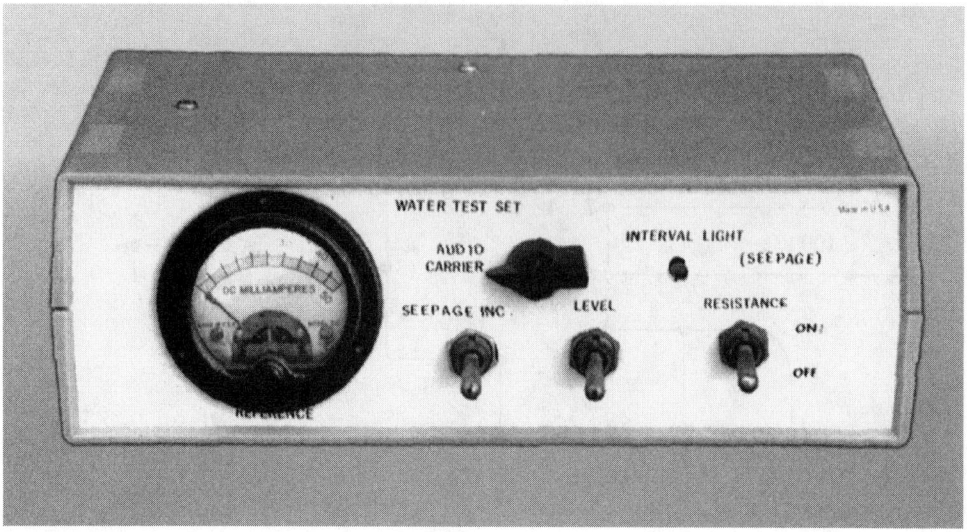

12-9 Completed water characteristics monitor.

three circuits for water level monitoring, seepage detecting, and resistance value monitoring is shown in Fig. 12-9.

Electromagnetic energy receiver

The electromagnetic energy receiver is a versatile and sensitive monitor for various kinds of electromagnetic signals. The basic breadboarded circuit is shown in

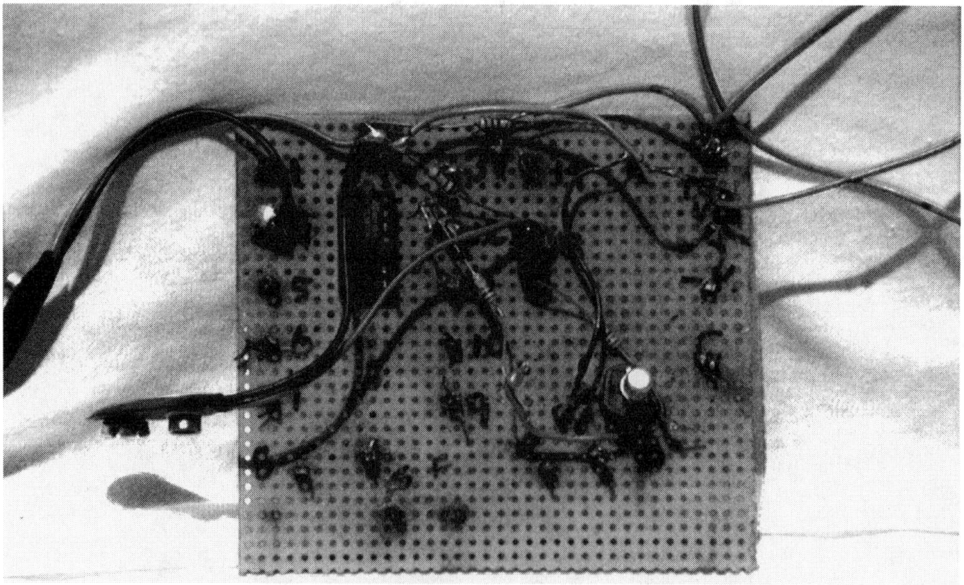

12-10 Breadboarded electromagnetic energy receiver.

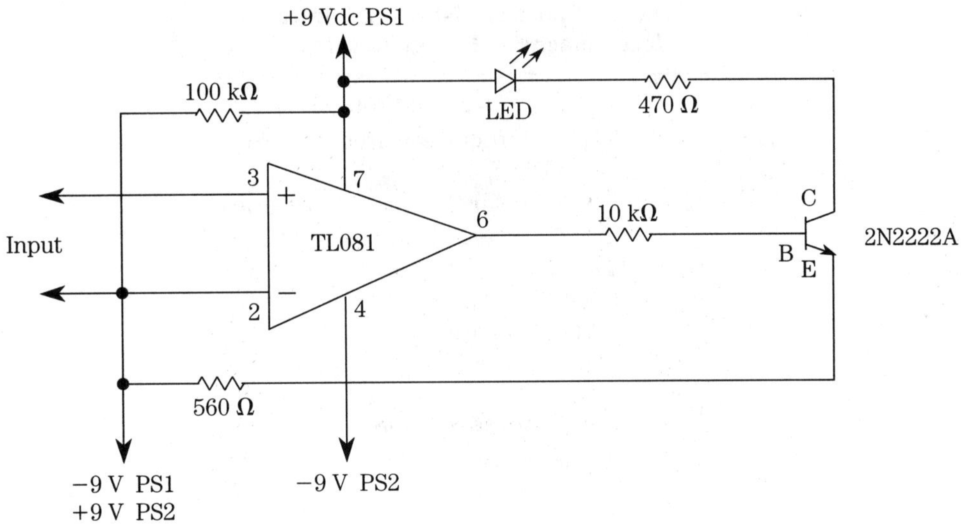

12-11 Basic schematic of the electromagnetic energy receiver.

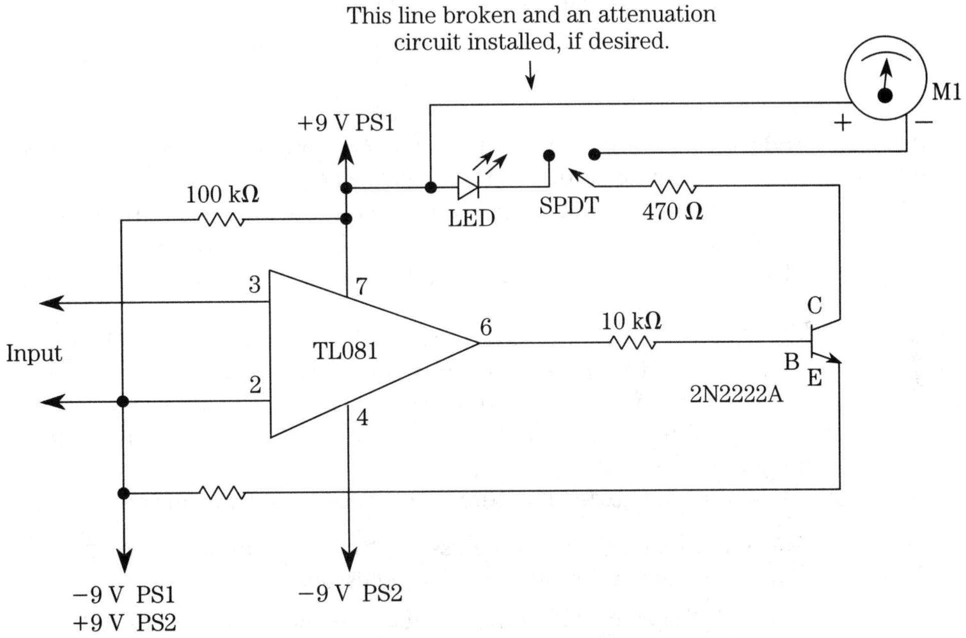

12-12 Basic receiver schematic with meter.

Fig. 12-10. Two versions, one of which has a meter, are shown in the schematics of Figs. 12-11 and 12-12. The parts list is provided in Table 12-2.

The schematics are oriented to the reception of seismic impulses, but, with suitable antennae or sensors, other kinds of signals can be received. These signals in-

Table 12-2. Parts List for the Basic Electromagnetic Energy Receiver

Part	Description
IC1	TL 081 Integrated Circuit and Socket
LED1	Red LED
PS1, PS2	9-Vdc Power Supplies or Batteries
Q1	2N2222A NPN Transistor
R1	470-Ω, ¼-W Resistor
R2	560-Ω, ¼-W Resistor
R3	10-kΩ, ¼-W Resistor
R4	100-kΩ, ¼-W Resistor
	For Metered Version:
M1	dc Meter (See Text)
S1	SPDT Switch
	For Attenuation Circuit, If Used:
S2	8-Position Rotary Switch
R5 to R12	Resistors (See Text)

clude those produced by the induction of electric equipment, emissions of energy from neon lights, broadcasting stations, static discharges, etc.

The basic unit of the electromagnetic energy receiver also serves as the basis for a seismic sensor-display scheme when it is constructed as an interferometer.

Figure 12-13 shows the schematic diagram for the receiver with signal amplifying and audio output sections.

In Fig. 12-12, the SPDT switch selects between the LED-flashing and the meter-reading circuits, for the choice of how to observe the impulses being received.

The line between pin 7 of the IC and the + terminal of the meter is broken by an eight-position resistor attenuation-control circuit switch installed, which is used if the meter is of a milliampere value.

The builder has a wide choice of meters that can be used with this circuit:

- A 0- to 1-Vdc full-scale meter can be used with the circuit, as shown in the diagram, and no attenuation resistors are necessary.
- A 200-μA meter can be used. With this value, the following ½-W resistors should be used in the attenuation circuit: 8.2 kΩ, 16 kΩ, 27 kΩ, 33 kΩ, 47 kΩ, 62 kΩ, 82 kΩ, and 100 kΩ.
- If a 1-mA meter is used, the recommended resistor values are as follows: 1 kΩ, 2.2 kΩ, 3.3 kΩ, 4.7 kΩ, 5.6 kΩ, 8.2 kΩ, and 10 kΩ, with one terminal direct. The highest sensitivity is achieved with the lowest resistance setting.

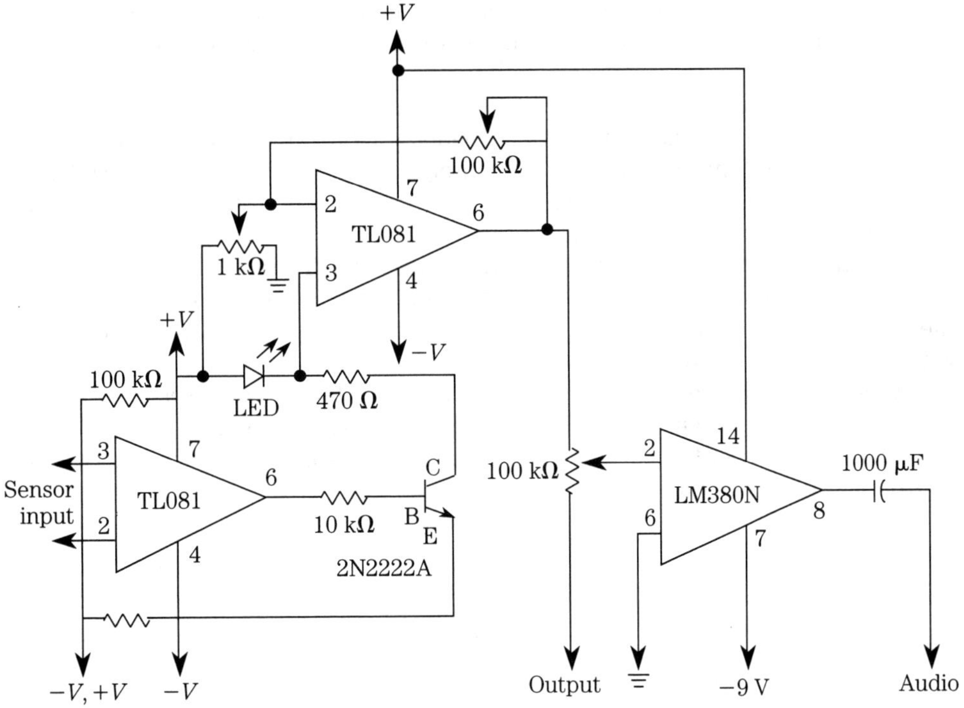

12-13 Electromagnetic energy receiver, with amplifier and audio sections.

The basic electromagnetic energy receiver can be constructed by following these steps:

1. Pin 2 of the IC connects to one end of a 100-kΩ resistor, and the other end of this resistor goes to pin 7.
2. Pin 2 is the terminal for the – lead of power supply 1, and the + lead of power supply 2, as well as one sensor input and one end of a 560-Ω resistor.
3. The other end of the 560-Ω resistor is connected to the emitter terminal of a 2N2222A transistor.
4. Pin 3 connects directly to one sensor input.
5. Pin 4 connects to the – of power supply 2.
6. Pin 6 is connected to one end of a 10-kΩ resistor, the other end of which goes to the base of the transistor.
7. Pin 7 goes to the anode (long lead) of the LED and to the + of power supply 1.
8. The cathode (short lead) of the LED goes to one end of a 470-Ω resistor, the other end of the resistor connects to the transistor's collector lead.
9. Pins 1, 5, and 8 of the IC are not used.

The meter, if used, is connected in place of the LED. If desired, an SPDT switch can be installed to select between the LED or the meter for signal display.

Interferometer

An *interferometer* is a device that divides a light ray into two or more parts, then brings them together to cause interference. The interferometer is used in astronomy, helping to determine direction, distance, and the size of the light source (star) being analyzed.

The interferometer can also be used to measure air pollution. It is capable of making the small measurements necessary to measure the refraction of the gases in the atmosphere. It can also be used to measure high temperatures, and to measure with great precision, like the threads of screws.

A very basic interferometer can be constructed for use as a seismic detector. The schematic diagram in Fig. 12-14 uses the basic circuit of the electromagnetic energy receiver. A super bright LED is used for the light source, along with a length of fiberoptic cable and a basic receiver.

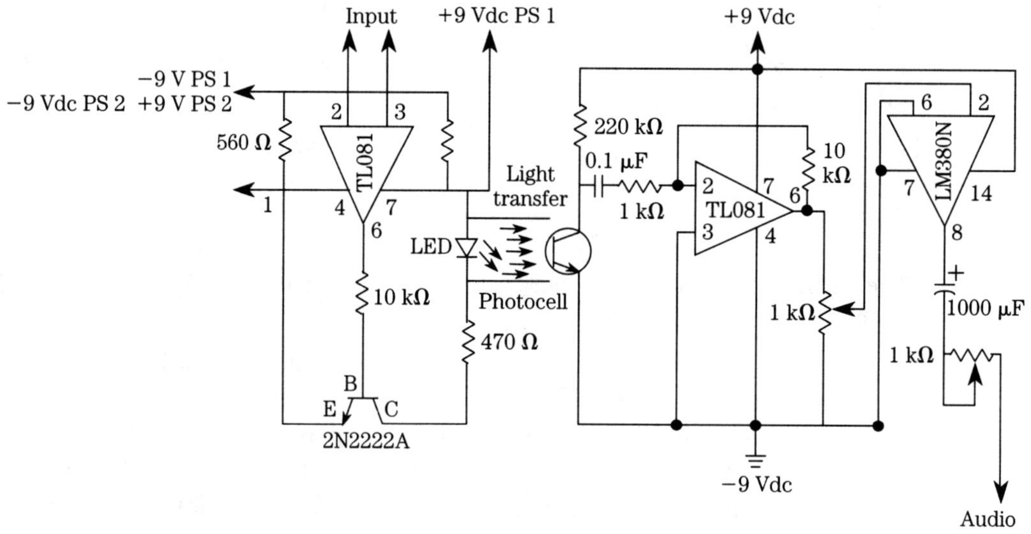

12-14 Interferometer schematic.

Lasers are often used in this application, and, in fact, the laser-operated principle is the same as that used here. This circuit, however, is simpler and far less expensive, yet it is quite effective.

The interferometer is basically an interference meter. The fiberoptic material takes the LED's coherent light, which varies in intensity, and breaks up the light. The vibrations inherent in the fiber contribute to the noise, and, therefore, its interference.

The receiving circuit uses a phototransistor to monitor the light output. When the fiber vibrates, the reflection of the light changes. This inconsistency is monitored.

Phototransistors are usually most sensitive to infrared light, so the use of an infrared emitting diode in place of the LED, along with the use of an infrared transistor in the receiver, can be a very effective system.

When a visible LED is used, the "super-bright" red is the best to use. Of course, the visible LED is the one to use when a visual indication of interference is desired.

An alternative system uses a bright white light, such as from a "grain of wheat" bulb, in the interferometer. Especially with this source, and somewhat less so with the superbright LED system, for a very short run of light from the emitter to the receiver, a plastic drinking straw can be used with surprising results!

The plastic optical fiber is the type used for seismic monitoring because interference is most prominent with this kind of fiber. This feature is put to good use because the phototransistor is positioned to monitor the light output, and the circuit is designed to monitor interference. The incoherency is processed through the audio amplifier as hisses, squeals, tonal changes, thrums, and growls, as some of the light waves arrive at the end of the transmitter before others. The sounds are similar to those produced by the seismic-flare receiving system, but the vibration periods are different.

The electromagnetic energy receiver and the associated interferometer are very sensitive to seismic vibrations, and settle very quickly, as can be seen when monitoring the meter output circuit.

Coilless fiberoptic-based seismic sensors

Sensors for the light-interruption and interferometer detection systems are fiberoptic based, in contrast to the electromagnetic seismographs used in the earlier systems.

The sensor in Fig. 12-15 uses a high-output light source, 10 to 15 feet of 1000 µM plastic fiber, an infrared phototransistor, and two plastic containers.

One of the containers is a five-gallon size, and it serves as the protective housing for the sensor. The other container will hold 2½ to 3 gallons, and the fiberoptic material must be wound around its outside. The hook-up is via coaxial cable to the amplifier and monitor.

In operation, vibrations received at the base of the container interrupt the coherency of the fiber, and cause interference to the light transfer. This interference is converted to display by the phototransistor and the receiver's electronics.

Figure 12-16 is an example of coilless sensor for use with the light-interruption/intensity detection systems.

The sensor does not require a magnet, just a weighted mass for the plumb. Movement of the light source over the receiving phototransistor changes the resistance, which is monitored by the receiver.

The model used a cadmium sulfide photocell with high resistance (12 kΩ minimum at 0.1 footcandle) in darkness, and low resistance, (250 ohm maximum at 100 footcandle) in light. A six-inch length of plastic straw was used for the light transfer.

The electromagnetic energy receivers operate with standard electromagnetic sensors, as well as with the light-interruption types. When using electromagnetic

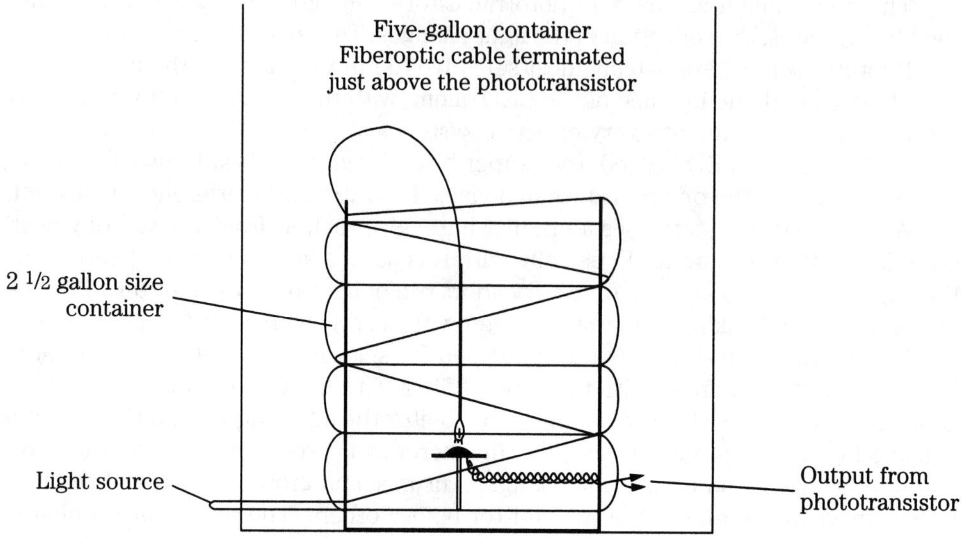

12-15 Coilless fiberoptic sensor.

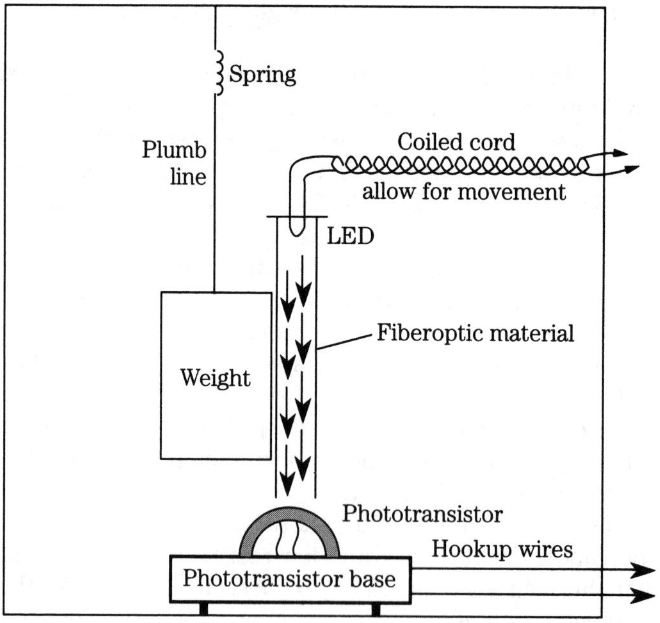

12-16 Coilless sensor.

sensors, commercial types and generic versions with the largest amounts of wire work best.

These receivers are perhaps the most sensitive of any described in this book. The meter circuit is capable of displaying even the most minute movements very rapidly and without saturation. The circuits might best be used to monitor microseisms, the possible precursors of earthquakes, and to trace or chart fault lines.

They can also be used to monitor ground conductivity, atmospheric electrical activity, and VLF/ULF radio propagation when used with suitable inputs. The outputs can be interfaced to computers, multimeters, and oscilloscopes for studies and comparisons. Although they work well with standard seismograph sensors, there are a number of other designs that can be used.

Interferometer-based sensors

Sensor A of Fig. 12-17 uses two light sources and two phototransistors for reception. The clear slide is attached directly to the weight, and the graticule pattern is selected for the desired sensitivity. If the circuit of the electromagnetic energy receiver is used as shown, the slide can be used clear, with no light interruption pat-

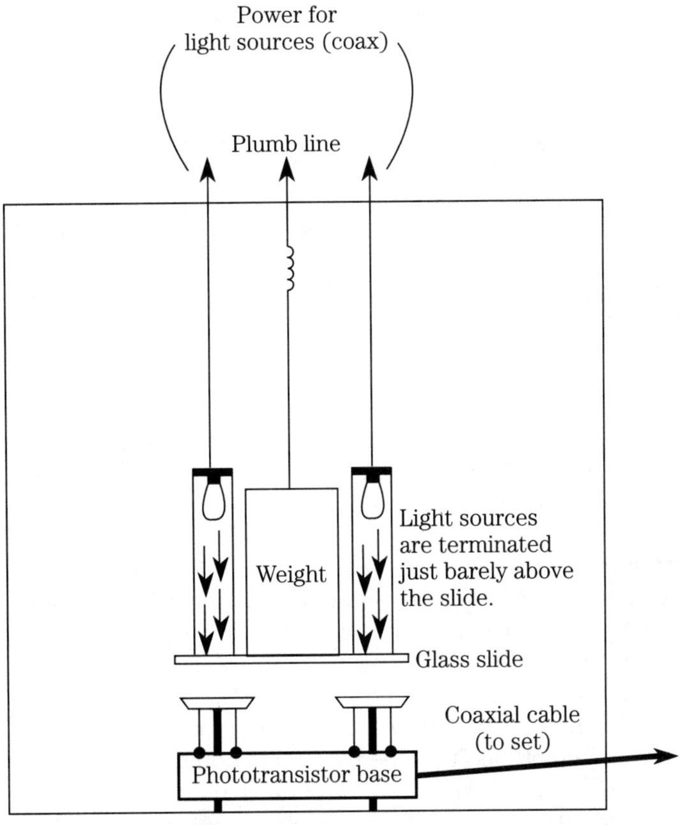

12-17 Interferometer-based sensor A.

tern, as the intensity of the movement governs the intensity and pulsation of the light. In this case, an infrared or visible LED is used as the light source. When a constant-intensity light source circuit is used, a graticule pattern must be selected.

The sensor of Fig. 12-18 uses the LED for the light source, but in a stabilized position. The phototransistor is attached to the plumb weight. A lens is used, positioned to focus the beam of light at the center of the phototransistor.

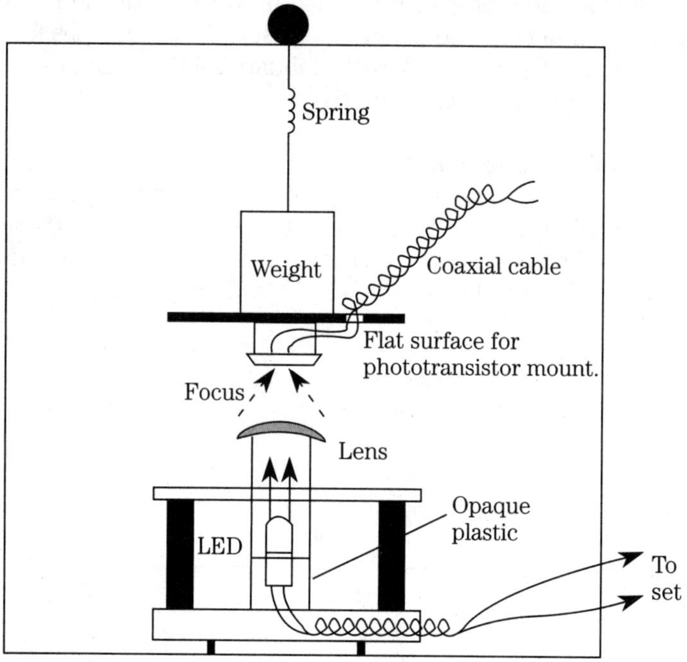

12-18 Interferometer-based sensor B.

Figure 12-19 uses a superbright red LED, or an infrared LED emitter positioned inside a PVC pipe fitting that is attached to the plumb weight. A second fitting holds the lens, and allows for the adjustment of the focus by the use of the threading of the pipe fittings. The phototransistor is positioned beneath the patterned slide.

Light intensity meter

A circuit variation of the interferometer is a light intensity meter. Any op-amp IC can be used. The circuit shown in Fig. 12-20 is capable of very sensitive readings, and using a high-gain, low-noise IC will provide the best results. The circuit described uses a TL 081 IC.

The fixed resistors must be *balanced*; they all must have the same value. Resistance values of 83 kΩ to 150 kΩ can be used, depending on the sensitivity requirements. Fixed resistors are ¼ W. VR1 is a 500-kΩ potentiometer used for circuit balance (values from 100 kΩ to 2 MΩ are usable), and VR2 is the potentiometer for meter null. Here, 1 kΩ to 10 kΩ can be selected. Potentiometers are ½ W.

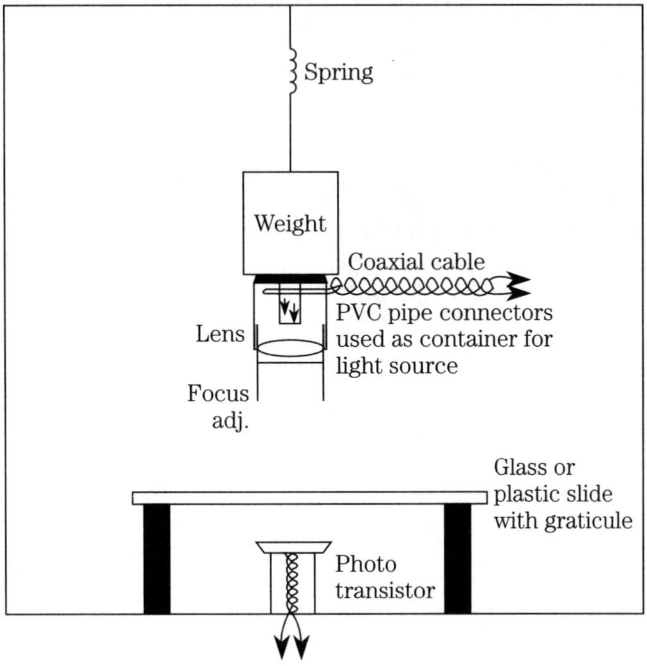

12-19 Interferometer-based sensor C.

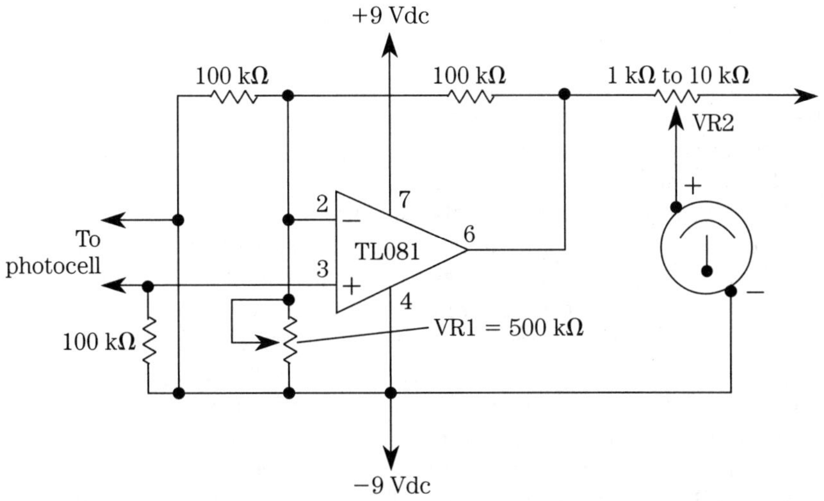

12-20 Light-intensity meter schematic.

When a 1-mA meter is used, fixed resistors will typically require values from 91 kΩ to 100 kΩ, VR1 will use a 500-kΩ potentiometer, and VR2 will need a 5-kΩ or a 10-kΩ potentiometer. Light intensity changes observed by the photocell can then be read on the meter.

Earthquake alarm

A variation of the light-interruption detection system uses transistor-based circuity. Essentially, the circuit is a pulse counter, using a photocell and a relay that can activate a device to indicate the interruption of a light beam. Figure 12-21 shows the schematic for such a circuit, and Table 12-3 is its parts list.

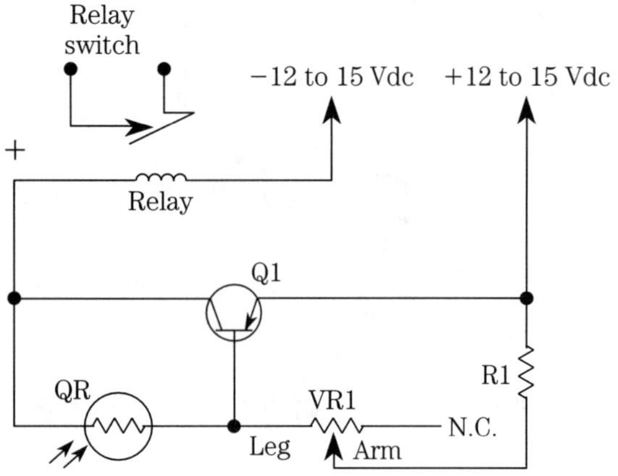

12-21
Schematic for the light-interruption switch: the earthquake alarm.

For seismic survey, using a counter as the device activated by the circuit would read the number of times that a light source, or the receiving photocell, has moved from their contact point. Indications would be either movement occurred or by the number of counts, the intensity of the movement would be set. The earthquake alarm would use a buzzer that would sound during earth movements.

Other uses are possible, including use as a burglar alarm; when a buzzer is used, it would be activated when unauthorized movement interrupts the light beam. A driveway annunciator is also an option.

Both the light meter and pulse-counter can use an already existing light source for the light beam, and, depending on intensity, the source can be located quite some distance away from the receiver. Using the lenses in a toy telescope can dramatically increase the system's range. A high-intensity, lens-focused infrared emitter or a laser beam can also be used. Variations in components allow for either day or night use.

As for the circuit, the pnp transistor is used as an amplifier. Its collector lead is connected to a relay. When used with a normally closed relay, light striking the photocell lowers the resistance, producing bias voltage to the base of the transistor. This is a negative voltage. Thus, with the transistor on, the relay switch is open. As soon as the light striking the photocell is interrupted, the resistance of the photocell goes "high." Then the bias of the resistor turns the transistor off, and the switch closes. The relay opens again when light hits the cell, switching the device off.

Table 12-3. Parts List for the Light Interruption Switch

Part	Description
QR	Photocell: Cadmium Sulfide (or equiv.)
Q1	2N3906 pnp Transistor
R1	100-Ω, ½-W Resistor
VR	1- to 10-kΩ, ½-W Potentiometer
PS	12- to 15-Vdc Power Supply
RY	Relay with Contacts to Suit Load

The potentiometer (VR) serves as a sensitivity control for the circuit. It can be adjusted so that the amount of light hitting the photocell must be above a certain intensity to operate the relay contacts. By the use of a normally "open" relay, light hitting the photocell will turn on the relay, as opposed to a normally "closed" relay, where a light interruption will turn it on. The VR serves the function of allowing different light conditions to operate the relay.

When using a small relay, the contacts can be used to operate a 6- or 9-V buzzer. Thus, it can be used to indicate earth movements. A counter can be used for pulse counting; in this way, any object (such as cars or people) can be counted.

It is very important to select a relay that is capable of handling the load connected to it or the circuit will be destroyed. To operate ac circuits, a relay with coil contacts of 6 to 12 Vdc must have switch contacts rated at 120 Vac and amperage to cover the requirements of the circuit.

Interferometer graticules

Some simple examples of graticules used with the interferometer are shown in Fig. 12-22, which shows samples of how a clear slide can be inked to disrupt the light source during movements. Opaque screen or wire mesh patterns can also be used. Wire mesh, available at many hardware stores, comes in a variety of standard sizes, is very durable, and works well in the system. Using ¼-inch wire (or a smaller gauge) is recommended. The higher the density of the pattern, the more sensitive the reception. With consistent, measured graduations (such as ¼", ⅛", 1/16", 1/32", for example), the interferometer can be used as a measuring tool.

Damping mechanisms

In electronics, *damping* is a process used to control spurious vibrations or oscillations. A damped oscillation, then, is a mechanical or electrical oscillation whose amplitude is sharply reduced in successive cycles, so that they gradually die out to zero amplitude.

A *damped balance* is a balance equipped with a method for damping the to-and-fro movements of the suspension, so that it comes to rest more rapidly. There are electrical and mechanical ways to accomplish damping. Among the electrical methods are the use of intricate biasing systems, resistor shunting, and relatively simple

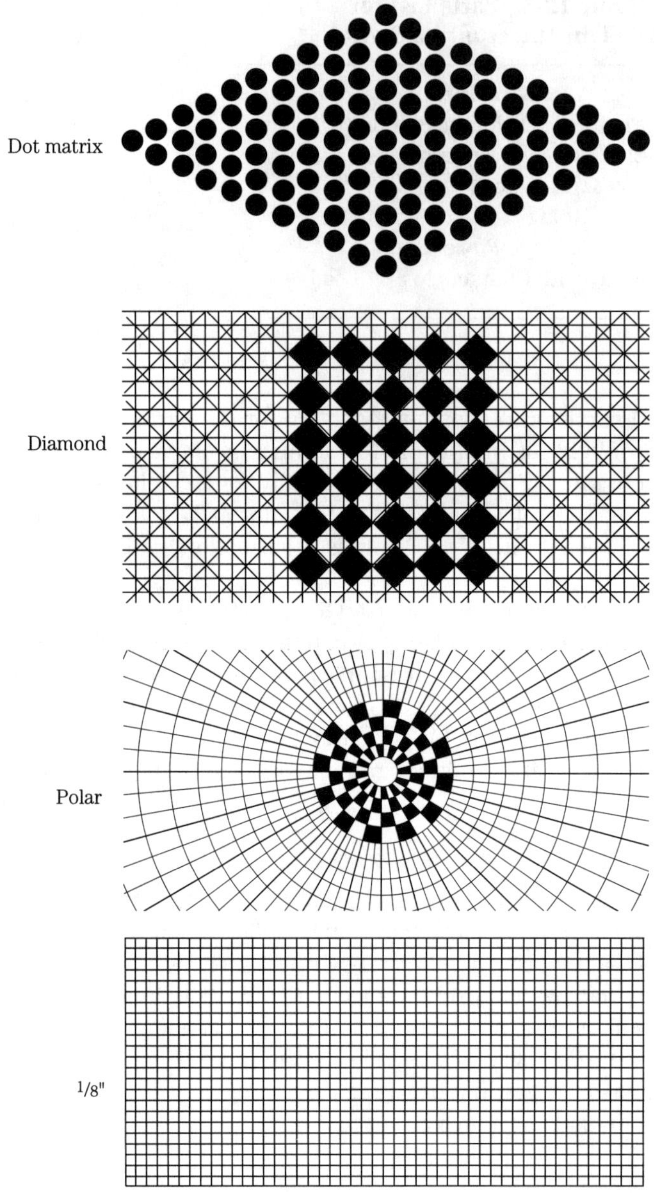

12-22 Sample interferometer graticules (reticles).

resistor and capacitor-based attenuation circuits. Some of the circuits described in this book incorporate one or more of these features as a part of their designs.

For the reception and transductance of seismic signals, mechanical damping can also be built into the sensors. The object is to detect movement, then allow the sensor to settle quickly, smoothing out the "knee-jerk" movements. One way of doing this can be implemented on plumbs using either magnets or weights.

When a weight is used, a rigid appendage is attached to the weighted mass, and is allowed to dip into the damping medium. The attachment can take the form of a needle or nail, and be bolted, wired, welded, soldered, etc., to the plumb.

Water, oil, and even alcohol can be used as the damping medium. The volume of liquid used, along with the specific gravity, influence the sensitivity and the amount of damping action provided.

An example of mechanical damping is shown in Fig. 12-23. If a magnet is used as the weight, wind the coil around the outside of the container. If the coil is part of the plumb, use magnets around the perimeter of a metal container. Mineral oil is the sample damping medium.

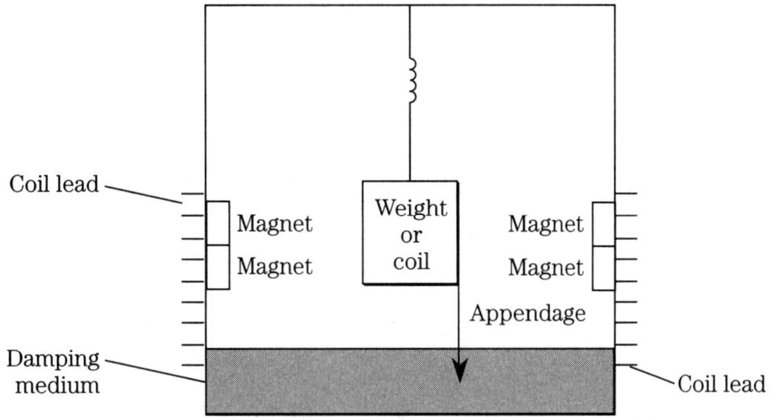

12-23 An example of mechanical damping.

Sensor mounting and testing

It is rather difficult to use mechanical damping in field installations where sensors are buried very deeply. However, a kind of "sand box" can be constructed for testing and for semi-permanent installations, allowing the "back yard" placement of seismic sensors.

Four pillars usually support the box. The pillars are driven into the ground to a depth of four to six feet. The inside dimensions need to be at least double the overall size of the sensor's container. A level wooden platform is built onto the pillars. This is then painted and weatherized. Then, a coating of hard plastic or metal, depending on the kind of sensor used, lines the bottom and sides of the box. The sensor is mounted onto the center of the box. Then the box is filled with sand or small-diameter gravel. The entire installation is then covered for protection. This final covering can be very simple (such as a tarp) or more elaborate (such as a doghouse style of construction).

The finished box can be used to test different sensors, or as a permanent home for a continuous-duty station. A sample construction diagram is shown in Fig. 12-24.

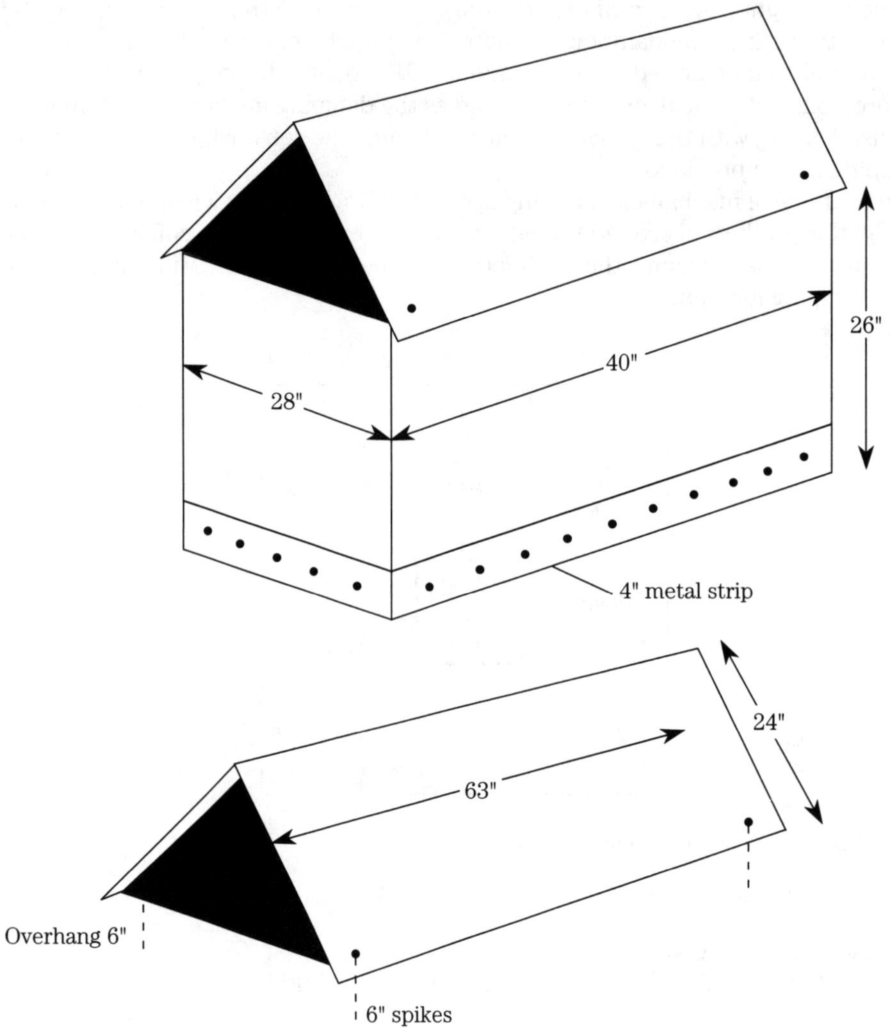

Sensor test box construction notes:

A. The roof is removable. This is accomplished by using four 2-×-2 posts, one at each corner of the inside box. Holes are drilled through the roof and into each of these supports, through which 6" long spikes are inserted, which can be removed when necessary.
B. The box base is secured to four 4-×-4 posts driven into the ground. The base is lined with weatherproof plastic (or metal) with a 4" wide metal strip around the base of the unit.
C. The coaxial cable hookup from the box to the electronics should be underground, inserted into a length of PVC pipe and sealed for protection. This keeps the cable out of the way and safe from most damage.

12-24 Sensor test box.

Geiger counter

A Geiger counter is a device used to detect radiation (i.e., radioactive emissions). It is a relatively simple instrument to build and use, and quite versatile.

When used to monitor cosmic rays, the Geiger counter can serve as a solar activity and flare receiver.

The Geiger tube is very fragile. Use great care in its handling. Also, the circuitry is of high voltage, so you must take great care in constructing it. The metal wall of the tube is the – electrode, the ground. A wire going through the tube is the + electrode. These electrodes are maintained at a high voltage, just below the value necessary for the breakdown of the gas contained in the tube. When any other ionizing influence is received by the tube, the gas ionizes, setting up a weak current that is processed by the circuit as meter deflections or audio clicks.

The Geiger counter circuit of Fig. 12-25 is very sensitive. When an "alpha"-type tube is used, it can detect even these "alpha" rays, which have the least penetrating power of all the radioactive emissions. Beta rays, with more power, also can be detected, and gamma rays, with the most penetrating power, are detected along with cosmic rays.

Cosmic rays are responsible for the random clicking or meter jumping of the Geiger counter. This occurs several times per minute every day of the year. Cosmic rays are from outer space, and are influenced by solar activity, flares, and the earth's magnetic field. The intensity of cosmic radiation varies with the sunspot cycle. Usually, during times of high solar activity, the observed cosmic radiation is low. During solar flares, cosmic radiation is often intensified. The earth's magnetic field affects the speed and direction of the rays.

Most Geiger tubes require a high voltage of 350 to 500 V for operation. In this project, you need a high-voltage supply, a two-transistor audio amplifier, and a meter-reading circuit. The design is for a benchtop counter and the main objective is to observe cosmic radiation.

With a few modifications, the unit could be adapted for field use. The main need for this would be an inverter transformer power supply that could step up battery voltage to the high voltage needed for the tube. The following paragraphs show how to construct a "brute-force" power-transformer supply.

You should enclose the tube in some kind of protective housing. Metal must be avoided because the frame of the tube is at high-voltage potential. Glass works very well, and a small bottle with a screw-type cap can easily be adapted to stabilize the tube. The window of the tube is positioned near the top of the bottle, and the cap can be removed when the counter is in use. The bottom of the bottle can be drilled out with a glass drill bit or it can be carefully punched out.

The bottle is positioned so that it is level with the top of the cabinet's housing, or just slightly below this level. Then, a door is affixed or a hole is drilled in the case to allow for the tube's window. The bottle's cap gives protection for the very fragile window of the Geiger tube. A grate can be placed over the opening of the case to prevent objects from falling on the tube's window.

The Geiger counter can be used to monitor cosmic rays. It can be used to detect radon gas, and to check for radioactivity in water, steel, concrete, etc.

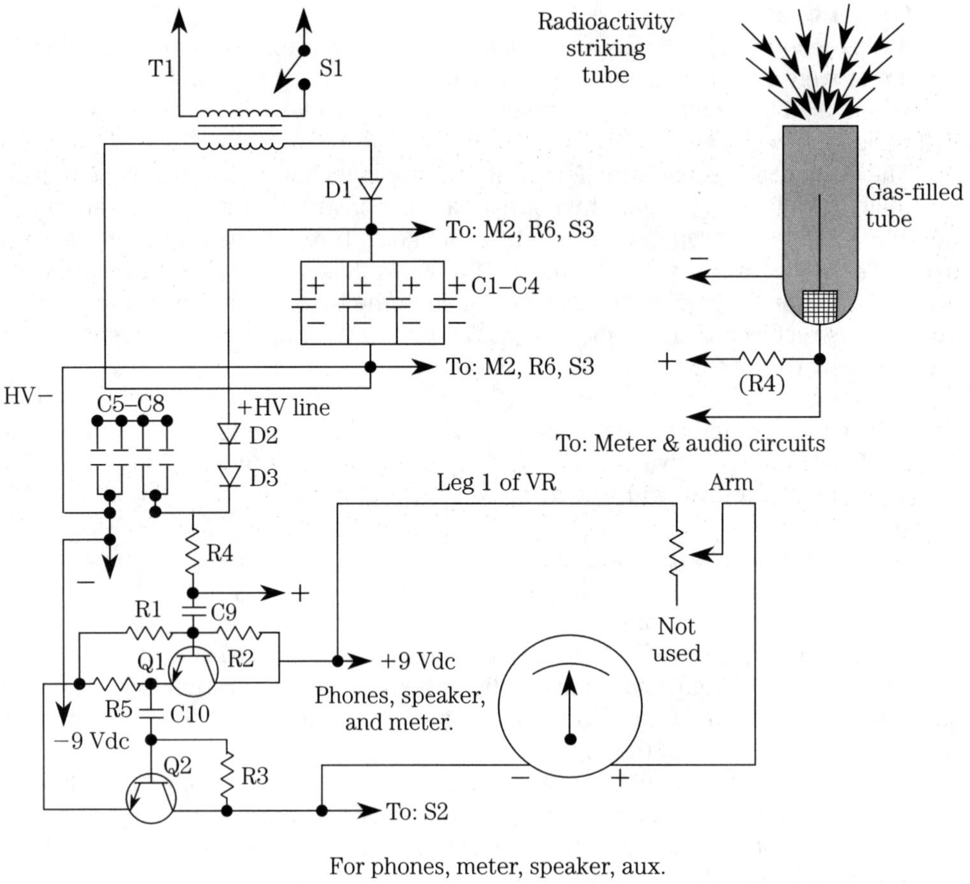

12-25 Geiger counter schematic diagram.

One seismic survey use of the Geiger counter is by setting the tube at the end of a length of coaxial cable and positioning it at the top of a well casing, or at a protrusion of the bedrock, to monitor the possible changes in radioactivity, or the emission of radon gas during times of seismic activity. The parts list is shown in Table 12-4, and a finished unit is shown in Fig. 12-26.

Construction of the Geiger counter begins with the selection of a power transformer. The transformer's secondary should be from 240 to 270 V, at a 40- to 90-mA current rating. Many old radios use a 250-0-250 V transformer in the power supply, and these are fine for use. The center tap is used in this case, because the secondary leads would put out too much voltage (over 500 V) for the circuit to operate properly.

Table 12-4. Geiger Counter Parts List

Part	Description
C1 to C4	100-µF, 450-Vdc Electrolytic Capacitors
C5 to C8	0.47-µF, 630-Vdc Capacitors
C9	0.0068-µF, 1600-Vdc Capacitor
C10	0.047-µF, 630-Vdc Capacitor
D1 to D3	1N5408 Diodes
Q1, Q2	2N3904 Transistors
R1 to R3	560-kΩ, ½-W Resistors
R4	1.5-MΩ, ½-W Resistor
R5	2.7-kΩ, ½-W Resistor
S1	SPST On/Off Switch
S2	Rotary Switch; 1-Pole, 4-Position, for Signal Out: Meter, Phones, Speaker, Auxiliary
S3	SPST Pushbutton Momentary Switch, Normally "Open"
R6	3-kΩ; 5- or 10-W Bleeder (Power) Resistor
M1	0 to 3 mA dc FS Meter
M2	0 to 500 Vdc FS Meter
VR	100-kΩ ½-W Potentiometer
VT	Geiger Tube
PS	9-Vdc Power Supply for Audio Circuit
T1	Power Transformer: 115 Vac Primary, with 260 Vac Secondary
Misc.	Case, Phones, Speaker, Terminal Posts, Knobs, Etc.

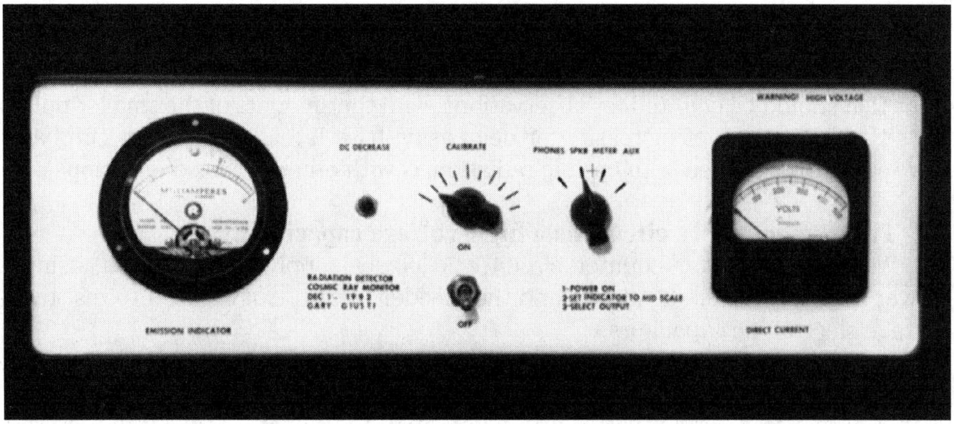

12-26 Geiger counter.

In the unit pictured, the transformer is rated at 260 V, using the center tap, but it puts out almost 300 V under no-load conditions. Upon rectification, the dc voltage available is about 410 V, and about 360 V can be measured at the Geiger tube. Large-value electrolytic capacitors are a must in this circuit, and 450-V electrolytics are used in the set.

An alternative power supply, not requiring such large capacitors, can be constructed from two or more smaller transformers, rectified and connected in series to produce the required voltage for the circuit. The last stage of such a circuit could be adjustable, such as the one used in the Harbor Master radio circuit, to really peak the power supply for the optimum performance.

Either way, because the circuit uses such a high voltage, the capacitors will retain the voltage for some time after the circuit is turned off. Therefore, a bleeder circuit is used to discharge the capacitors.

The bleeder is connected as follows: A 500-Vdc meter is connected between the + and – of the power supply, observing polarity. Hook one end of a 3-kΩ, 5- or 10-W power resistor to the – lead. The other end of this resistor goes to one terminal of a normally open SPST pushbutton switch, with the other switch contact connected to the + lead of the electrolytic capacitors. Use this circuit after the power to the unit is turned off, and observe the draining of the dc voltage from the capacitors.

The meter in the counter is a 0- to 3-mA, FS dc, like that used in several earlier projects. In operation, it is adjusted to a mid-scale reading, or to where the maximum sensitivity to radioactive emissions is obtained. The – terminal of the meter connects to the collector of transistor Q2, and the + of the meter connects to the arm of a 100-kΩ potentiometer. One lead of this potentiometer is then connected to the collector lead of transistor Q1.

A note regarding interpreting the readings and counting the emissions: this Geiger counter is very sensitive, more sensitive than most low to moderately priced commercial units, and you can be able to detect smaller and more subtle radiation bursts than with other units.

As this Geiger counter is operated, some deflections, and the subsequent audio readings, will be at a distinctly higher level than others. The larger pulses are the ones that should be counted for observations, with the amounts of the smaller pulses noted as indications of increasing or decreasing base levels of radiation (probably not so much associated with cosmic radiation as with other localized emissions).

The power supply circuit as a high-voltage capacitor tester

Without the Geiger counter circuit, the power supply configuration, using a power transformer, diode, meter, and the bleeder circuit, can be used to test high-voltage electrolytic capacitors.

For the tester, two pushbutton, momentary (normally off) switches are used. One is placed in series with the incoming ac line, the other is used in the bleeder circuit. A power transformer with a secondary of 60 to 90 V at 30 to 100 mA is used, along with a 150-Vdc meter.

The electrolytic capacitors are omitted, and at their connection points (+ and –) two terminal test posts are installed, along with the meter and the bleeder circuits.

A capacitor under test is hooked to the test posts, and is first charged by pressing the power switch, then it is discharged by pressing the bleeder switch.

A good capacitor will charge, rectify the voltage, and then will recover some of the voltage after the bleeder switch is pressed momentarily and released.

As always, be sure to install electrolytic capacitors correctly (according to the proper polarity), and also, to fully discharge the capacitor before disconnecting and handling it.

This circuit will work with any capacitor having a voltage rating higher than the rectified output voltage of the power supply. In this example, any capacitor with a voltage rating of 150 WVdc or higher can be tested.

13 CHAPTER

Digitizing the output

When used with some of the seismometer circuits previously described, a digital multimeter can provide a precise, rapid-reading display of seismic activity.

The readout displays deviations from a referenced quiescent-state observation value. In this use, the seismometer circuitry is used as a signal receiver/converter and as an amplifier.

Three circuits are particularly well-suited for this purpose: the LF 356-based seismometer, the magnetic field disturbance detector, and the magnetic transducer detector.

Figure 13-1 shows the basic LF 356-based seismometer hooked up to a BK-388-HD digital multimeter. The multimeter is connected with the correct polarity to the A output of the seismometer. For this use, the seismometer must be built with the larger-value milliampere meter or one of at least a 0- to 3-mA or a 0- to 5-mA full-scale meter value. The seismometer is set up and calibrated to a 1-mA reading on the meter, as was previously done. The multimeter is set to the 0- to 200-mV range. The quiescent reading on the multimeter should be about 16 to 20 mV. Small vibrations will cause dips of 0.1 mV to 0.2 mV, and large quakes can cause vibrations from under 10 mV to over 20 mV.

The multimeter's 0- to 200-Ω scale range can also be used. In this setting, the seismometer's meter reading will jump to 2 mA, which is the reason for the use of the larger value meter. The 2-mA reading is a normal condition, and it should not be readjusted. Quiescent readings will be about 33 Ω, and oscillations from 20 to 40 Ω can be observed during quakes.

The 2-kΩ frequency range setting of the multimeter can also be used, and with excellent results. Quiescent readings will be near 0.000, and the readings can oscillate to about 120 Hz during seismic activity.

The quiescent stated here are not absolute values, but will vary somewhat because of variations in set and sensor construction, particularly with the selection of the power supply used to power the set. These figures are to be used as a guide.

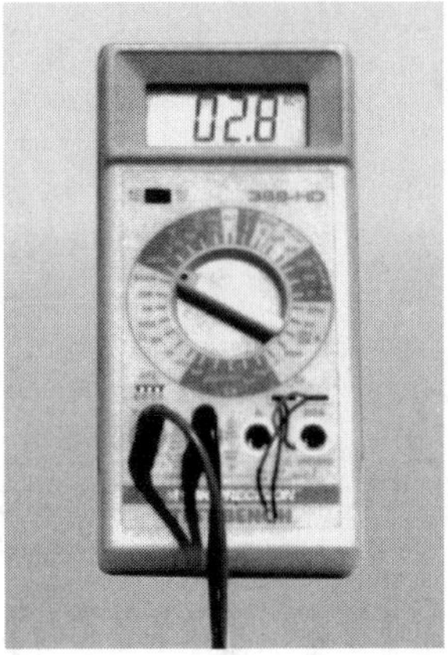
Maxtec International Corporation

13-1
The BK 388 HD multimeter.

For testing, wind a solenoid-type radio coil with about 100 turns of #24 magnet wire around a 1½ inch form. Hook up the ends of the coil to the seismometer's input, adjust it for a 1-mA reading, hook up the multimeter, set it to the 0- to 200-mV range, and, while observing the multimeter, wave a magnet back and forth over the coil. You should observe wide fluctuations of values on the multimeter.

When using the magnetic field disturbance detector as the converter/amplifier for the digital readout, the multimeter connection is again made to the 1-mA meter connections.

Using the 0- to 200-mV range setting, the quiescent value for this example should be about 43 mV. Readings from 20 to 70 mV might be observed during earthquakes. Unless a larger value meter is used in set construction, the multimeter's ohms scale should not be used because it will pin the meter. The 0- to 2-kΩ scale can be used, but the sensitivity of the display will be very low.

The 0- to 2-kHz frequency range can be used, with quiescent values being about 60 Hz. During quakes, readings from 20 Hz to 60 Hz will be displayed.

With the magnetic transducer detector, the multimeter connection is again made in parallel with the meter's output. With a 200-μA meter in the set, the unit is adjusted for its mid-scale observation position of about 100 μA.

The multimeter, set to the 0-to 200-mV range, should read about 43 mV, and can range from 35 to 50 mV during times of seismic activity. Again, because of the low value of the meter used in the set's construction, the ohms scale of the multimeter should not be used.

Perhaps the greatest efficiency in using this circuit will be obtained by using the frequency ranges of the multimeter. Either the 20 kHz or the 200 kHz range can be used. Quiescent values will be about 19 kHz, near the top of the multimeter's 20-kHz range. During periods of seismic activity, oscillations down to nearly 0 will be recorded. In all cases, and with each of the circuits used for the digital readout, the quiescent readings observed on the multimeter can be adjusted with the detector's sensitivity adjustment. Figure 13-2 shows the later version of the BK-388-HD multimeter, the model 389.

Figure 13-3 shows a complete analog/digital seismic wave display unit. On the

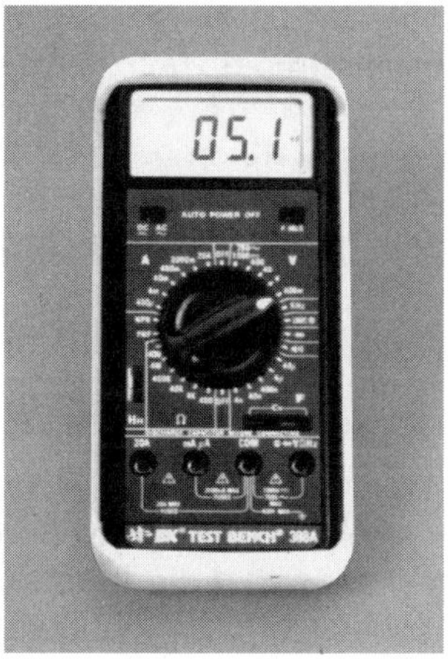

Maxtec International

13-2
The BK 389 multimeter.

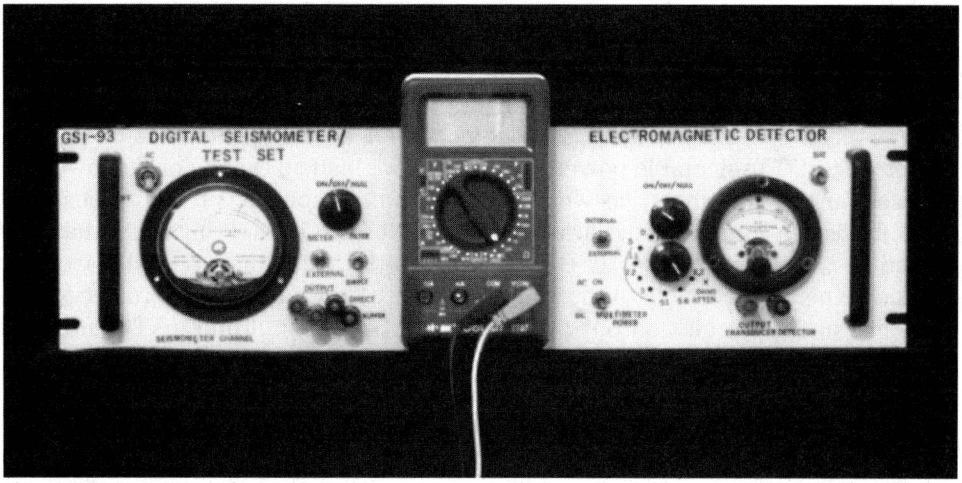

13-3 Analog/digital seismometer.

left is a single-channel basic seismometer circuit and on the right is a magnetic transducer detector circuit.

The seismometer section utilizes a 0- to 3-mA meter, wired for both A and B outputs. A switch selects between meter reading only or meter plus output terminals. A resistor attenuation network is not used, and only a single 1-kΩ resistor is used in series with one sensor input lead, as a filter. This resistor can be switched in or out of the circuit.

The magnetic transducer detector uses a 0- to 200-μA meter, with the output terminals connected in parallel with the meter. An SPDT switch selects either meter or output. In this section, filtering resistors are connected in series with one sensor input lead. Here, an eight-position rotary switch selects between one direct position, and one of seven other resistors, varying in values from 500 Ω to 8.2 kΩ. The resistors provide electrical damping for the output circuits, helping to steady the meter deflections.

Each circuit contains an off/on switch with the 100-kΩ, ½-W potentiometers used in the null/offset sensitivity controls.

Power requirements are very low; an ac adaptor was disassembled so that its transformer could be mounted directly on the two units' common circuit board. The supply was rated for 8.3 V at 10 mA, and both circuits, plus the multimeter, can be operated simultaneously.

In parallel with the power inputs to both of the display circuits, a 9-V battery snap is routed to the outside of the cabinet, where a battery can be installed for portable use. A panel mounted SPDT switch selects from the ac or dc power options, with another switch selecting between the battery and off positions.

The multimeter was mounted to the cabinet with two bolts driven through the back of the meter's case. The meter's power supply, again a 9-V battery, was also routed to the back of the cabinet, where another battery snap was mounted for the multimeter's power supply.

The multimeter can be set up independently, with no modifications to its power supply. However, an experimental power hook-up is used in this set: The negative lead of the multimeter's power supply is connected to the same negative lead as the two display circuits, and the positive lead is connected to one lead of a switch (SPDT) selectable power supply positive lead so that ac or dc power can be used for the multimeter, as well. If this kind of a hook-up is tried, be sure that all dc leads are connected in parallel, not in series. An interesting thing that happened in the prototype set is that the magnetic transducer detector's circuit wound up being accessible to the multimeter with only a one-lead connection: the plus side of the circuit's output to the plus lead of the multimeter. The negative lead is not necessary.

Here, the multimeter's circuits are not altered, so all multimeter functions can be used for other testing as well. A 3½-digit meter is excellent for digitizing outputs, and is capable of excellent resolution and sensitivity. Four- or 4½-digit meters are not recommended. They are too sensitive; so many numbers flash that readings can easily become confused.

When used in radio propagation and solar flare study, a minigraph chart recorder can be used in place of the multimeter. A suitable receiver is used as the sensor input. Figure 13-4 shows my Rycom 2174A, a tube-type receiver from the early 1950s. This military-commercial set has a frequency range of from 300 cycles to 420 kHz in six bands. It has been used for solar flare monitoring, atmospherics and lightning-strike monitoring, seismic wave theory tests, and for the calibration and testing of the receiver projects in this book. The Rycom 2174A is an excellent moderately priced professional-duty receiver, often available from surplus outlets.

Figure 13-5 shows an excellent, easily made electromagnetic sensor that can be used for fixed installations in areas of high seismic activity, or for detailed readings of large intensity, close-proximity earthquakes. It can also be used in portable installations after large quakes for mapping and testing faults and fault zones.

13-4 The Rycom 2174-A military-commercial VLF receiver.

13-5
"Water bottle" electromagnetic seismic sensor.

The case is a plastic, five-gallon water bottle 19" in height. An opening was cut into one side, and a "U" bracket-mount radio transformer (power or filament) was bolted to the bottom center of the bottle.

The secondaries of the transformer were cut off, and the primary leads were routed out of a bottle through a hole drilled near the top. These leads are connected to the display unit's input. The plumb line uses two compression springs; one at the top and one at the bottom. The line is 15-lb. test fishing line, the length of which must be carefully measured so that the magnet can move near the transformer iron, but not so near that the metal attracts the magnet.

A rectangular magnet is used, with pole pieces that are bolted together. The magnet assembly measures 2¾" in length by ⅝" wide. The end of the bottom spring is held by this bolt. A hole was drilled through the neck of the bottle, ¾" from the top and a 3" long bolt installed here. This bolt holds the end of the top spring. During transport, this bolt is removed, allowing the entire suspension line to be removed without disassembly. In this way, quick and easy installation, and portability of the sensor is accomplished.

The physical size of the transformer is of some importance. A very small transformer sacrifices efficiency and sensitivity. A very large transformer compromises the period swings of the suspension line.

The transformer used in the sensor example is 3.8" W × 2.5" D × 2.5" H. This is a good size, and any transformer near these dimensions should work well.

A pair of secondary wires could also be used to run another receiving unit, if desired, or another coil could be wound around the outside of the container after the holes are taped and sealed with waterproof sealant. Make sure everything is waterproof, and strap-mount to a cement or wooden post for a quasi-permanent installation. For portable use, bury the sensor half-deep at a selected observation site.

This kind of sensor, along with the analog/digital seismometer, has been used in a fixed installation with an oscilloscope as a seismoscope. It is not as sensitive to microseismic activity as other electromagnetic sensors, but it has been used as a portable unit for measurements and tests along the San Andreas and other nearby faults. Figures 13-6 through 13-10 are "field-photos" of the units in actual survey work. Figure 13-10 shows the sensory units in a novel arrangement, allowing for some particularly interesting readings. The commercial vertical sensor was hooked "in series" with the generic "long-period" sensor, which was oriented in a northwest-reading direction. Resulting deflections contained both long and short period, and vertical and horizontal data.

13-6
At the south end of the Calaveras fault.

13-7 Near the San Andreas fault, northwest of Hollister.

13-8 Surveying near the Pinnacles State monument.

13-9
Near the San Benito river, 20 miles southwest of Hollister.

13-10 Testing near Tres Pinas, near the Diablo mountains.

14

CHAPTER

Meteorological applications

Independent studies of unusual elements require extensive local observations, and, if networking is involved, speedy dissemination of information. As wide a geographical coverage as possible, along with homogeneity of data and of monitoring systems, are also important considerations. A computerized network would be ideal for this effort, allowing for the rapid transmission and detailed analyses of data.

This chapter covers some theories and tools for some special investigations of atmospherics as they pertain to weather.

Solar radiation study fits in very nicely with atmospherics and weather study because solar radiation, being the only source of the earth's energy, is the ultimate control of weather.

The study is not a simple one, however, because one of nature's laws is that a body, upon receiving radiant energy, itself begins to radiate energy. The earth's surface receives energy radiated from clouds and from atmospheric particles, as well as from the sun, then radiates energy back to the atmosphere, and through into space.

The distribution of particulate matter in the atmosphere is an interesting study, particularly when the matter consists of particles around which water droplets can condense. Variations in the concentrations of solid particles in the atmosphere should obviously have some effects in the distribution of clouds and precipitation.

The distribution of radioactive matter in the atmosphere is also of interest. The atmosphere is naturally somewhat radioactive because of the decay of naturally occurring isotopes on the earth's surface, and also in response to the cosmic radiation from the sun. Atmospheric radiation measurement is a valuable tool for studying the mixing and movement of air masses. Local average measurements should be recorded over long periods of time so that they can be compared to the values recorded during storms, frontal passages, and areas of jet-stream influences. The atmospheric radiation differences among the various storm routes are interesting to study.

Apparatus

For radiation measurements, the Geiger counter project in chapter 12 is ideally suited for measuring atmospheric radiation. Normal "counts" per minute of atmospheric radiation can range from about 10 counts to about 40 counts per minute. Values are quite variable because of the local composition of the earth at the observer's location, the prevailing air mass, the area where a storm originates and accumulates moisture, and the nature of the sun, as the amount of radiation emitted is slightly and irregularly variable. The solar energy constant fluctuates by one or two percent. Some variations are periodic, such as those associated with sunspots and solar storms. Also, the presence of the atmosphere weakens the intensity of the incoming solar radiation. Solar radiation also is scattered by small particles of dust, of the air itself, and by water vapor. Some of the energy is even absorbed by the water vapor. Ozone and oxygen also absorb solar radiation.

For measuring solar radiation, a very simple integrator is shown in Fig. 14-1. This is simply a solar energy panel. This model has 16 cells, each 3 inches in diameter, that are wired to give an energy rating of 6 V at 500 mA under load. The no-load value of the panel is nearly 10 V in full sunlight. This integrator is called a *solarimeter*.

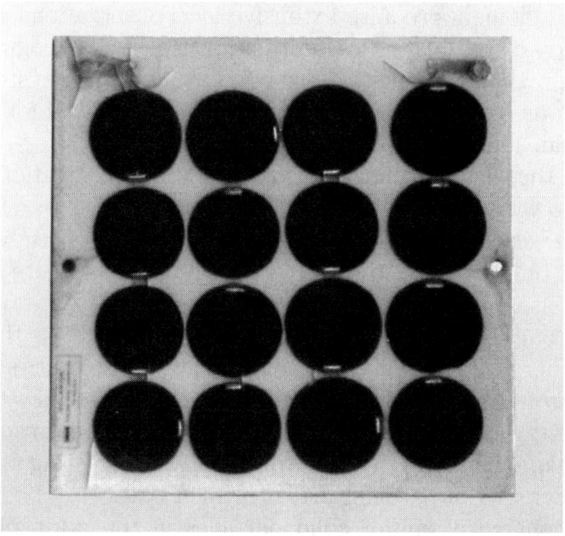

14-1
Basic solarimeter (panel).

The key to this kind of observation is to mark the "trend" of the variation in readings. Any kind of solar panel can be used for a local observation, but if studies are to be coordinated over wide areas, panels, ratings, and components must be similar.

The solar panel is positioned as a rain gauge would be: on a horizontal surface, away from all obstructions. In this case, it is also situated away from anything that can cast a shadow on the panel in the daylight hours, except, of course, clouds. A dc voltmeter is connected, with an on/off SPST switch to the panel, like in Fig. 14-2. A 100-Ω, 1-W resistor is used for load readings (Fig. 14-3). This circuit uses a normally "open" pushbutton switch, with the connections made across the terminals of the

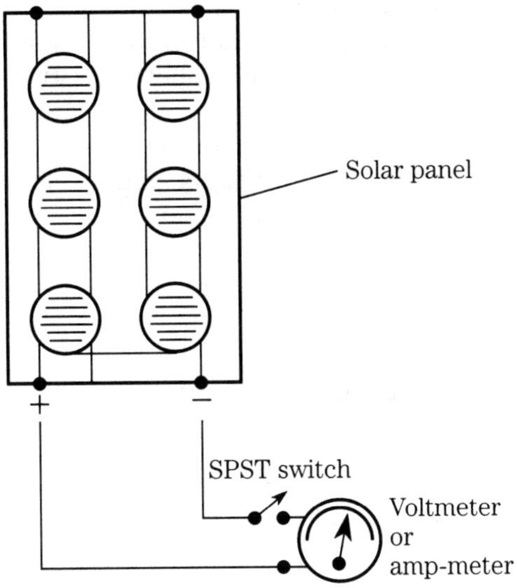

14-2
Solar energy integrator.

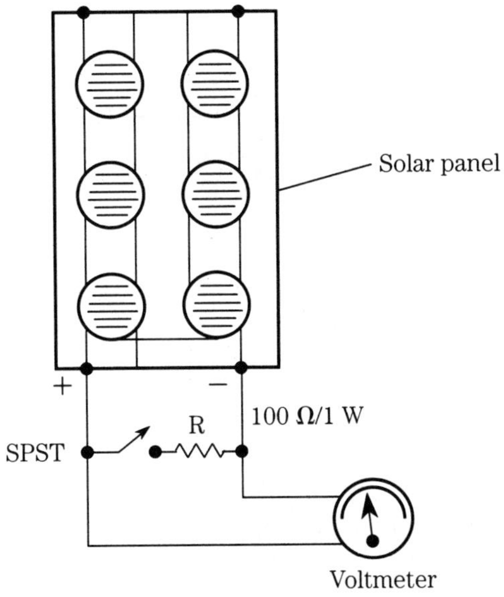

14-3
Solar energy integrator with the "load" resistor.

panel. For current strength, the voltmeter of Fig. 14-2 is replaced with an appropriately valued milliampere or ammeter.

To use the integrator, a predetermined time of day is calculated for the observation; usually mid-day, being the midpoint between local sunrise and sunset. A five-minute average of voltage potential is observed, a 10-second push of the load potential switch is made, and a 30-second average current reading is taken.

Turbidity, reflecting the total amount of particulate matter in the atmosphere, scatters solar radiation. Turbidity variations are important in measuring the radiation and heat balances at the earth's surface. Very simple photo-resistive measurements can be made to monitor this condition.

By using the sky as a giant "graticule," the resistance of the atmosphere to solar radiation can be very easily monitored. Figure 14-4 is the schematic diagram of just such an integrator for this, using a photocell and an ohmmeter.

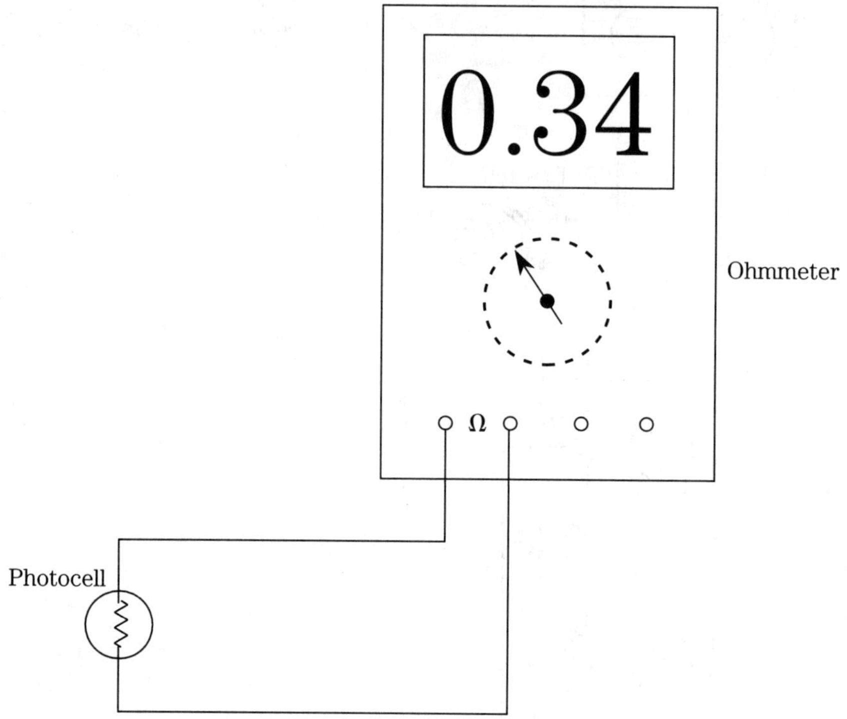

14-4 Solar/atmospheric resistance measurement hook-up.

Take a few considerations into account when building the equipment for monitoring the intensity and characteristics of incoming solar radiation.

One of the variables in building the solar resistance integrator is the characteristics of the photocell used—whether the cell is designed to respond more to visible light or to the infrared light spectrum.

So many different types of solar power cells and panels are available that selection of a panel could be confusing. A common solarimeter configuration uses three solar cells connected in series, along with a 2-Vdc meter or a multimeter set to a 0- to 2-V scale for reading the direct amount of solar power produced by the cells. Other small panels use several cells to provide various outputs of voltage and current ratings. Any type of panel can be used, along with the appropriate meter for observing the readings. Be sure to use a weatherized panel or provide a clear, hermetically sealed, ventilated enclosure for the panel.

A level surface is important to mount the solar cells and panels, and for the photocell used to measure solar resistance. The plastic spools used to hold magnet and other hook-up wire are excellent for mounting the photocell. Cut a piece of plastic to fit over the top of the empty spool, then run the wires through the center of the spool, and down to terminals installed in the sides of the spool for hook-up. The bottom of the spool can be screwed or bolted onto the mounting base for a secure installation.

Diffuse sky radiation can vary greatly even within small distances. Local variations caused by man-made pollutants can be very great, particularly in industrial areas and in heavy traffic areas when an inversion layer is present. Cloudiness affects the resistance. Also, the increase of upper-level water vapor associated with approaching storms, even when the sky is seemingly and visually clear, affects the resistance values.

Atmospheric resistance measurements should be taken in two steps:

1. By measurement with the photocell on a horizontal surface.
2. By orienting the photocell directly at the sun.

Using the basic solarimeter, the load and current reading circuits, and the photoresistive integrator, we have constructed a simple, efficient "Pyrheliometer."

Observation of sky color

The color of the sky results from the atmosphere around the earth. When sunlight passes through the gasses and particulates there, molecules scatter different colors. The shorter the light's wavelength, the more it is scattered. Blue is scattered the most, red is scattered the least. Therefore, the clearer the air, the more blue the sky appears. The shade of blue observed in a cloudless sky depends on the prevailing air mass. When the cloudless sky is not a deep blue, but is somewhat diffuse, the effects of particulate scattering are exhibited. The light from the clear sky is polarized, the degree of polarization varies considerably, depending on the amounts and sizes of scattering particles in the atmosphere. Volcanic dust is especially important. The polarization also depends on the position of the sun in the sky.

Red or pink colors often occur at near sunrise and sunset because of the light passing through so much more atmosphere. Particulates in the atmosphere cause scattering in varying degrees at other times. Water vapor in the air is invisible. The amount of water vapor (moisture) is not constant. Warm air can hold more moisture than can cold air. Various other sky color conditions can be observed, as is covered in the following sections.

Whiteout

Typical of polar regions, *whiteout* occurs in clean air. The main factor necessary for whiteout is the presence of a snow surface with a high reflecting power for visible light.

Rainbows

Rainbows are arcs of color produced on sheets of water droplets.

Halos

Halos are caused by refraction and reflection of sunlight or moonlight by ice crystals. Different types of halo can be observed, depending on the size, shape, density, and orientation of the ice crystals in the atmosphere, and whether or not they vibrate in winds while falling.

Coronae

Coronae are sets of colored rings produced by diffraction when the sun or moon shines through relatively thin clouds of water or ice. In this case, the particles are about the same size as the wavelength of light. Coronae are usually much smaller than halos. In general, the smaller the particulates, the larger the diameter of the coronae. Because they are small in diameter, they are seen more often around the moon than around the sun.

Apparent colors

The *apparent color* of the sun or moon is an important indicator of what kind of particulate matter is in the atmosphere. *Blue sun* is caused by the scattering and absorption of light by particulates that are relatively large and uniform in size. *Purple light* appearing soon after sunset at a point just above the horizon where the sun has set is caused by fine particles of volcanic ash in the atmosphere. *Green light* on the sun's disk as the sun disappears at sunset can occur when the atmosphere is very clean.

The precise mechanism that connects solar activity and weather is not known. However, there are definite connections between solar events and terrestrial weather and climate. It is important to measure the monthly and yearly means of solar radiation received at the earth's surface because these amounts vary considerably. Dust and haze, particulates from fires and volcanic debris, and water vapor content are all important in measuring the resistive factor of the atmosphere to sunlight.

Again, volcanic dust in the atmosphere is particularly important because it can markedly decrease direct solar radiation at the ground by up to 20 percent. Offsetting this somewhat, volcanic dust increases the diffuse radiation because the dust acts as a scattering medium.

The earth's magnetic fields reflect the sunspot cycle, varying with the values of ultraviolet radiation, which causes ionization of the upper atmosphere.

Days of maximum particle emissions from the sun are generally followed by rises in atmospheric pressures in high latitudes, and a slight fall in pressures in the lower mid-latitudes. This results in less weather activity in polar regions and increased storminess in the mid-latitudes, about 5 and 14 days after the solar disturbance. Blocking high-pressure ridges often develop inland just east of the Gulf of Alaska, which consequently lower the jet-stream to over the continental United States.

Eruptions of solar flares is associated with storm development in the atmosphere above the mid-Pacific Ocean. Storms often develop 6 to 10 days after the flare. Stratospheric warming occurs, more water vapor is absorbed, and clouds form.

Data and applications

This section covers a specific case study. Table 14-1 lists the solar power and resistance indices received at Fresno, California during March 1993. Figures 14-5,

Table 14-1. Solar Electrical Indices for March, 1993

	Voltage		Amperage	
Date	Open circuit	Load[a]	(mA)	Resistance[b]
1	8.91	7.79	400	48.0
2	8.87	7.73	397	46.5
3	9.14	8.10	423	43.0
4	9.10	8.05	357	42.3
5	9.30	8.10	423	42.5
6	9.03	7.90	432	42.0
7	9.19	7.45	434	41.5
8	9.10	8.05	437	40.3
9	8.70	6.80	208	62.6
10	9.10	7.90	423	41.5
11	8.57	6.96	190	67.2
12	9.00	7.95	426	39.8
13	9.10	7.92	388	40.4
14	8.62	7.30	159	54.5
15	8.90	7.80	235	48.9
16	9.17	8.10	389	41.1
17	9.10	8.20	434	37.5
18	9.30	8.35	438	39.5
19	9.10	8.10	377	40.3
20	9.10	8.20	407	43.3
21	9.06	8.12	418	40.0
22	9.18	8.24	421	39.4
23	8.97	7.60	262	53.6
24	8.88	7.70	204	56.2
25	8.78	7.50	169	57.4
26	9.25	8.44	573	35.4
27	8.83	7.70	192	56.5
28	9.37	8.46	480	38.0
29	8.85	7.68	172	55.2
30	8.46	6.43	104	70.0
31	9.22	8.28	390	41.6

Note: Data for this table was gathered with the solar panel pictured in Fig. 14-1.

a Load reading is taken using a 100-Ω resistor.

b Resistance value is in ohms, taken with the sensor positioned on a horizontal surface.

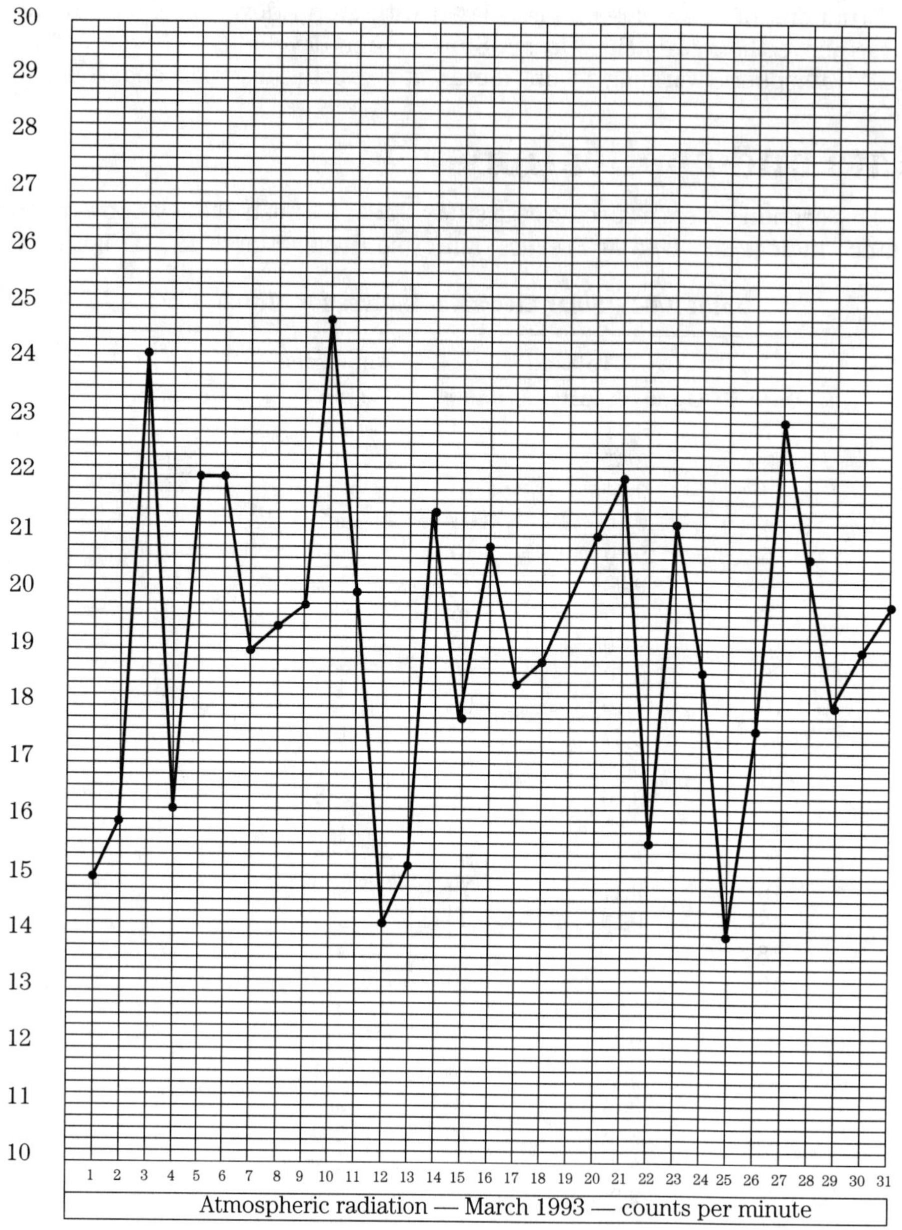

14-5 March 1993, graph of atmospheric radiation.

14-6, and 14-7 are graphs of the atmospheric radiation counts, the solar flux, and the geomagnetic field activity, respectively, for the month, and Figs. 14-8 and 14-9 are graphs of the solar flux and the geomagnetic field activity for April 1993. Table 14-2 lists the significant events in the weather, seismic, and solar conditions for March 1993.

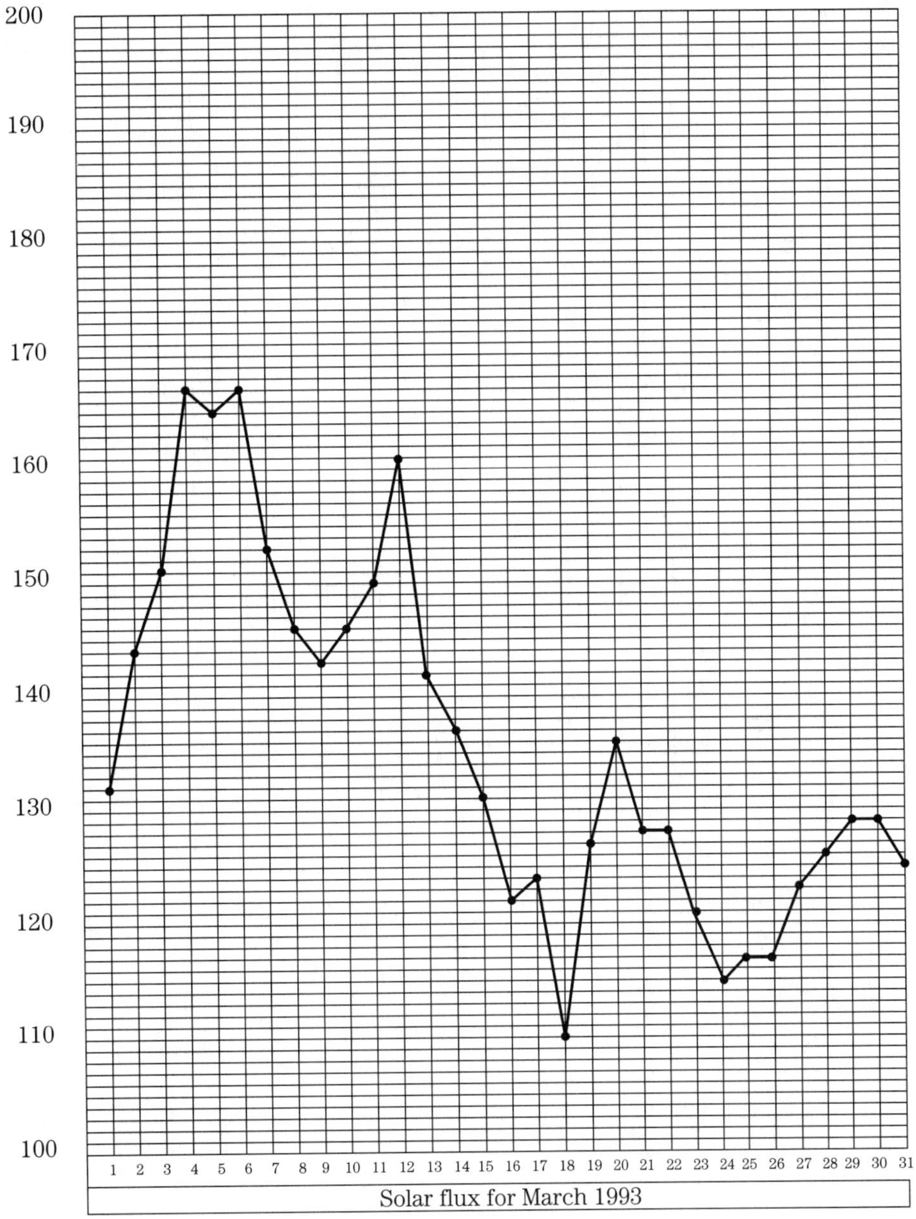

14-6 Graph of the solar flux for March 1993.

Studying an ENSO event

An *ENSO event*, short for the *El Nino Southern Oscillation*, is a valuable entity to study because it has the capability to affect weather around the globe. Here, this study is detailed as to some of the causes of the event, and how it can be studied with the apparatus presented in this book.

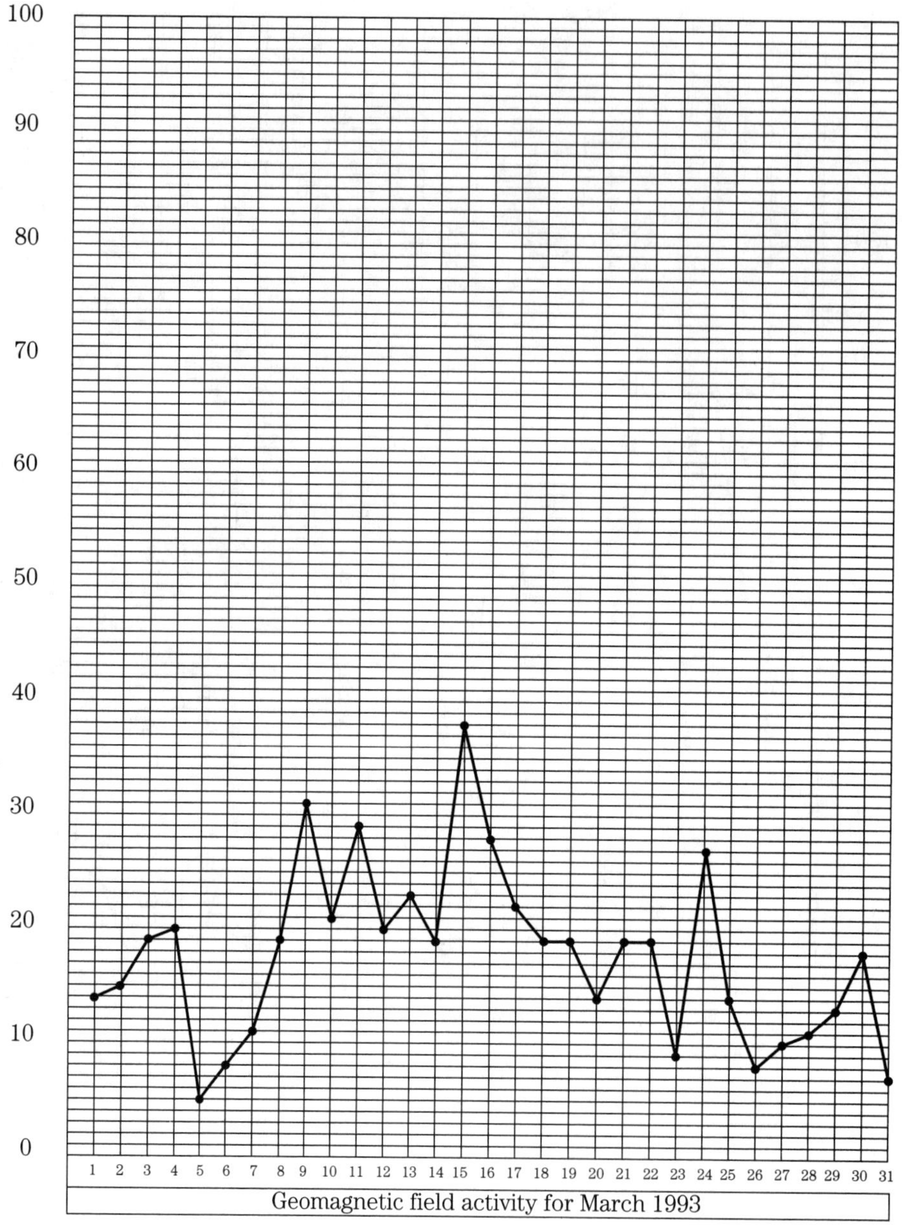

14-7 Graph of the March 1993 geomagnetic field activity.

The Pacific Ocean's vast size, along with its active tropical weather, makes it a very important contributing factor for much of the world's weather.

In the late spring and summer, the Intertropical Convergence Zone is most active. Figures 14-10 through 14-12 are depictions of the eastern tropical Pacific, from

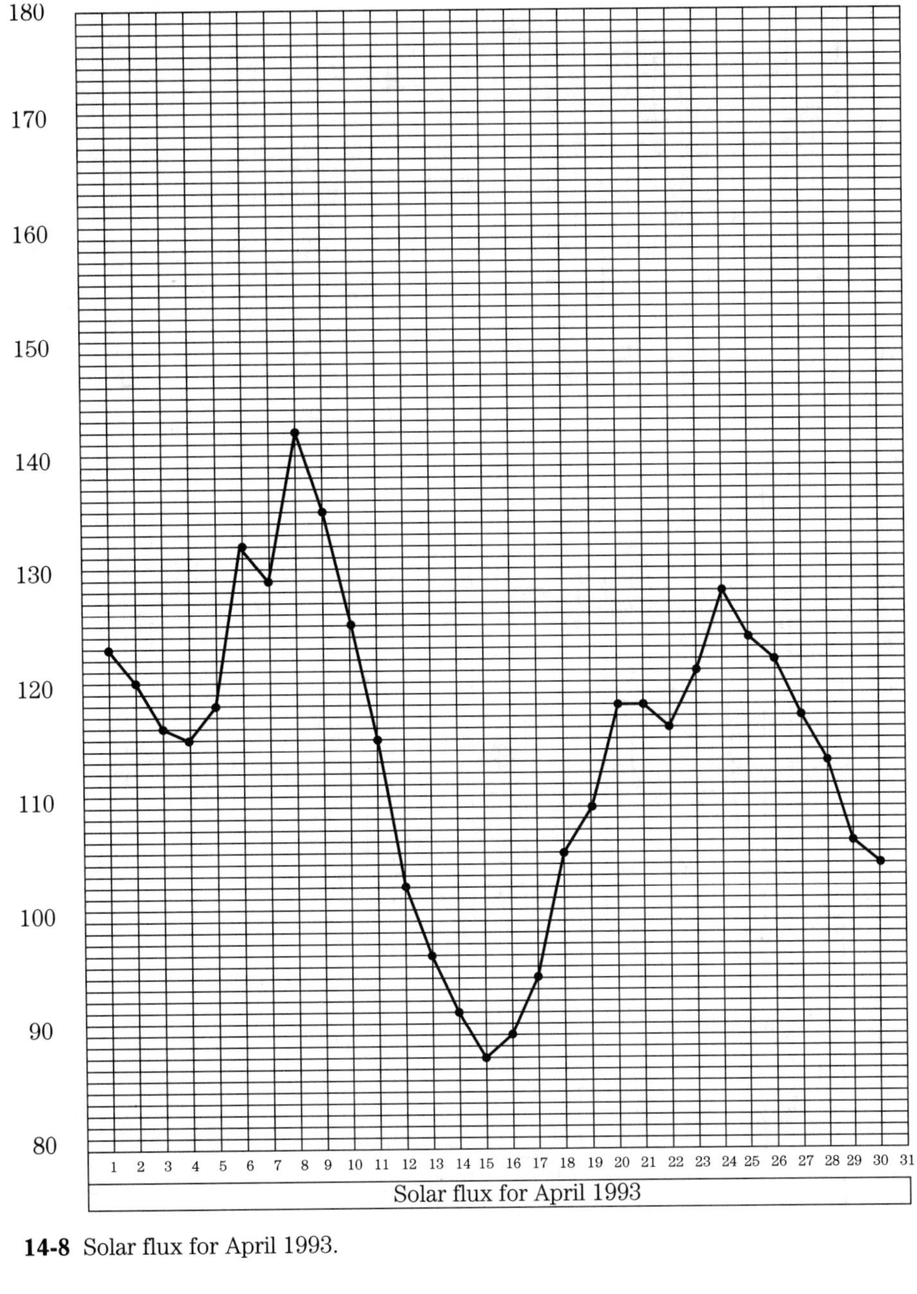

14-8 Solar flux for April 1993.

radiotelegraph and radiotelephone data received on long and shortwave using the Harbor Master and Navigator radios. The maps show the thunderstorm activity for 1200 UTC on the 3rd, 5th, and 10th of March 1993, as an example of this product.

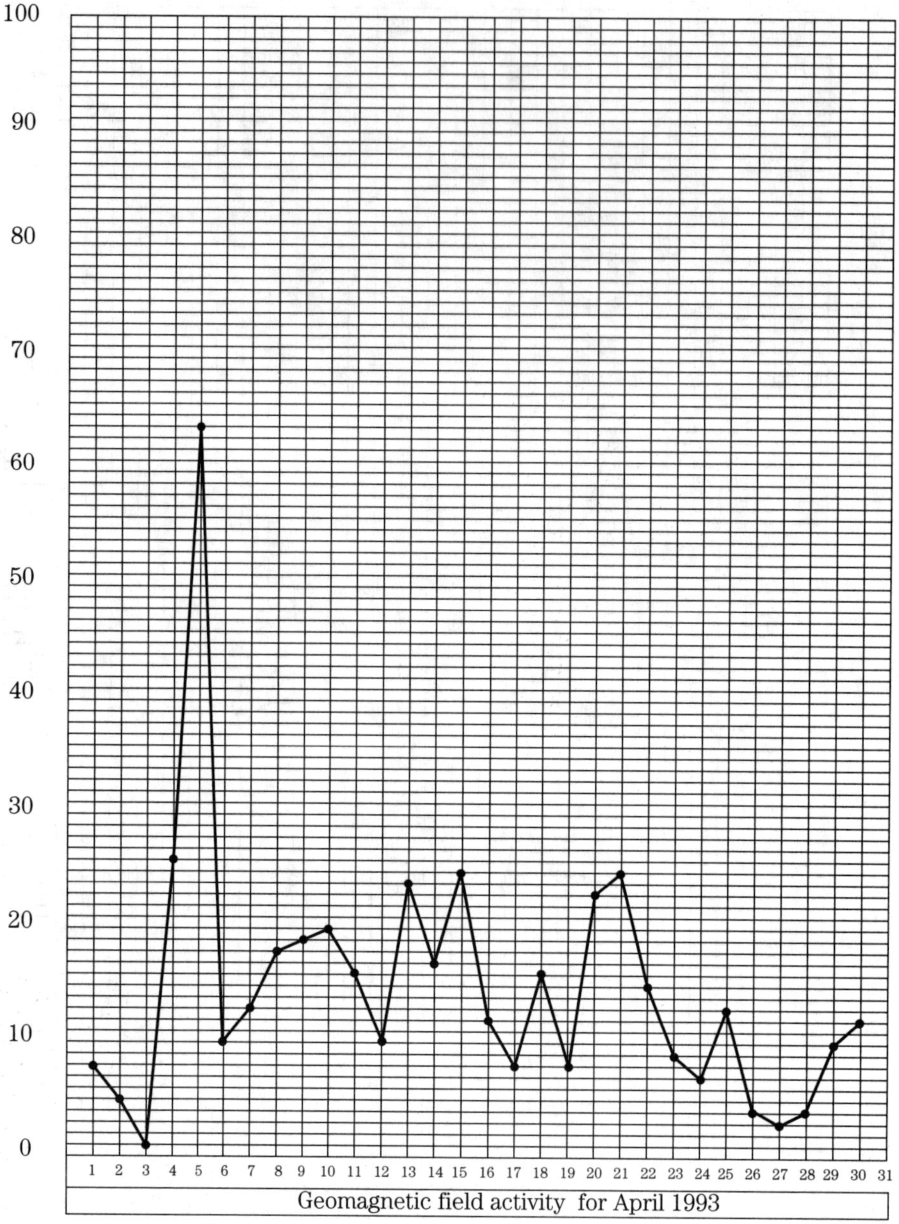

14-9 Geomagnetic field activity for April 1993.

Eastern Pacific tropical storms and hurricanes develop in the warm waters southwest of Mexico. Most of these storms die out at sea as they move over colder waters, but some exceptions do occur. One factor that influences this development is the ENSO phenomenon.

Table 14-2. Anomalies for March, 1993

Date	Event
2	Major solar flare at 2149 UTC
4	Satellite observed proton event
6	Major solar flare at 2049 UTC
8	Major geomagnetic storm
11	Major solar flare, minor geomagnetic storm. Flooding in Nebraska and Arizona. Storm beginning to build into the southeastern U. S. Volcanic eruption in the Philippines. After a two-week period of very low seismic activity, swarms of earthquakes hit in many places. Iran had a major quake, swarms occurred on Raoul Island, 550 miles northeast of New Zealand, and the California zones were active.
15	Minor geomagnetic storm. The "Storm of the Century" hits in the southern and eastern U. S.
25	Thunderstorms occur in parts of California, and a major earthquake hit near Portland, Oregon. Seismic activity in Chile remained high after a damaging 5.9 magnitude quake there the previous week.

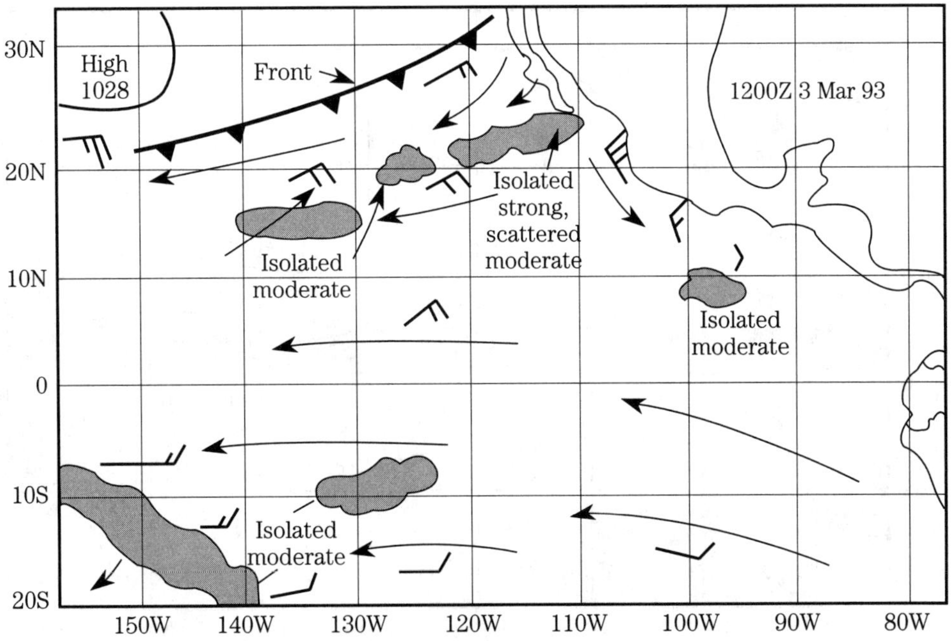

14-10 Tropical analysis for March 3, 1993.

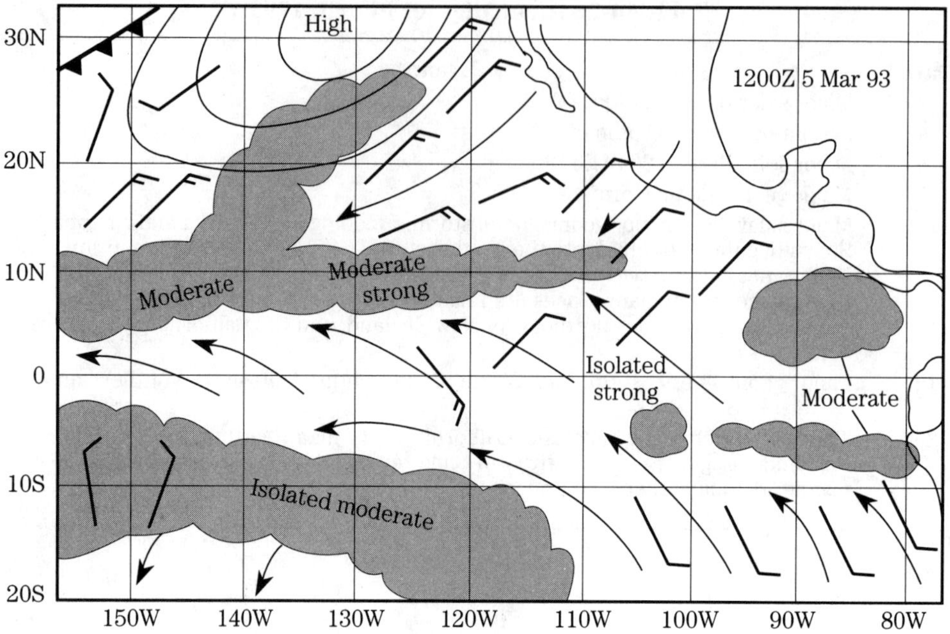

14-11 Tropical analysis for March 5, 1993.

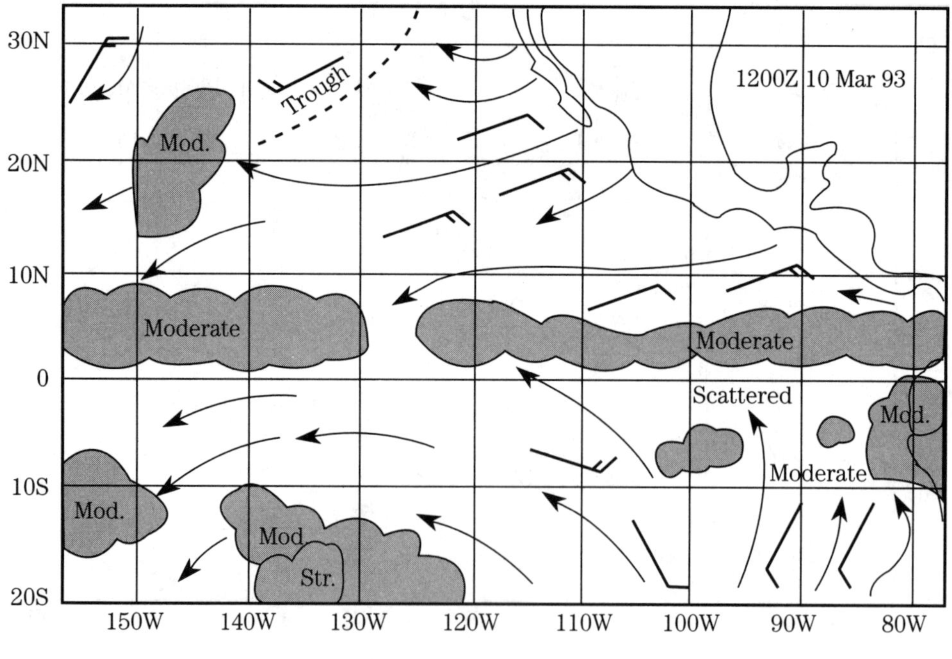

14-12 Tropical analysis for March 10, 1993.

El Nino is an annual Christmas-time appearance of warm water off the coastline of Ecuador and Peru that disrupts local fishing. The El Nino event, however, occurs roughly every two to seven years as a large pool of relatively warm ocean water in the tropical western Pacific moves eastward along the equator and replaces the normally cooler waters of the central and eastern Pacific tropical areas. The warm surface water affects the atmosphere by creating convection clouds in the normally arid region.

Figures 14-13 through 14-16 are GOES weather satellite products from August 1992 that show tropical storm and related weather activity from the various Pacific Ocean areas; the Western Pacific, the Eastern North Pacific, the North American continent, and the area south and west of Baja California. Several hurricanes and typhoons are shown in the various views, along with thunderstorm activity in the intertropical convergence zone.

Solar activity comes into play as the ionization associated with flares leads to stratospheric warming. The convective activity of the clouds heat the upper atmosphere, altering the path of the jet streams, which are belts of extremely strong winds at high altitudes that drive weather systems around the globe. Jet streams are found both north and south of the equator, between 30° and 60° latitudes. The jet streams often have split flows, as shown on the map of Fig. 14-17. Both the polar and the sub-tropical jet flows are shown. Figures 14-18 through 14-20 are streamlined maps for March 3, 7, and 10, 1993, respectively.

The pattern of the ENSO event tends to persist longer than the actual currents of above-average temperatures because the ocean's temperatures change relatively slowly. Figure 14-21 is an Eastern North Pacific Surface Analysis Map.

The scientific use of the term *ENSO* is now usually reserved for the more intense events, like those that occurred in 1972-73, 1976-77, 1982-83, and 1991-92. The most intense ENSO event recorded was the one in 1982-83, when sea-surface temperatures off of Peru warmed as much as 15° above normal, with thermal anomalies noted along the West Coast of North America. The southern oscillation portion of the

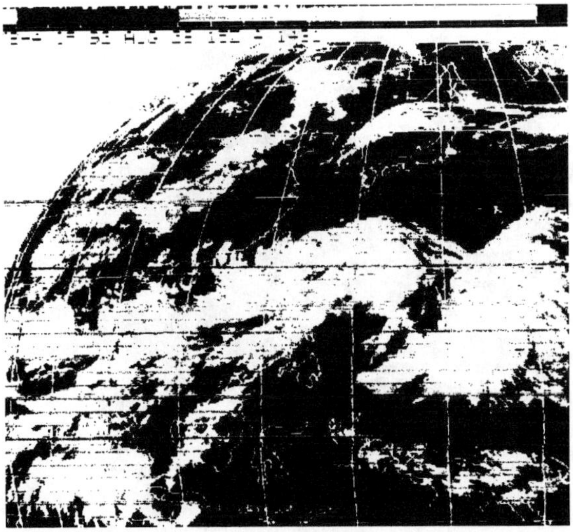

14-13
Sample GOES satellite product of the Western Pacific.

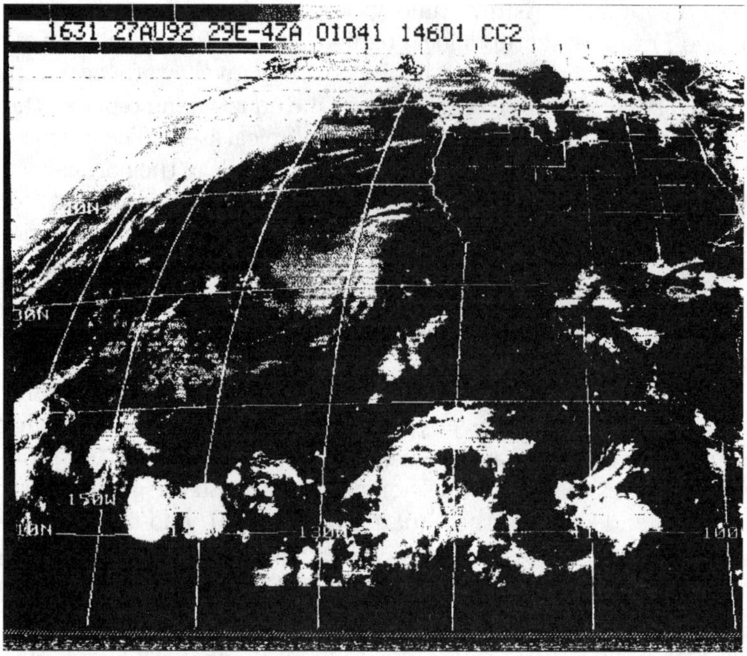

14-14 Sample GOES satellite product of the Eastern North Pacific.

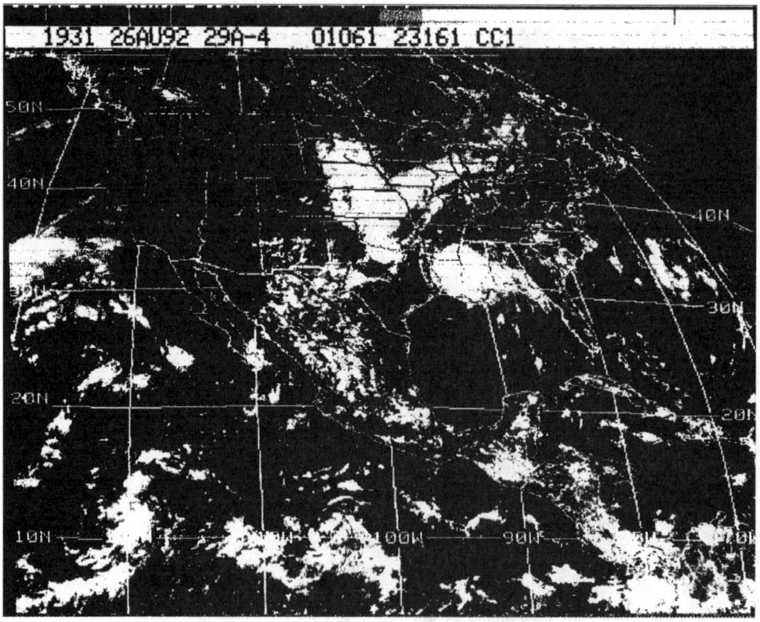

14-15 Sample GOES satellite product of North America.

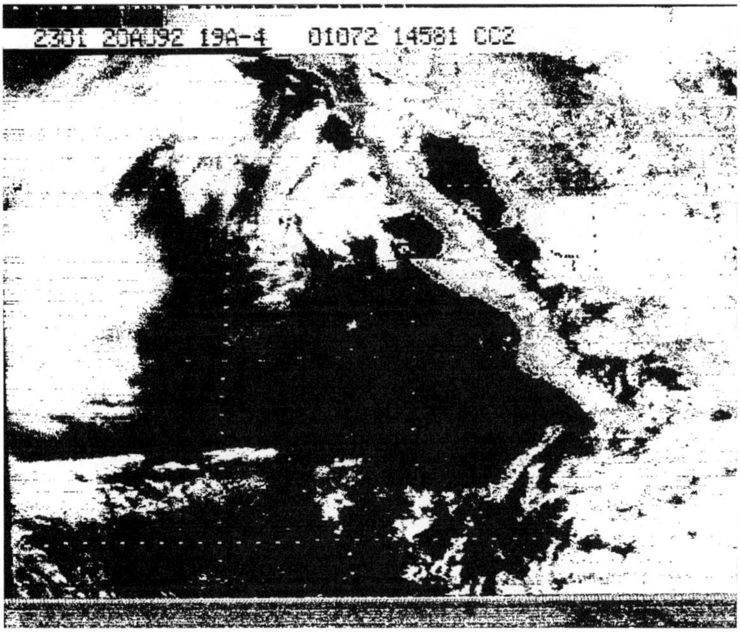

14-16 GOES satellite product of the areas south and west of Baja California.

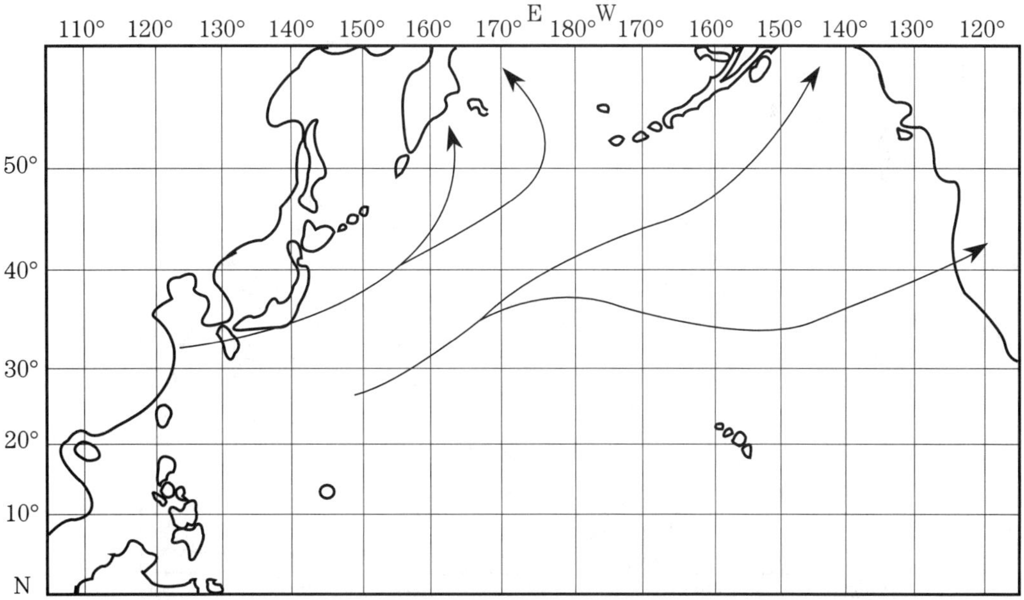

14-17 Jet streams and storm tracks for March 22 through 28, 1993.

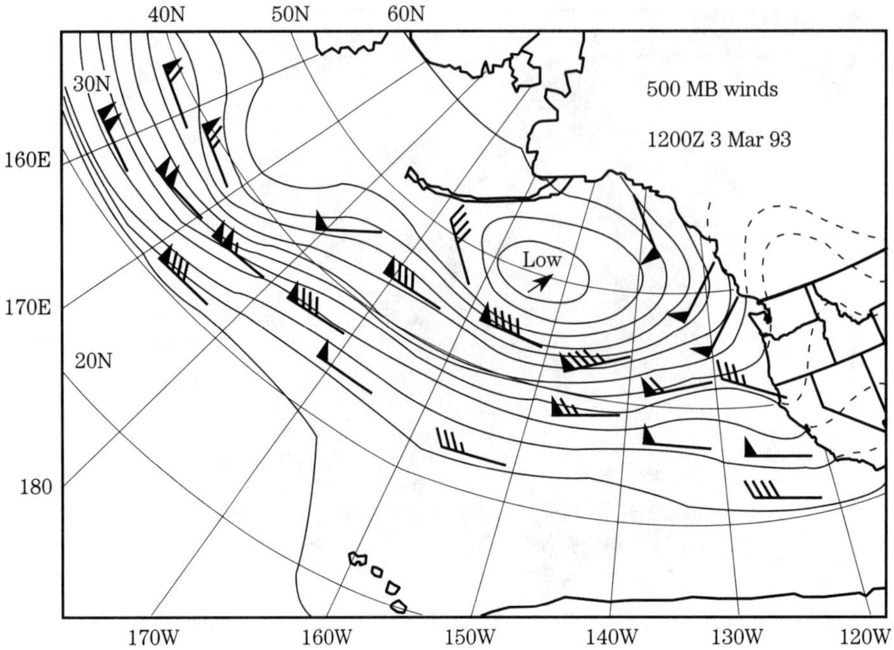

14-18 Stream line (upper air) map for March 3, 1993.

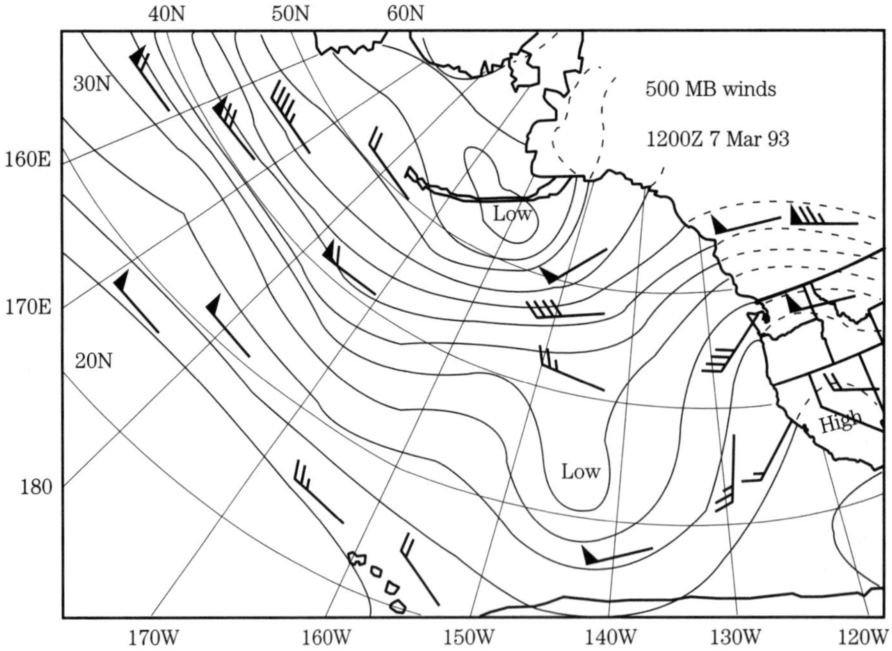

14-19 Stream line (upper air) map for March 7, 1993.

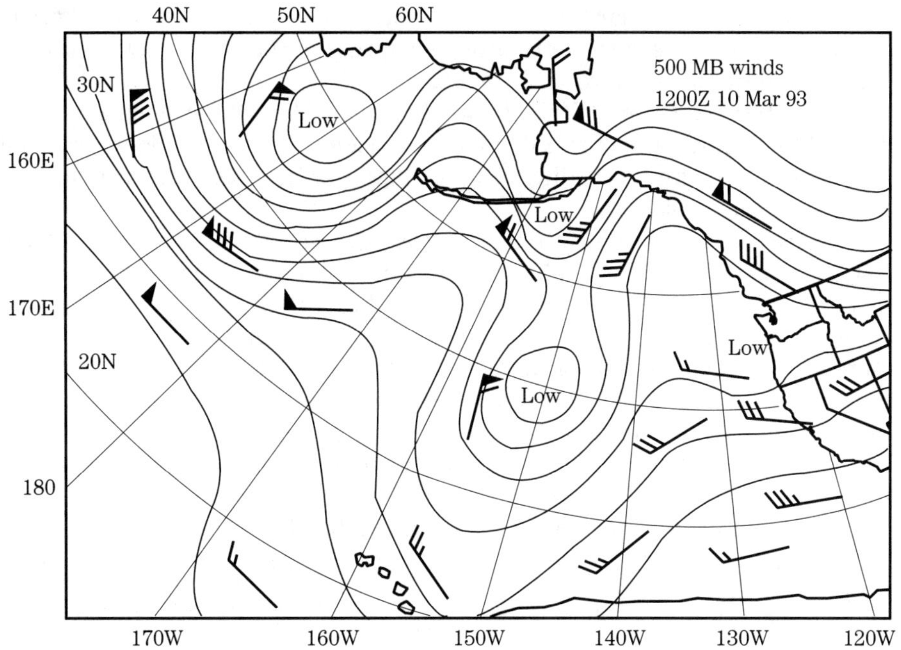

14-20 Stream line (upper air) map for March 10, 1993.

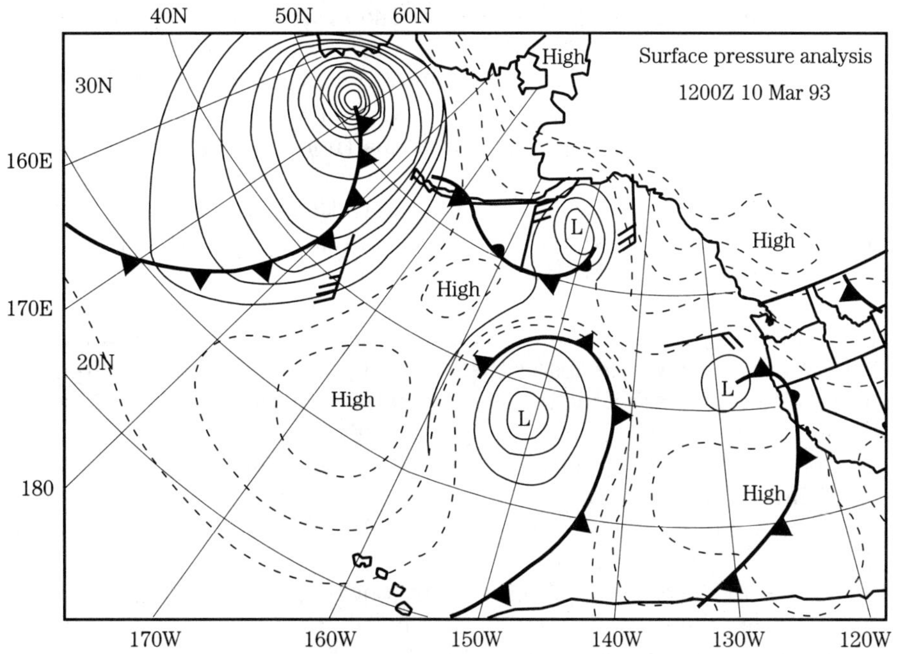

14-21 Surface analysis of the Eastern North Pacific.

ENSO is responsible in large part for the more intense events. The Southern Oscillation is a massive see-sawing of barometric pressure gradients between the southeastern and the western tropical Pacific.

The 1991-92 ENSO, although weaker than the 1982-83 episode, has been blamed for record rainfall and flooding in Arizona, Texas, and Southern California, and is credited with alleviating, at least temporarily, a six-year drought in the remainder of California.

Typically, in an ENSO episode, the tropical storm season in the Atlantic Ocean, Caribbean Sea, and the Gulf of Mexico is quieter than average, in terms of numbers of storms. However, destructive storms, like Hurricane Alicia in 1983, and Hurricane Andrew, in 1992, can form. The opposite effect is often noted, with the number of storms increased, in the tropical Pacific. In fact, the 1992 season produced a record number of named tropical storms in the Pacific region. The increased sea-surface temperatures were a primary cause.

By areas; during an ENSO event the waters of the Central Pacific are warmer than normal and spawn an increased number of tropical cyclones from August to November north of the Equator. Storms tend to head in a northwest direction past Guam and toward Japan, then curve northeast into the North Pacific to the Gulf of Alaska.

Hawaii sees enhanced tropical storm activity because of the warm waters to the south of the islands. In the late fall and early winter, the jet streams out of Japan become stronger, splitting off toward Canada and Mexico, from a point just northeast of Hawaii. The jet stream steers strong storms from Japan to the Gulf of Alaska, while the northeast Pacific high-pressure area weakens and drifts south of Hawaii, which causes the dry "Kona" winds and are from a southwesterly direction.

In California, the strong North Pacific storms from near Hawaii send numerous frontal waves to the west coast. The sub-tropical jet stream over Baja helps steer storms southwest of Los Angeles, which track northeastward and bring heavy rains to Southern and Central California. Santa Ana (northeasterly) winds often precede and follow these storms.

In the Gulf of Mexico and the Southeastern United States in the winter season; the jet stream steers over the Gulf as deep low pressure areas are spawned that track across the Gulf and into the Atlantic regions.

In South America; the Southern Hemisphere summer (December to March) is characterized by heavy sea-swell conditions near the coasts of Ecuador, Peru, and Chile. Warm ocean waters foster the development of strong convection, and these storms often inundate Ecuador and northern Peru with floods. Off the coasts of southern Peru and Chile, the normally strong high-pressure area weakens, and deep lows form in the latitudes between New Zealand and Chile.

Figure 14-22 is an area map of the world that shows areas affected by a typical ENSO event. The reference numbers are listed in Table 14-3, along with descriptions of associated weather anomalies.

Studies of the ENSO phenomenon are ongoing and have been successful. The probability of a moderate-to-strong ENSO for the 1991-92 season was correctly forecast with about a six-month advisory. The 1991-92 ENSO coincided with many global weather extremes. There was drought in Hawaii, Western Europe, South Africa, the

14-22 ENSO event area map.

Philippines, and the Pacific Northwest of the United States. Floods occurred in Arizona, California, Nebraska, and Texas, and in Australia, Mexico, and Ecuador. The Southeast and Atlantic regions of the United States were pounded by a powerful storm called the "Storm of the Century." Winter weather was unusually warm in the central U. S. and very cold in the Middle East.

Table 14-3. ENSO Reference Areas, and Associated Weather

Area	Characteristics
1	Floods from January to May. Heavy seas.
2	Hotter from December to March. Light winds and seas.
3	More tropical storms.
4	More tropical storms south of Hawaii, a lack of trade winds, and more dry southwest winds.
5	Heavy seas from October through March. Frequent intervals of strong storms.
6	Drier and warmer weather.
7	Colder winter weather.
8	Storm potential greater from December to March. Decreased hurricane activity August through November.
9	Dry weather: often drought conditions.
10	Relatively mild winter weather.
11	More typhoon activity.
12	Drier weather.
13	Weak monsoon season.
14	Wetter, colder weather.
15	Hot and dry weather.
16	More tropical cyclones.

Many possible factors for these weather events remain to be studied and analyzed to determine which of these weather events are associated with the ENSO event.

Weather events, such as the 1982-83 and 1991-92 ENSO events, Hurricane Andrew and the March 1993 "Storm of the Century" can cause billions of dollars in damages as well as catastrophic losses in life and property. Weather enthusiasts from amateur to professional can benefit and contribute to the study.

ENSO and the March 1993 "storm of the century"

The storm that socked the Southeastern and Atlantic regions of the United States in the middle of March 1993, was one of the fiercest storms on record. It was so named the "Storm of the Century" because it struck on the anniversary of another of the worst storms on record; the blizzard of 1888, which dumped at least four feet of snow in some places during March 12 through 14 and killed hundreds of people.

Did the 1991-92 ENSO event cause this storm in 1993? Probably not. However, the ENSO event, which carried over into 1993, was probably indirectly involved because it allowed conditions that put more moisture into the air.

In the formation of this storm, during March 9 and 10, warm water upwelling and the bases of convection were evident in the Gulf. The jet stream dropped down from Canada along with frigid Arctic air. The cold air mixed with the warm, moisture-laden air from the Gulf of Mexico forming the storm on the 11th and 12th, which then tracked along the eastern seaboard for the next three days. Low pressures and high winds recorded during the storm brought it comparisons with moderate- to large-sized hurricanes. By March 13, the storm covered nearly the entire eastern one-third of the United States. The southern extent of the storm ravaged parts of Cuba. Mexico, near the Texas border, was also hit.

Another probable factor, though, again an indirect one, was the eruption of Mt. Pinatubo in the Philippines, which blocked sunlight and caused global cooling. Exactly what caused the jet stream to dip so far south to Florida is not known. Figure 14-23 is a map showing the factors of the formation of this storm.

Along with other speculation on what causes ENSO episodes is the recent discovery of active volcanoes in the depths of the ocean around Easter Island, in the South Pacific.

Thus far, over 1100 seamounts have been discovered in the area, some almost 7000' tall, yet still submerged 2500 to 5000 feet below the sea surface. It is estimated that two or three of these volcanoes could be erupting at any given moment. The density of these seamounts is such that no other area has been discovered with as high of a concentration of activity undersea or on land.

If the undersea volcanoes influence the ENSO episodes, it might be as follows:

Volcanic activity provides huge amounts of heat into the ocean, changing water temperatures enough to affect weather patterns in the Pacific. Periods of extreme volcanic activity underwater (perhaps when more than a dozen volcanoes are active at the same time) might be one of the mechanisms that triggers the ENSO, which occurs when the semi-permanent high-pressure cell centered near Easter Island drops slightly in pressure, and alters ocean currents and temperatures.

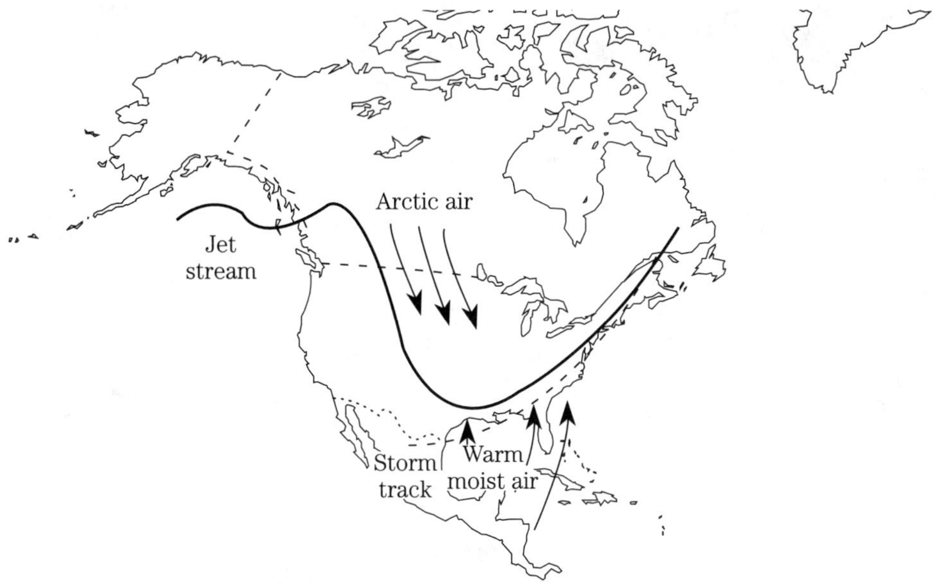

14-23 "Schematic" of the March 1993 storm in the United States.

Although it is uncertain whether the ENSO event directly influenced the March 1993 storm, the possibility does exist that the high solar activity of February and March 1993 was in some measure responsible for some of the month's storminess.

Refer to Table 14-2. Major solar flares occurred on March 2, 6, and 11. Figures 14-10 through 14-12 indicate how the tropical Pacific's convection built through the period of March 3 to 10; and Figs. 14-18 through 14-20 show the upper-air winds, along with storm development in the Pacific area from March 5 through March 10.

As a result of the solar activity, February and March 1993 were characterized by long periods of stratospheric warming. Stratospheric warming in the high latitude regions of the winter hemisphere of the earth is associated with gross distortions of the normal circulation patterns associated with the winter season.

Along with providing important information about the solar activity and the geomagnetic field, radio station WWV also provides important weather information. Marine weather reports, with information about major storms in the Atlantic Ocean and the Eastern North Pacific are broadcast at 8, 9, and 10 minutes past each hour. Similar storm warnings for the Eastern and Central North Pacific are broadcast from WWVH at 48, 49, and 50 minutes past each hour.

Figure 14-24 shows a completed pyrheliometer with meters, and the connections for the digital multimeter, ohms reading, and dc output. A voltage regulator circuit could be added, which would allow for the use of the solar panel to power other equipment in the field when the pyrheliometer is not being used for observations. The completed unit is pictured in Fig. 14-25 with a solar panel and photoresistor on top, in a portable configuration of the set.

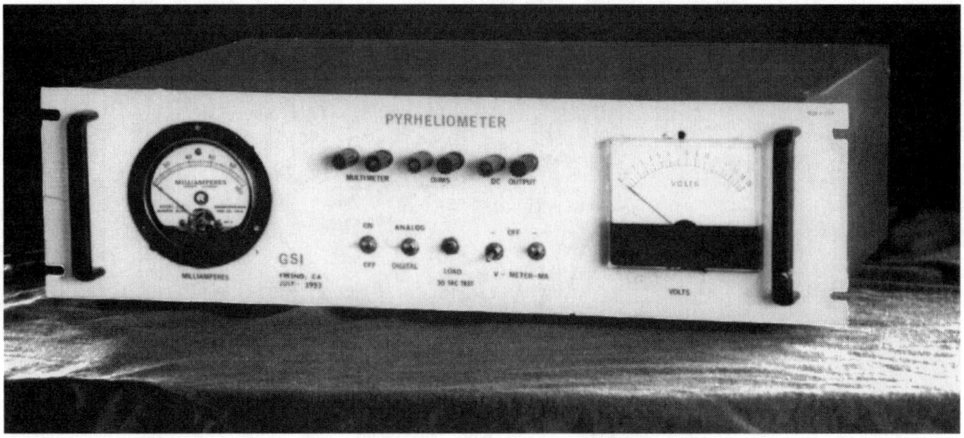

14-24 Completed pyrheliometer.

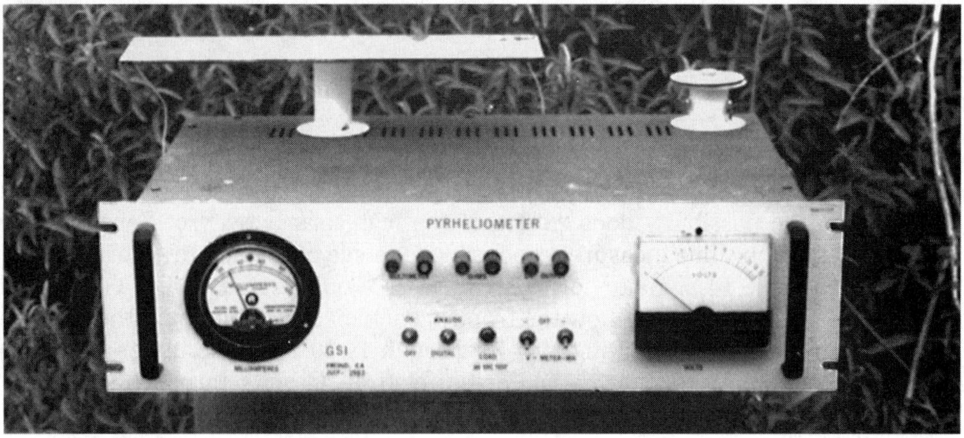

14-25 Pyrheliometer set up for portable use.

Wiring of the set is very simple. The load switch is the push-button switch in the center of the line of switches. All of the toggle switches are single-pole, double-throw (SPDT on/on) types. These switches are a good choice for many circuits because they can be used as simple on/off switches, or for circuit- or tap-selection switches.

Figure 14-26 depicts a screen view of a ULF radio signal input into a computer-generated strip-chart program. Here, the VIC-20, with only a 16K expansion slot, provides the display—proving that the exotic is often not needed for results! The display was being monitored mainly for seismic-wave study, but, as pointed out many times in this book, the signals are capable of providing propagation, atmospheric, and solar data, as well.

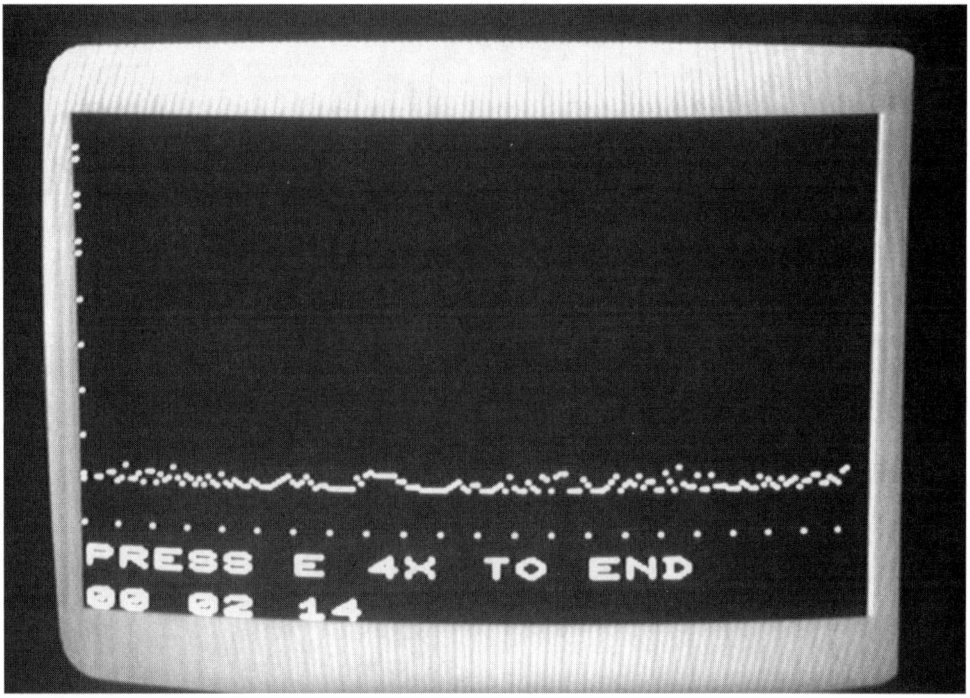

14-26 Sample screen of a ULF signal input into a computer strip-chart program.

15
CHAPTER
The Northridge earthquake

At 4:31 A.M. PST, on January 17, 1994, a violent earthquake measuring at least 6.8 on the Richter scale struck 11 miles beneath Northridge, California, just northwest of downtown Los Angeles. Composed of two main shocks occurring about five seconds apart, the earthquake contained complex components of horizontal and (especially) vertical movements that caused the violent movements and resulted in great damage.

The apparent percentage of energy released from this earthquake was much higher than normal for a quake of 6.8 magnitude. The powerful up-and-down motion from the rupture of an 11-mile-deep "thrust fault" caused a vertical displacement of about 15 feet at the epicenter (Fig. 15-1).

The complex waves underground were unusual. Crests and troughs of seismic waves rippled through many kinds of rocks. Some of the waves probably bounced off the San Gabriel and Santa Susana Mountains. Where wave crests intersected, shock waves were concentrated and magnified, causing the great damage at the surface.

Thrust faults often leave ruptures a mile or more in depth, and leave little surface evidence of an earthquake. This is precisely what happened at Northridge, and aftershock data will be the key to mapping the fault responsible for this earthquake.

Thrust faulting, where giant pieces of rock are thrust at a sharp angle above another piece of bedrock, releases great amounts of energy at the surface. This process causes the formation of hills and mountains. Because the ruptures often do not reach the surface, earthquakes that result from thrust faults often are difficult for scientists to study, as opposed to strike-slip faults, such as the San Andreas, which leave visible traces on the surface and are composed primarily of horizontal movements. Thrust faulting is a sign of compression of the earth's crust. The resulting "squeezing" upward motion is responsible for the violent damage capabilities of earthquakes. Local fault zones are shown in Fig. 15-2.

As a result of the Northridge earthquake, power and water were knocked out to at least 700,000 homes; freeways and overpasses buckled (eight of which crumbled and collapsed); and 61 people suffered quake-related deaths, 10,000 were injured,

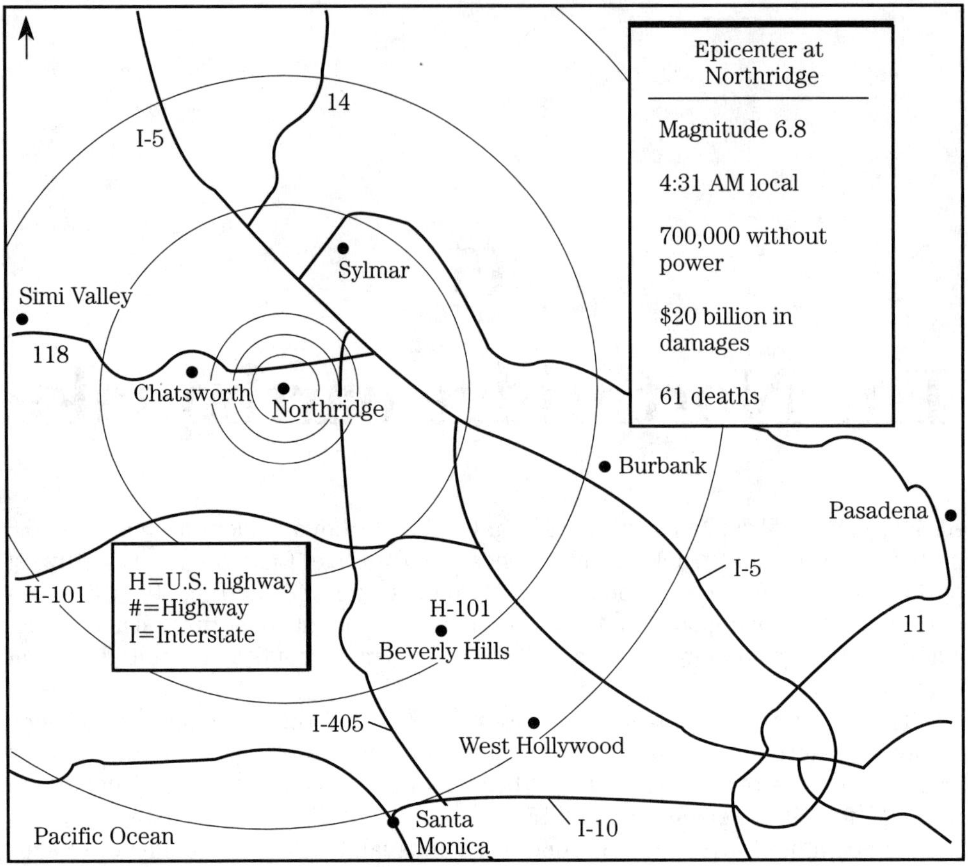

15-1 Epicenter of the Northridge earthquake in relation to the surrounding communities and the highway system.

and hundreds were left homeless. Tent cities were necessary to temporarily shelter quake-displaced people.

One week after the earthquake, electrical service was restored to all but about 1000 people, but over 25,000 still were without running water. Many more were advised to continue boiling water before drinking it. Damage estimates were set at about $20 billion, which makes this disaster at least the third largest, and possibly the worst, disaster in U. S. history.

Were there precursors?

Several interesting observations were made in the days prior to the earthquake.

On January 9 and 10, a swarm of significant small earthquakes rattled the Los Angeles area. A 3.7-magnitude quake struck at 3:01 P.M. (L) on January 9, followed by a 3.0 at 10:12 P.M. These earthquakes were centered two miles south of Santa

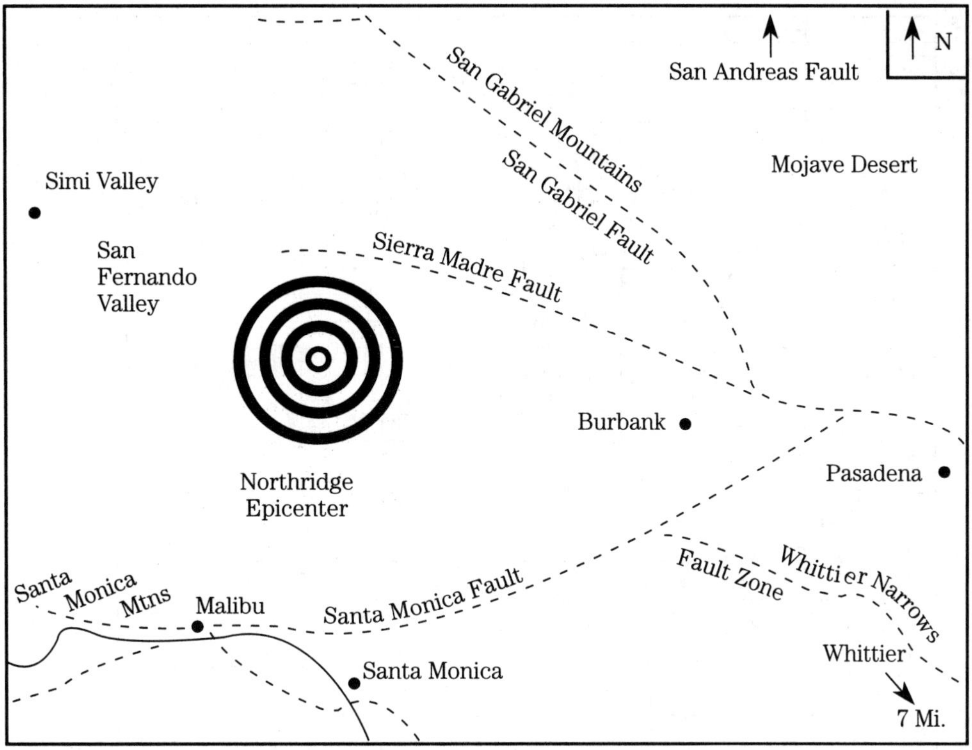

15-2 Local fault zones.

Monica and were felt by residents from Malibu, south along the shoreline to southern Santa Monica Bay.

Two more quakes rattled the area on January 10. They appeared to be in the Whittier Narrows earthquake zone, which is part of the Elysian Park fault system running from Whittier, through Los Angeles, and along the Santa Monica Mountains to the ocean.

Short-range reception of VLF and ELF radio signals, running stable for the first few days of January, became unstable and more intense by January 7. Frequencies of 300 Hz, 6 kHz, and 25 kHz were charted for solar activity as well as for seismic study. Figure 15-3 shows a graph of the 25-kHz activity. Signals are referenced to an "S Meter" reading of average daily intensity on a scale of 1 to 10. The average ambient noise level reading during quiet periods is from 1.5 to 4.0, using a constant intensity audio amplitude and the same longwire antenna. The signal intensity reading was taken via the SEA method. As a note, the 300-Hz and 6-kHz readings, while different in values, showed the same activity patterns. Table 15-1 lists solar, geomagnetic field, and seismic events for January through April 1994. Figures 15-4 and 15-5 are solar and A index charts for January and February. More reference information through April 1994 is provided in appendix F.

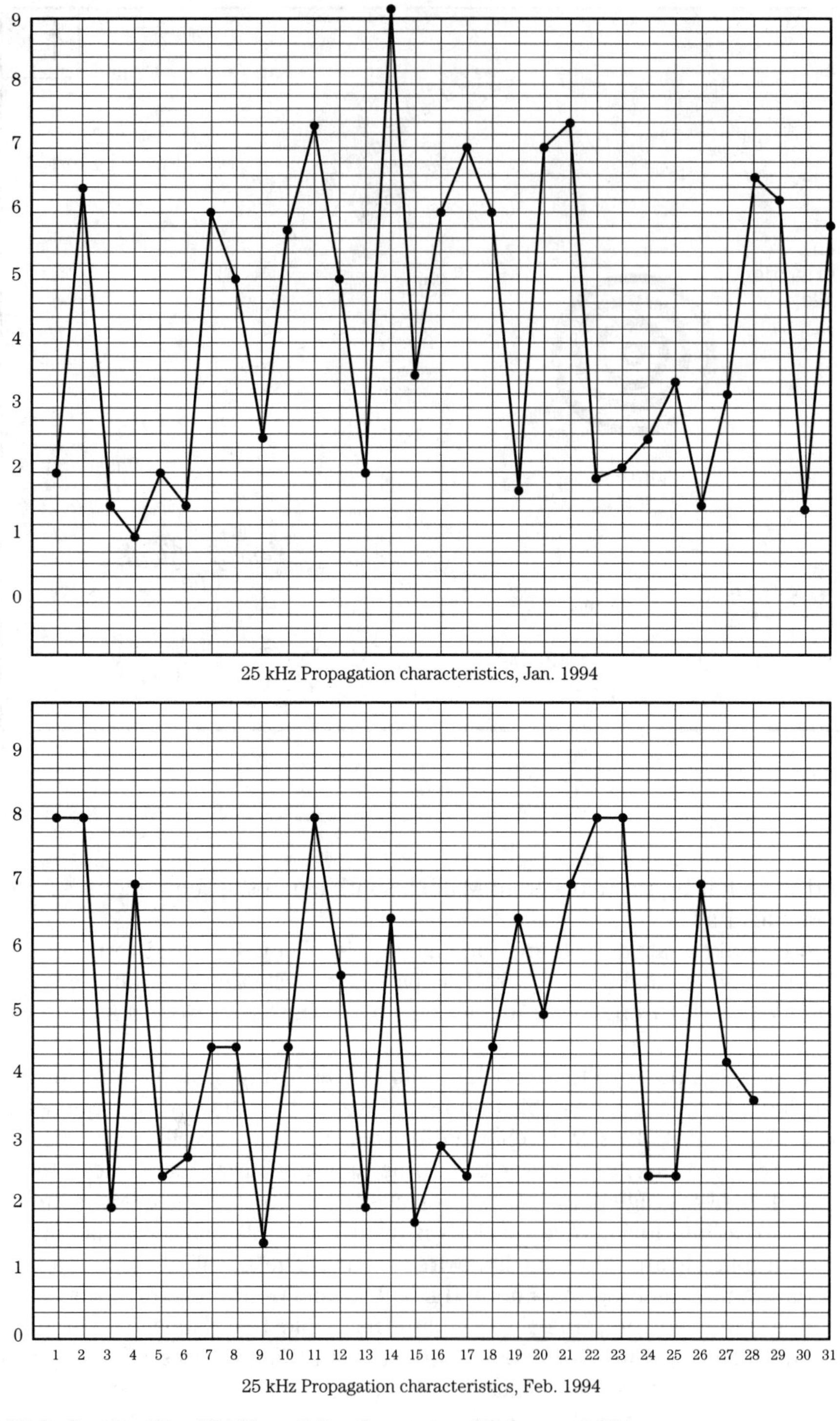

15-3 Graph of the 25-kHz activity; January and February 1994.

Table 15-1. Geomagnetic, Seismic, and Solar Anomalies for January through April 1994

January	
1	Earthquake; 2.7 at 1912L, 10 miles west of San Francisco, CA, on the San Andreas Fault.
2	Earthquake; 2.5, located same as above.
3	Solar Flare at 22:56Z.
9	Earthquakes; 3.7 at 15:01L and 3.0 at 22:12L, two miles south of Santa Monica, CA.
10	Earthquakes; two more located as above, also one estimated at 2.9 in the San Francisco Bay area.
11	Earthquake; 4.2 at 03:24L, 5 miles south of Watsonville, CA. Stratospheric warming episode ends.
16	Solar Flare at 23:25Z
17	Earthquakes; 6.8 at Northridge (See Text) Also, earthquakes on the Green River Fault (largest at 3.5), and in other areas in northern California (Largest 4.2)
21	Significant aftershock episode to the Northridge quake with five shocks from 4.0 to 4.6 recorded from 10:39 to 10:57L.
26	Geomagnetic Storm Conditions
28-29	Significant aftershock episode to the Northridge quake; five measuring 3.5 to 4.0 late Friday, and nine over 3.0 early Saturday. The largest measured at 5.0.
February	
1	Earthquakes; 4.3 10 miles southeast of Bishop, CA, a 3.6 aftershock to Northridge, a 5.2 in Hawaii.
2	Earthquakes of 4.5 and 5.8 near the Idaho/Wyoming border, followed by another quake in the area of 5.2, plus 27 aftershocks greater than 3.5 Magnitude in the first 12 hours following the initial quakes.
5	Earthquake; 4.2 at 1453Z 15 miles southwest of Marion, Illinois.
6	Significant aftershock swarm to the Northridge quake, including 4.1 and 3.6 magnitude shocks just after 05:00L.
6-13	Geomagnetic storm conditions.
10	Stratospheric warming episode begins.
15	Earthquake; @ 7.0, in Indonesia.
16	Stratospheric warming episode ends.
16-18	Major aftershocks to the Northridge earthquake; including 3.3 and 2.6 on the 16th and 3.7 and 3.1 on the 18th.
20-22	Satellite observed Proton event, began 0115Z on the 20th.
21-23	Geomagnetic storm conditions
25	Significant aftershock event to the Northridge quake, including; 4.0 at 04:59, 3.2 at 05:11, and 3.6 at 05:56L.
March	
1-31	Stratospheric Warming episode.

Table 15-1. Continued

2	Polar Cap disturbance began at 00:01Z.
3	Geomagnetic storm conditions.
5	Severe and violent thunderstorm activity in central and southern California
5	Significant aftershock swarm to the Northridge quake; including 3.0 and 3.1 aftershocks.
7-10	Minor geomagnetic storm conditions.
12	Severe thunderstorms in northern California.
12-13	Major geomagnetic storm.
15-16	Geomagnetic storm lessens intensity, but continues.
20	Major earthquake; 5.3, 1 mile WNW of Panorama City, in southern California; near, but not in, the Northridge aftershock zone.
23	5.4 Earthquake 33 miles east of Ensenada, Mexico, at 02:59Z. Also, 3.6 at the same location at 03:10Z.
27	Erratic readings from the "Coalinga Gripper" sensor.
30	Pronounced drop of observed ULF radio propagation as observed by the "Coalinga gripper" sensor.
31	Two moderate earthquakes strike five miles from Coalinga, California; 4.4 at 11:59 A.M. and 4.2 at 12:01 P.M. (L)
April	
2-11	Minor geomagnetic storm conditions.
6	Earthquakes; 3.8 at 00:26L and 4.8 at 12:01, at Lake Arrowhead in the Yucca Valley area of California. Also; 3.5 at 21:45L, followed by a 2.8 and several more 1.5 to 2.5 aftershocks to the Northridge quake.
8	Seismic sensor readings high and unsteady. Several small earthquakes observed scattered around southern California.
9	Several small earthquakes in Ventura County and in the San Bernardino Mountains. The strongest was a 3.6 at Fillmore at 10:53L. The center of this quake was out of the Northridge aftershock zone. The San Bernardino shocks; one at 2.9 at 16:57L, were aftershocks of the April 6 quake at Lake Arrowhead. With this activity, southern California is currently being affected by three aftershock patterns: From Northridge (1994), Landers (1992), and Lake Arrowhead (1994). South central California is still feeling aftershocks from the Coalinga earthquake of May 1983 and the San Francisco Bay area is still experiencing the effects of the 7.1 Loma Prieta earthquake of 1989.
13	Earthquake: 170 miles southeast of Manokwari, Indonesia, at 23:22Z.
16	Coalinga Gripper sensor readings are unstable. A severe geomagnetic storm began at 20:30Z and ended at 1500Z on April 17.
21	Earthquake: 4.4 near Coalinga, California at 09:37L.
27	Significant swarm of aftershocks to the Northridge earthquake, with the largest at about magnitude 4.5.
30	Earthquake: Kyushu, Japan. Magnitude about 6.8. Violent thunderstorm activity in central and southern California.

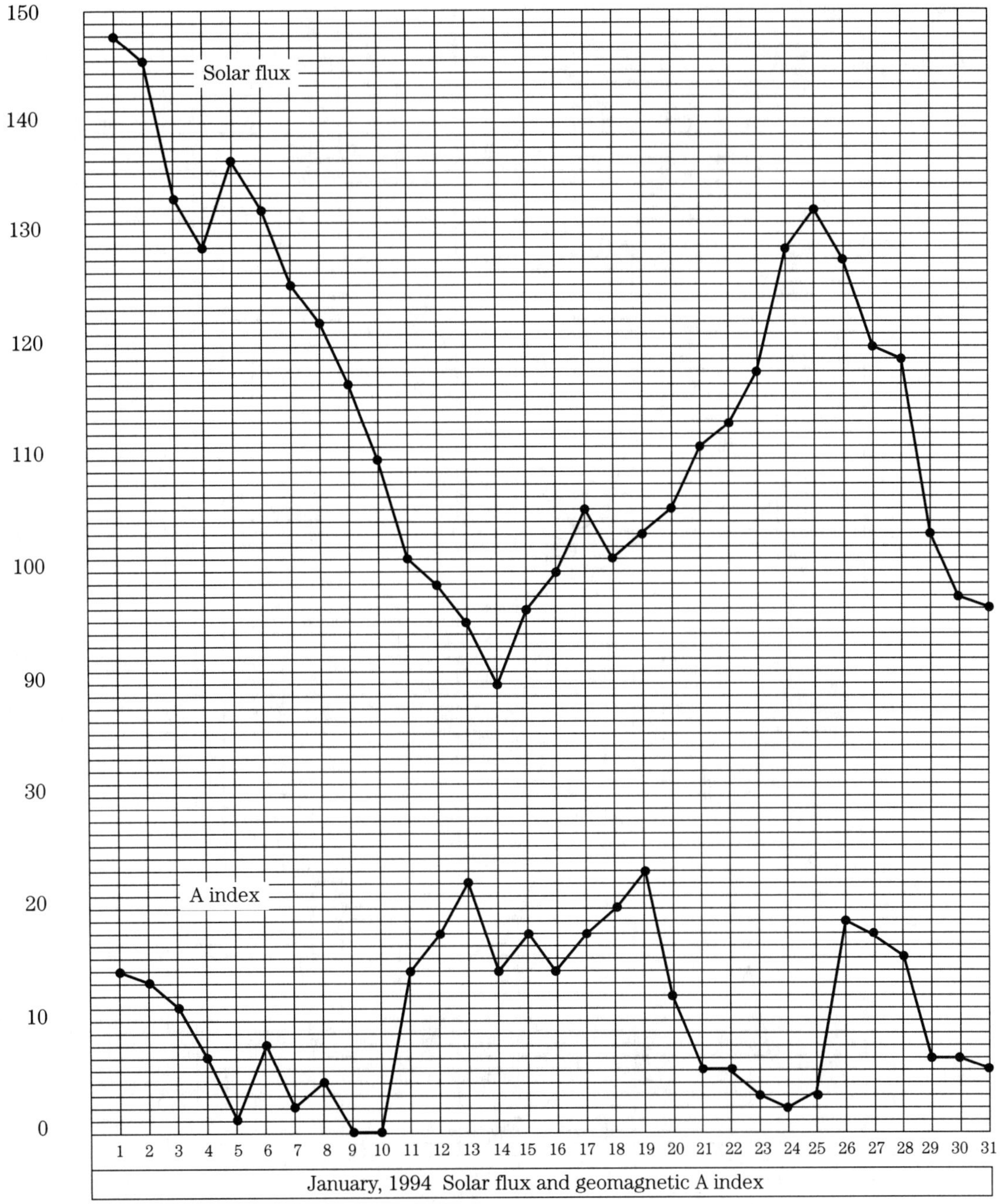

15-4 Solar and geomagnetic A index values for January 1994.

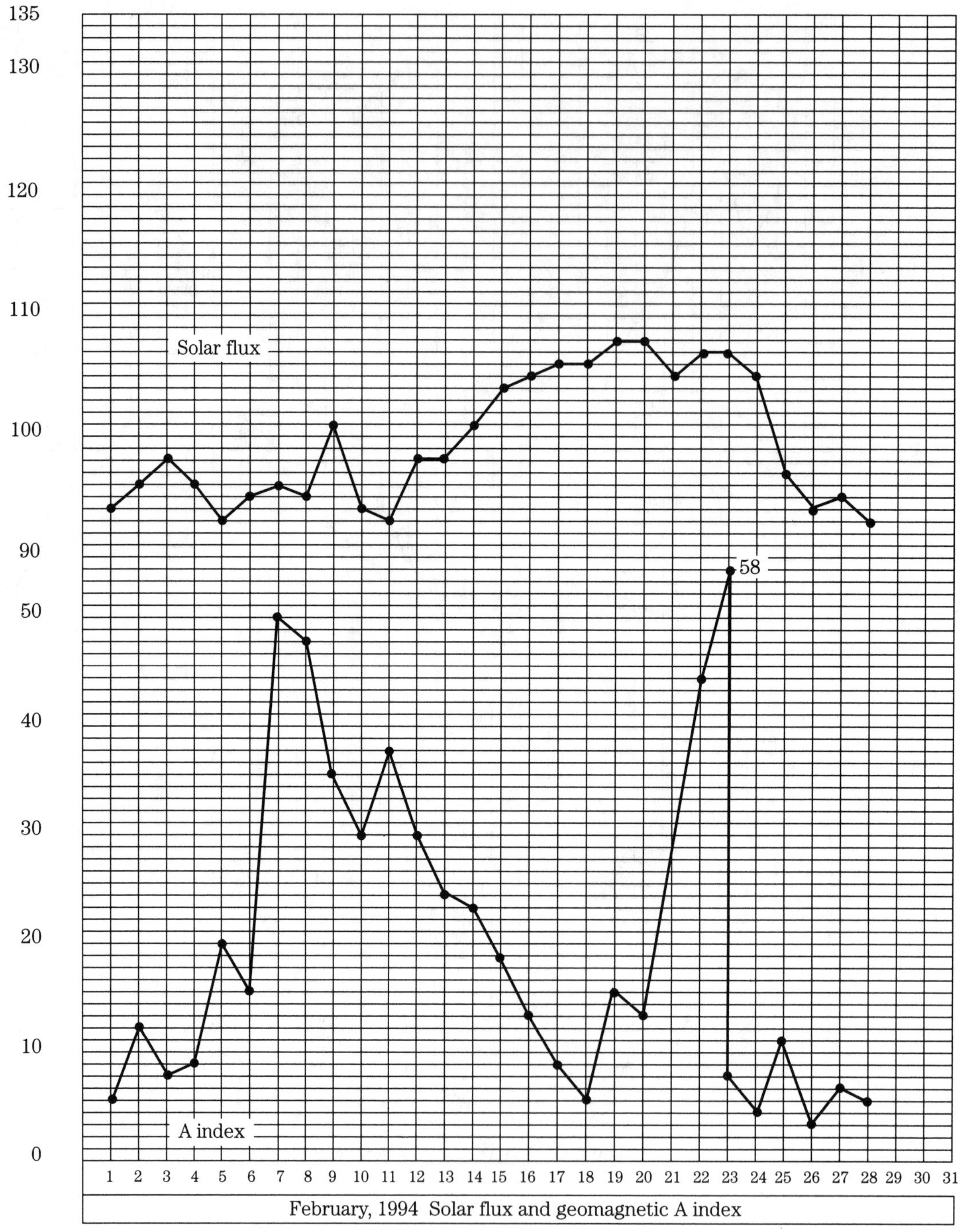

15-5 Solar and geomagnetic A index values for February 1994.

Catching the waves

The electromagnetic energy receiver (EER), which has an LED that is keyed by a buzzer alarm circuit, recorded the Northridge earthquake from Fresno a few seconds after the quake. The 9-V "smoke alarm" buzzer produced a loud report that definitely would have alerted anyone in its vicinity.

The monitoring apparatus was kept simple. Input to the EER was from an "extra-deep" sensor, and the signals from the EER were input into a basic LF 356 single-channel seismometer. Output A was input into a strip-chart recorder. Indicating the high gain of the system, the chart recorder had to be set to its 100-mV operating range to avoid pinning the movement. Because a sensitivity of 1 mV is available, you can readily see just how responsive a simple system can be. The sensor characteristics for the "extra-deep" sensor are listed in Table 15-2.

Table 15-2. Northridge Earthquake Sensor Characteristics

Northridge sensor	
Type	Dual Sensor - Two-Circuit Transformer type, and Electromagnetic type.
Transformer	Power Transformer, 115 Vac: 18 VCT/2 A
Receiver Used	Electromagnetic Energy Receiver
Container	"Extra-Deep" Type; 24 Inches in Height
Coil Circuit	6000 Feet of #24 Magnet Wire
Coil Form	4 Inches in Diameter PVC
Form Length	7 Inches
Suspension Line	30 Pound Test Fishing Line
Spring	Compression Type
Magnet	1 Used; 2¾ Inches Long, with Pole Pieces
Amplifier Used	Basic LF 356 Seismometer
Damping	Provided inductively by the transformer-type sensor installed in the same container, at the bottom center of the coil form.

Figure 15-6 shows a seismogram from a 5.0 aftershock recorded from Fresno that contains both horizontal and vertical movements. Amplitude measurements and comparisons of relative strength can easily be made with the system.

Using an oscilloscope with a 1-mV sensitivity as a seismoscope, very intricate and detailed deflections can be seen on the screen. A VCR would have been handy to record some of these views.

After the earthquake, a sensor that used mechanical damping was transported closer to the epicenter, and it was hooked to the EER. A 9-Vdc counter was installed in place of the buzzer and alcohol was used as the damping medium, in an attempt to "count" significant aftershocks.

After a little adjusting, aftershocks over 3.7 were automatically counted. The alcohol worked very well. Spurious counts were virtually nonexistent, and, as planned,

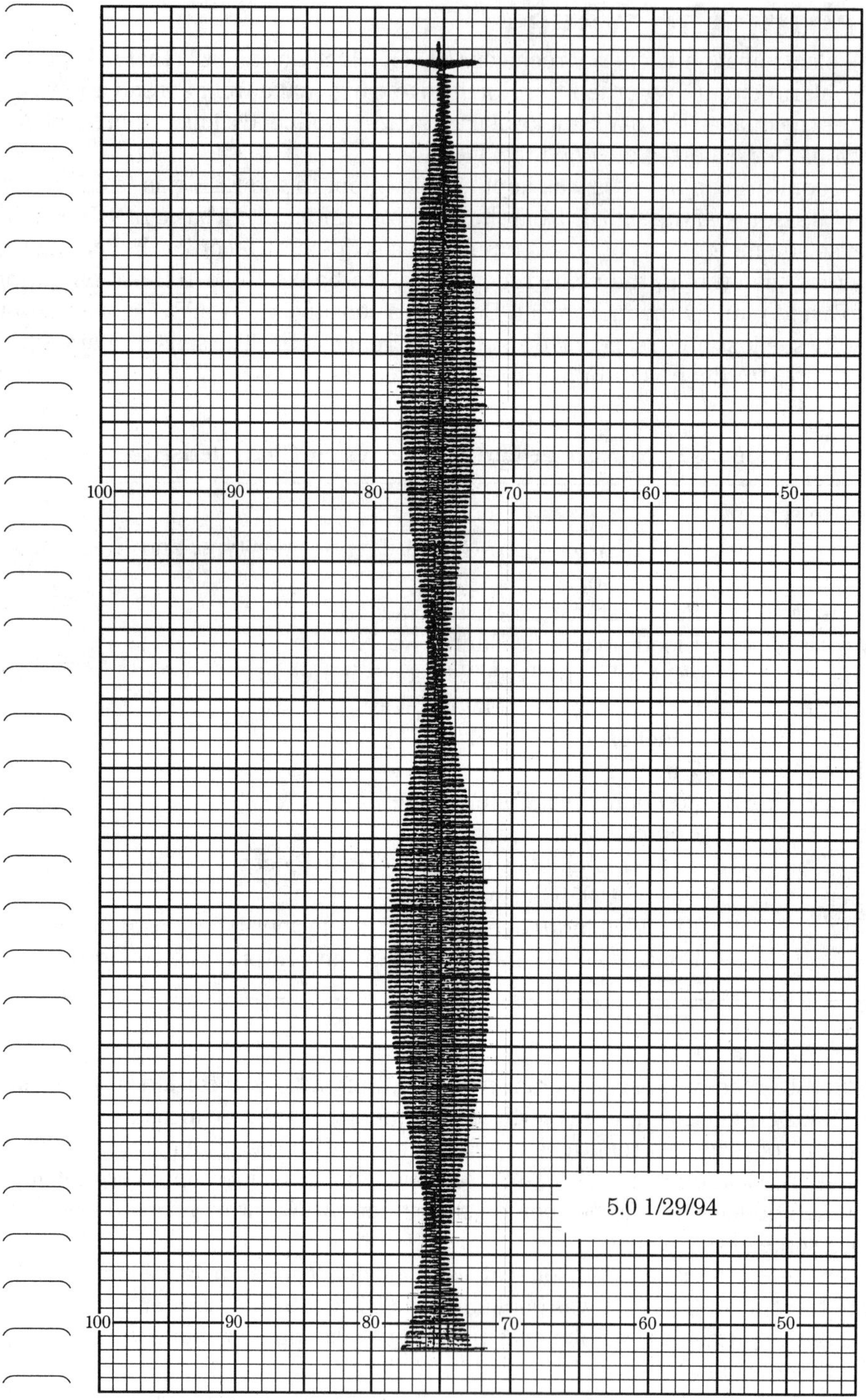

15-6 Seismogram from a 5.0 aftershock.

sensitivity was restricted to the bigger aftershocks because smaller ones were absorbed by the damping medium.

About a week later in Fresno, the EER was hooked to a commercial HS-10-1 sensor to see how the meter and LED circuits were reading the lower-level activity. Both showed the almost constant movement generated by the great number of aftershocks.

Officially, in the first month after the January 17 earthquake, over 3600 aftershocks occurred, By mid-April, about 7000 had occurred, and seven measured at least 5.0. Ultimately, over 25,000 aftershocks are expected by the end of 1996. With the major 5.3 aftershock at 13:20 L on March 20, at least one more major aftershock over 5.0 is likely by the end of 1995.

Circuit adjustments

Many of the events of major earthquakes and aftershocks in California during the early part of 1994 showed interesting correlations to the observed activity in the VLF radio bands.

Some important circuit adjustments and observations were made that could greatly affect how the theory of monitoring these low frequencies is used for earthquake observation and prediction.

For the 25-kHz reception, the navigator's TRF circuit was used. The radio was used with the basic seismometer and an oscilloscope for readings. The radio circuit was adequate. However, a couple of problems surfaced that had to be dealt with quickly.

The first problem dealt with the selectivity and sensitivity of the four-tube TRF circuit. A simple triode-tube RF stage was added at the front end of the receiver. This addition gave the new five-tube version dramatically better performance with more gain and better rejection of unwanted signals. The addition (Fig. 15-7) requires only a 1G4 tube, a variable capacitor, coil, and a tube socket. Connections out of this stage are made to the first 1N5 tube, and to the same power-supply connections, as in the original circuit.

The two-winding, solenoid-style coils, wound for reception down to 24 kHz, are identical in construction, using #32 magnet wire on empty paper-towel tubes. All tuning capacitors were changed to have the same value (365 pF) for easier tuning. If coil (band) switching is desired, use tube bases for the plug-ins and attach these to PVC shields that the coils fit into. Tube sockets are wired into the set's coil connections. This way, the five-tube version can accommodate other bands as long as all coils are wound identically.

The second problem involved the antenna input. A longwire, amateur-radio beam, and a vertical were tried. All seemed to either swamp and overload the receiver, or attenuate signals. Either way, efficiency was reduced. What seemed to work well for lightning-strike monitoring, solar flare study, and for observing teleseismic events, was not working very well for the nearby activity.

A large loop was tried. The low-gain and excellent null capability produced very good results. The directionality of the loop helped to focus into more specific areas. The lower gain helped to cut down signals from other areas further away.

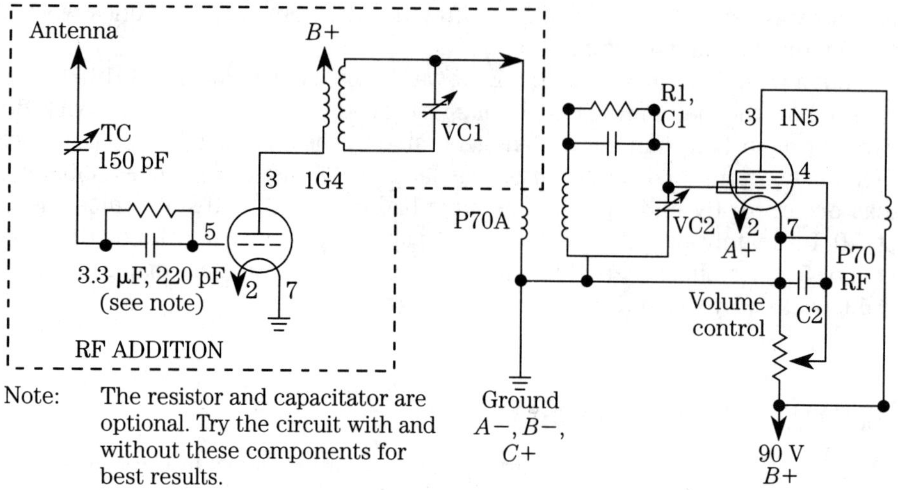

15-7 RF addition to the navigator.

This scheme has worked well for large earthquakes within 250 miles of the receiving station, as well as for receiving primary radio waves related to precursor and aftershock activity. Surprisingly, with the five-tube TRF circuit, a 36-inch-long telescoping whip antenna (trimmed with a 150-pF variable capacitor) and a good earth ground has also provided excellent reception.

This makes sense, as very-low frequency signals can, and often do, propagate great distances. Often, less antenna gain is better than more when trying to pin down signals from specific areas.

Preliminary indications are that this type of observation might be particularly suited to monitoring thrust-fault activity because of the nature of the subsurface movements; displacement of ground water, along with the changes in conductivity produced, and the friction associated with the vertical component of the movement; and the resulting electrical influences produced.

This is important because of the great damage potential of earthquakes stemming from thrust-fault activity, and because thrust-faults probably exist in far greater quantities than were previously thought. In California, they probably run perpendicular to the San Andreas Fault, intersecting or nearly intersecting the San Andreas, then extending to other parallel-running strike-slip faults. Earthquake activity in thrust-faults could then influence activity in various segments of strike-slip faults, thereby increasing the potential for great damage.

It is probable that such a dangerous potential exists in the San Francisco Bay area, between the Hayward Fault and the San Andreas. Around California, thrust-faults seem to run from the various smaller mountain ranges to the major known strike-slip fault lines.

History

As far as the observation of radio waves relates to the Northridge earthquake and to prediction studies, the 5.9-magnitude earthquake in 1987 on the Whittier Narrows Fault provides some precedent for the theory.

Another example of thrust-fault activity, this significant quake was preceded by several small (plus one moderate earthquake) earthquakes in the week prior to the 5.9 event.

Although this earthquake occurred before seismic ELF-ULF studies, low-frequency solar-flare studies were being done. Low-frequency propagation of signals below 100 kHz was noted as being enhanced at times, beginning eight days before the quake.

Since 1987, there have been several earthquakes with a magnitude greater than 5.0 in the northern edge of the Los Angeles basin. All of these earthquakes resulted from thrust-fault activity. The patterns of possible precursor activity (both of small earthquake swarms, and of low-frequency propagation anomalies), have been very similar for most, if not all of these events.

Attention needs to be given to the San Francisco Bay area as well. Beginning especially with the 7.1 Loma Prieta earthquake in 1989 and the seismic activity since, the more frequent microseismic swarms, and moderately larger (3.0 to 4.0) earthquakes indicate that an observation network needs to be set up for this area as soon as possible. The Loma Prieta earthquake occurred in a more remote region than the Northridge earthquake, and the aftershock pattern is much different. Although aftershocks from Loma Prieta continue to this day, there have already been more intense aftershock events, both in magnitude and in quantity, from Northridge. Still, the 7.1 Loma Prieta quake did not release the amount of energy that could be expected from a quake of this magnitude, and it is very possible another similar quake could occur in the area—on its own, or influenced by activity on the Hayward Fault, or on the probable thrust-fault that exists between the Hayward and the San Andreas faults.

One note regarding using computers for prediction studies: computers are excellent for recording historical data, making strength and energy comparisons, and calculating mathematical probabilities. However, events in nature very often do not follow computer projections. Variables exist that are not fully understood by scientists, therefore cannot be input into the computer. Random events in nature do not seem to be "random," and, in the case of earthquakes, computing projected times of future earthquakes by comparing time factors of historical quakes and expected stress accumulations and subsequent releases has not successfully predicted earthquakes. Computers, then, need to be used as a tool to acquire and analyze current conditions and anomalies, along with the historical data, to attempt to more precisely predict future events. In other words; time between earthquakes is important, but events leading up to an earthquake are more important. Computers can be most successfully used in projecting the probability of significant aftershocks because these events seem to follow more conventional "seismic gap" and stress-release theories.

Other observations

Obviously, for observers located at great distances from fault lines and earthquake zones, the most sensitive and responsive apparatus needs to be used for detection and monitoring earthquake activity. Although this is at least interesting to observers located on or near active areas, for prediction efforts, it seems that sensitive instrumentation designed to monitor locally active areas is more important.

Regarding the theory of monitoring radio waves to forecast earthquake activity: although direct seismic signals are subaudible using frequencies from 1.5 Hz to about 20 Hz, there seem to be features of harmonic activity at much higher frequencies. This is why frequencies of 300 Hz, 6 to 7 kHz, and 25 kHz, monitored primarily for solar activity, showed propagation anomalies in the seismic studies. Virtually devoid of man-made signals, various natural radio emissions are easily studied here via the "SEA" method.

As with electromagnetic emissions from thunderstorms, there are probably many more frequencies, from VLF through microwave, that could propagate seismic signals for short distances, either directly or harmonically. Further studies should include sensitive receivers, located at or very near fault lines, monitoring otherwise clear-channel frequencies in the upper shortwave, VHF, FM, and UHF bands.

A private short-range "SES" station

For those living in or very near, or those studying fault zones and seismically active areas, another monitoring procedure involves installing low-power beacon transmitters at various points along faults.

Transmitters would use low power (100 mW to about 1 W) and transmit a narrow-band tone or Morse-code signal. An omnidirectional radiating vertical antenna would propagate the signal. Signals would be very limited in range and researchers would try to correlate changes in the propagation characteristics of the signal to changes influenced by local seismic activity. Signals might travel further, the noise content might change, or signals might be attenuated as ground conductivity changes or as the bedrock becomes more electrically resonant before an earthquake. In other words, changes in natural conditions would be monitored by observing their effects on the man-made signal. Stable oscillators and receivers, and quiet, constant power supplies are musts for this study.

As an alternative set-up, BFO (beat-frequency oscillator) signal-processor units could be installed in remote locations, along with an FM transmitter to send the signals to nearby receiving stations. There are many commercially available low-power FM transmitters, some with as much as a three-mile range. Very sensitive remote-operating stations could easily be installed, using solar power and batteries for the power. Monitoring can be done via receiver S-meter readings, seismometer/receivers, chart recorders, audio recordings, computer data, or oscilloscope readings. The information could then be analyzed for study and possible warning forecasts.

This procedure is similar to how the U.S.G.S. monitors its radio link of seismometers. Various sensory units in the field use low-power transmitters to relay ground movements to receiving stations that use computers to analyze and catalog the movements.

The main difference in this system is that we are trying to observe, along with the earthquakes themselves, the precursors and influences *prior* to the actual physical seismic activity. We are, therefore, observing for prediction efforts, rather than to historically catalog temblor locations, intensities, and effects.

If an FM transmitter is not available or cannot be used, a short-range station can be built using a transmitter/oscillator, and building the LM 324 receiver to hear the signals.

The LM 324 receiver, using fixed capacitors and inductors, can provide stable reception for one or more channels in the VLF range very efficiently. The receiver has low power requirements, runs "cool," is stable in operation, and the output can drive any analyzing medium that you choose. When used in conjunction with a transmitter that has a range of, for example, 300 yards, the signals would be far less susceptible to influences from solar and long-distance atmospheric anomalies. Seismic influences might be more recognizable with the only other major influence being that of local thunderstorm activity. Of course, the emission patterns of these are much easier to identify than the longer-range atmospheric or solar-flare influences.

It would be best to build the receiver first, then construct the transmitter(s) for operation in the receiver's fixed channel ranges. For a 5- to 6-kHz channel; the fixed capacitor in the tuning circuit should be from 0.83 to 1.0 μF, and the inductor can be solid-state, or wound with 75 to 100 turns of #32 magnet wire around a 4½-inch form and tuned with a trimmer capacitor, if necessary. Be sure that the frequency used for the station is an otherwise clear frequency. However, in this range, finding clear frequencies is usually not a problem.

Obviously, this station would be most useful when installed in the immediate vicinity of a fault. The best situation would be placing the transmitter on one side of a fault and the receiver on the other side. Any of the seismic radio circuits could be used to monitor movements as they happen, either locally or by an FM transmitter. Remember that the propagation characteristics of VLF signals are influenced by, and contain impulses from, solar flare and atmospherics activity.

The "Coalinga gripper"

The "Coalinga gripper" might be the most interesting sensor/receiver yet. Most of this book has featured equipment to study earthquakes, solar activity, and low-frequency radio propagation characteristics. It really does seem that both solar activity and seismic activity affect radio propagation, but at different frequencies and in different ways.

Electricity does funny things under certain conditions. Low voltages are easy to work with and the higher voltages are pretty straightforward, too, until you reach about 500 V, then strange things start to happen. Electricity starts traveling on the outsides of the wires (and even through the air at high frequencies), such as in plasma bulbs, Jacob's ladders, Tesla coils, etc.

Magnetism is inherently strange. How it works and how to use it is simple. But, why it works defies simple explanations.

So, what does that mean for us? With a lot of luck, the "Coalinga gripper" was built on March 22, 1994, as a substitute for a commercial sensor in a permanent installation. Its combination of electromagnetism and induction has led to some strange observations.

Luck was involved in using just the right amount of wire for the inductor, the fortunate use of two magnets, adjusting the weight correctly and increasing the strength of its receiving potential, using just the right spring for the weight, and attaching it to a good mount. Specifications for this sensor are given in Table 15-3, and its schematic is shown in Fig. 15-8.

Table 15-3. Coalinga Gripper Sensor Characteristics

Nominal Frequency	0.6 Hz
Frequency Range	0.5 Hz to 1.5 Hz
Resistance at Rest	2500 Ω
Maximum Motional Resistance	250,000 Ω
Maximum Motional dc Voltage	400 mV
Frequency Response	100 Hz
Spring Type	Compression
Spring Length (No Weights)	2⅜ Inches
Spring Length (Installed)	2¾ Inches
Spring Positioning	Mid-Suspension Line
Wire Type	#33 AWG Magnet Wire
Length of Wire on Coil	11,835 Feet (Approx.)
Turns	10,000
Mean Turn Length	14.2 Inches
Type of Winding	Jumble Wound
Coil Form Material	Hard Cardboard
Form Diameter	4 Inches
Form Length	11 Inches
Magnets	2 Magnets Used; Each 2¾ Inches Long, with Set of Pole Pieces for Each
Suspension Line	30 lb. Test Fishing Line
Length	7 Inches
Termination Point	Mid-point Between the Two Magnet Assemblies
Container	Five-Gallon Plastic Drum with Lid
Coaxial Cable	RG-58
Mounting	Concrete, Permanent Installation

The sensor got its name after being on duty for 9 days, when two earthquakes hit Coalinga, California, 55 miles away.

More than just a seismic sensor for local small to moderate and distant large earthquakes, the Coalinga gripper is proving to be perhaps the first successful precursor sensor. Signals from the sensor were being input into a single-channel basic LF 356 seismometer, with output A going into a chart recorder. Again, the gain of the system

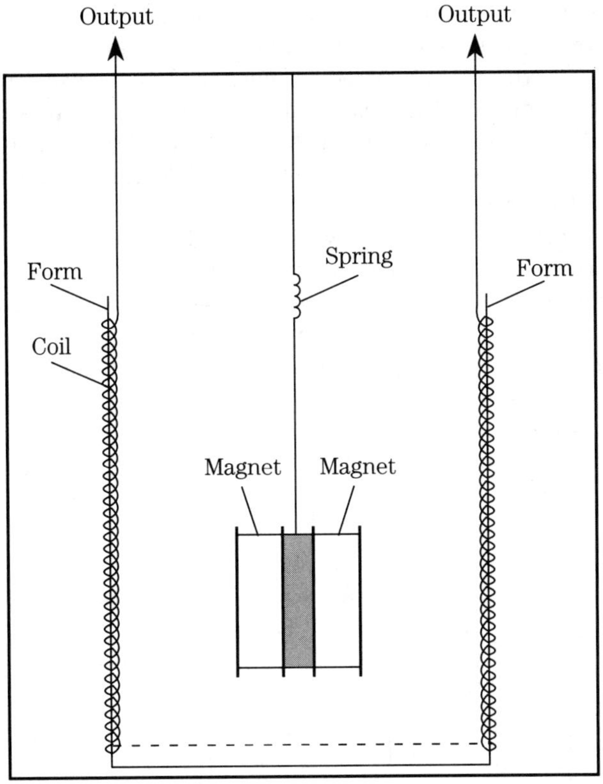

15-8
Coalinga gripper sensor schematic.

was such that the recorder's 100-mV scale was used. Note that a 0.5-mA seismometer setting allows the greatest sensitivity because a 10-mV scale can then be used.

For the first week, recordings showed very little movement, except for the common small local microseismic activity. However, in addition to this, and perhaps more importantly, the sensor was sensing something else: day vs. night. During the day, regardless of weather, the chart recordings show continuous small oscillations, punctuated at times by the local seismicity. At night, the same seismic activity would show up, but there are no oscillations. Essentially, a straight line would appear every day, from about five minutes after local sunrise to about five minutes before local sunset. Oscillations are always a little stronger just after sunrise and just before sunset than during the rest of the day. The action is like a "reverse" SEA monitoring station.

Three days before the March 31, 1994 Coalinga quakes, the paper-saving speed of the chart recorder, running at 0.2 inch/min. had to be increased to 0.1 inch/min. because the oscillations were so intense that the pen cut through the paper at the slower speed. After a day, the readings returned to normal, but with increased seismic activity noted. On the day before the earthquakes, the daytime readings were the same as the typical nighttime readings, a straight line. The following day, the oscillations, though erratic, were back, and the quakes hit around noon.

This is important because it confirms other researchers' claims of localized ULF reception anomalies. As a precursor sensor, a maximum range of 65 to 75 miles seems about right because the earthquake occurred 55 miles from the sensor. After a portable unit was built and taken into the field near the Northridge aftershock zone, the aftershock pattern there was reflected in the same kind of readings from the sensor. The "Coalinga gripper" name came from the fact that Coalinga is the only major seismic area in range of the prototype permanent sensor.

So, the Coalinga gripper serves as a local-range seismic detector and an extremely ULF receiving inductor and propagation monitor, as well as being a short-range precursor sensor. It will be interesting to see how the sensor reacts to solar flares. As a noise receiver, the sensor's output is fed into a basic seismometer, and output A is input into an audio amplifier (Fig. 15-9). This isn't a particularly "low-noise" audio amp, but it does work in the circuit. Listen just after sunrise and just before sunset to hear the big difference between normal daytime and nighttime readings. The parts list for this ULF noise receiver (amp section only) is listed in Table 15-4.

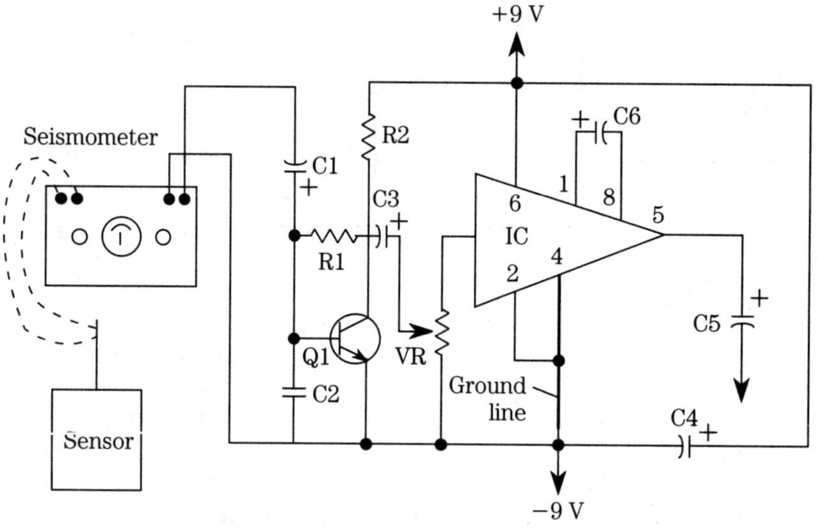

15-9 ULF noise receiver schematic.

Also, another mounting design can be used (or with any other sensor for that matter) that can increase the sensitivity to earthquakes. A permanent enclosure must be constructed that will be used only to house the sensor. A wooden-frame enclosure would be excellent. Particularly important is a wooden floor because it is easier to cut a hole for the sensor mount in wood than in any other material. This strut mount is shown in Fig. 15-10.

Here, one-inch-diameter pipe, fittings, and adapters were used, and the important thing to note is that the pipe mount, set in concrete to a depth of four feet beneath the ground level, never touches the frame or the floor of the enclosure. This keeps vibra-

Table 15-4. Coalinga Gripper Based ULF Noise Receiver Parts List

Part	Description
Sensor	Coalinga Gripper Seismic Sensor
Seismometer	Single Channel LF 356-Based Seismometer
Cable	Coaxial Cable for Hook-Up
Receiver:	
C1, C3	1-µF, 50-V Electrolytic Capacitors
C2	0.01-µF, 50-V Capacitor
C4	100-µF, 50-V Electrolytic Capacitor
C5	22 to 47-µF, 50-V Electrolytic Capacitor
C6	4.7-µF, 50-V Electrolytic Capacitor
IC	LM 386 Integrated Circuit
VR	1-kΩ to 5-kΩ, ½-W Potentiometer
Q1	2N3904 NPN Transistor
R1	33-kΩ, ¼-W Resistor
R2	10-kΩ, ¼-W Resistor
Speaker	8 Ω, 2-W Speaker
PS	9-V Power Supply

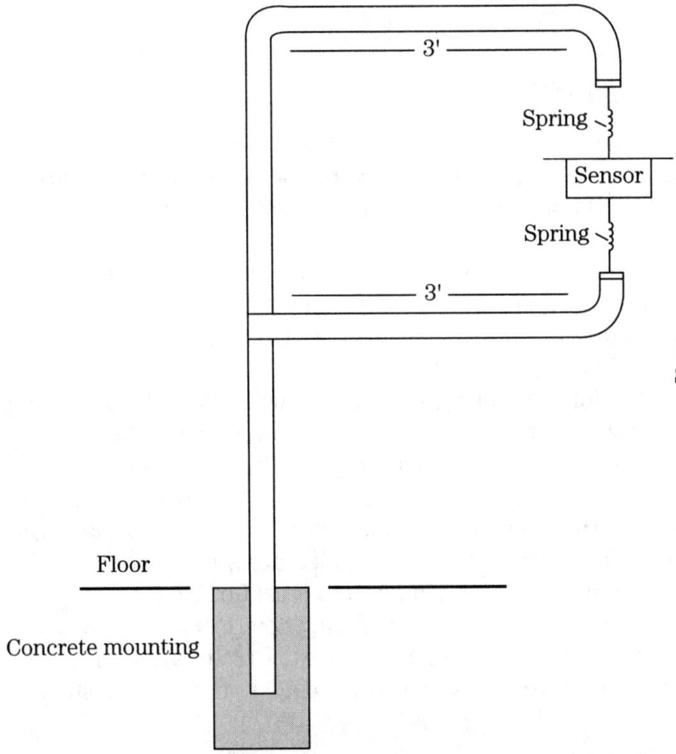

15-10
Strut-mounting for sensors.

tions of the frame and floor away from the sensor, effectively isolating it from accidental surface vibrations. The pipe extends to a height of 6 feet, and the sensor is suspended between two arms by strong fishing line and two strong springs that are adjusted so that the suspended sensor container just slightly stretches the springs.

This sensor also is sensitive to atmospheric disturbances such as thunderstorms and developing weather activity. On April 22, during hot and dry weather in central California, the sensor began to indicate atmospheric activity above normal intensities of oscillations on the chart recorder.

Similar to intense low-level seismic activity, the pattern persisted throughout the day, but the nighttime readings were normal for the system. The following evening was punctuated by moderate thunderstorm activity and the first measurable rain in weeks. Although the oscillations from the sensor looked like seismic activity, the readings were unmistakable when compared to ULF and VLF readings during thunderstorm activity. So, the Coalinga gripper also has a use as a short-range weather sensor.

A note on the construction of the coil. It was wound by hand around a hard cardboard form that was the same size as the oatmeal box sensor. 10,000 turns of wire were wound tightly in "jumble-winding" fashion. That is a lot of winding, but I was able to wind about 1000 turns per 12 to 15 minutes. Take a break (rest), each 1000 turns, re-tape the coil, and go on. If the wire breaks, don't panic; clean off the insulation, solder the wires together, insulate them with tape, and go on. This happened on the prototype coil, at turn number 6997. Test a spliced coil with an ohmmeter to be sure of the continuity.

Weather observations

Although the Northridge earthquake was the major seismic event in the United States in January, the natural forces of weather were pounding the eastern part of the country with extremely cold weather, including blizzard conditions and ice storms.

The jet-stream pattern brought arctic air and plenty of snow and ice to the region, while California and the southwest U.S. were dry and warm. The last half of December and the first half of January were so warm and dry in most of California that drought conditions were already being forecast.

The meteorological interest of monitoring VLF signals was rewarded beginning in late December, as solar activity increased and stratospheric warming began. The major solar flare on January 16 was followed by a drop southward of the polar jetstream. The eastern U.S. warmed, California cooled, and the first major storms in more than a month approached the West Coast. By January 22, heavy storms began pounding the state. In a couple of days, the Los Angeles area, devastated by the earthquake, had to deal with heavy rains, mudslides, and local flooding.

Stratospheric warming extended into February. Rain, snow, thunderstorms, and hail hit all over California after the middle of the month. Precipitation amounts, which had lagged far behind averages all season, finally began to approach historical averages.

However, the weather was very erratic, and hopes for average or better rainfall in California dwindled rapidly. Although the Northeast had 18 major winter storms and many cities set record extremes for cold while the "Siberian Express" occupied the region most of the winter, California remained mostly dry and warm.

A blocking high-pressure ridge pushed most storms to the north of the state; the polar jet-stream carried storms north and east of the West Coast.

The only significant storms in California during March came suddenly and violently. On March 5, severe thunderstorms developed west of the state and moved over central California during the afternoon. 55 MPH winds and 1⅛-inch-diameter hail, along with downbursts and funnel clouds caused the uprooting of trees, structural damages, downing of power lines, and about $7 million in agricultural losses to budding fruit and nut trees. One week later, northern California was hit by a similar barrage of severe thunderstorms.

During the period of March 22 to 24, an intense storm developed in the north Pacific, and it began to move southward along the coast from Vancouver Island. This was possibly the last legitimate chance for significant rainfall for the dry areas of California, and for the mountains to get large amounts of snow for summer runoff. Again, hopes were high as the storm looked impressive on satellite pictures. Hopes were that the storm would turn eastward over northern and central California and boost the state's rainfall, now lagging at about 65% of normal.

The storm finally moved east on the 24th, but over southern California. Central California stations received generally less than one-quarter of an inch from the once-promising storm.

The recurring, much publicized drought in the region was not totally unexpected. March signalled the fifth consecutive month of near-normal sea surface temperatures. ENSO anomalies were near zero, and the Southern Oscillation was stable and had been at near-normal conditions since late October 1993. The ENSO event of 1991-92, which had carried over into 1993, was over. There were no solar flares, solar activity was low, and nothing else persuaded the polar jet-stream to dip southward over California. The sub-tropical jet-stream stayed south, and the dry and warm conditions equalled drought for most of California, again.

However, mother nature had another springtime surprise for California. The stratospheric warming that existed throughout March 1994 increased thunderstorm potential, and the almost constant geomagnetic storm conditions adversely affected circulation patterns over the Pacific Ocean.

Such late-winter and early spring conditions would need solar radiation in the form of sunlight to provide the energy for storm development.

Beginning in late April, a series of cut-off low-pressure cells began to develop, then stalled just west of the Pacific Coast, from offshore central Oregon to southern California. The increased moisture from the Pacific was provided during the stratospheric warming "Episode." The cold, unstable air mixed in as a result of the unusual circulation patterns and a more electrically "charged" atmosphere led to a series of very intense, mostly unpredictable thunderstorm events in most of California during April and May.

Severe thunderstorms moving through south-central to northern California in the first week of May provided record daily precipitation totals in many areas. Rain, hail, and tornadoes struck much of the state from May 4 to May 7. Agriculture suffered greatly as rain and hail damaged many crops.

Monitoring stratospheric warming, polar absorption, and solar radiation can help in forecasting these strange conditions. The thunderstorm conditions were so severe during the early morning hours of May 6 that the Coalinga gripper sensor recorded the activity for nearly two hours.

Some selected stream-line maps are provided in appendix F for those who want to compare the conditions in future studies.

A
APPENDIX

Basic electrical laws

Resistances

Ohms Law: where I = amperes, E = volts, and R = ohms

$$I = E/R, \text{ or } E = IR, \text{ or } R = E/I$$

$$\text{Watts } (P) = \text{volts} \times \text{amps}; P = EI, P = E^2/R, \text{ or } P = I^2R$$

For calculating the values of resistors in series; the total resistance of a circuit with a number of resistors in series equals the sum of the individual resistances:

$$R_{Total} = R_1 + R_2 + R_3 \ldots$$

For resistors in a parallel connection; the total resistance is always less than that of the lowest value resistor present:

$$R = \frac{1}{\frac{1}{R_1} + \frac{1}{R_2} + \frac{1}{R_3} \cdots}$$

or, for only two resistors in parallel, the formula is :

$$R_{total} = \frac{R_1 \times R_2}{R_1 + R_2}$$

Capacitances

When capacitors are connected in parallel, the total capacitance of the group equals the sum of the individual capacitors: $C_{total} = C_1 + C_2 + C_3 \ldots$

If two or more capacitors are connected in series, the total capacitance is less than that of the smallest capacitor present:

$$C_{total} = \frac{1}{\frac{1}{C_1} + \frac{1}{C_2} + \frac{1}{C_3} \ldots}$$

Or, for only two capacitors connected in series; the formula is:

$$C_{total} = \frac{C_1 \times C_2}{C_1 + C_2}$$

The same units of measure must be used throughout the equation; pF or µF, with no mixing.

Inductances

For those experimenters and builders wanting to wind their own RF or choke coils; refer to the wire table and use the following formula:

$$\frac{L\ (\mu H)}{9a + 10b} = a^2 n^2$$

for the approximate inductance of single-layer air-core coils, where:

L = Inductance in microhenrys

a = Coil radius in inches

b = Coil length in inches

n = Number of turns

For inductors in series: When two or more inductors are connected in series, the total inductance is equal to the sum of the individual inductors, provided that the coils are separated so that no coil is in the magnetic field of another:

$$L_{total} = L_1 + L_2 + L_3 \ldots$$

For inductors in parallel, use the following formula:

$$L_{total} = \frac{1}{\frac{1}{L_1} + \frac{1}{L_2} + \frac{1}{L_3} \ldots}$$

And for only two inductors in parallel, use:

$$L = \frac{L_1 \times L_2}{L_1 + L_2}$$

B
APPENDIX

Electrical schematic symbols

- ammeter
- amplifier (operational)
- AND gate
- antenna (balanced, dipole)
- antenna (general)
- antenna (loop, shielded)
- antenna (loop, unshielded)
- antenna (unbalanced)
- antenna (whip, or telescopic)
- attenuator (or resistor, fixed)
- attenuator (or resistor, variable)
- battery
- capacitor (feedthrough)
- capacitor (fixed, nonpolarized)
- capacitor (fixed, polarized)
- capacitor (ganged, variable)
- capacitor (single, variable)
- capacitor (split-rotor, variable)
- capacitor (split-stator, variable)

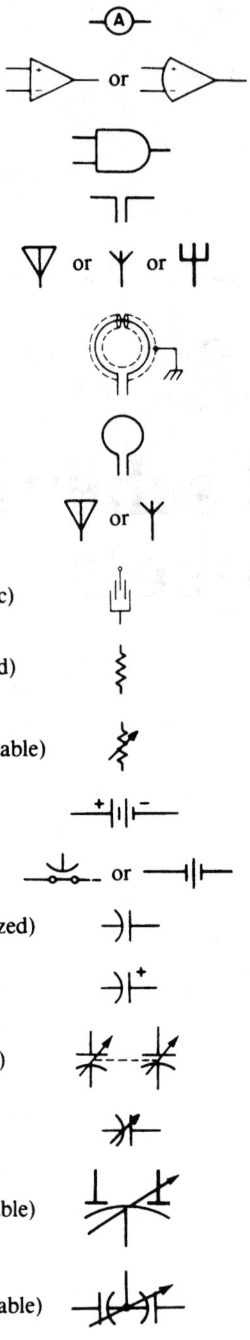

- cathode (directly heated)
- cathode (indirectly heated)
- cathode (cold)
- cavity resonator
- cell
- circuit breaker
- coaxial cable
- coaxial cable (grounded shield)
- crystal (piezoelectric)
- delay line
- diode (field-effect)
- diode (general)
- diode (Gunn)
- diode (light-emitting)
- diode (photosensitive)
- diode (photovoltaic)
- diode (pin)
- diode (Schottky)
- diode (tunnel)
- diode (varactor)
- diode (zener)
- directional coupler (or wattmeter)
- exclusive-OR gate

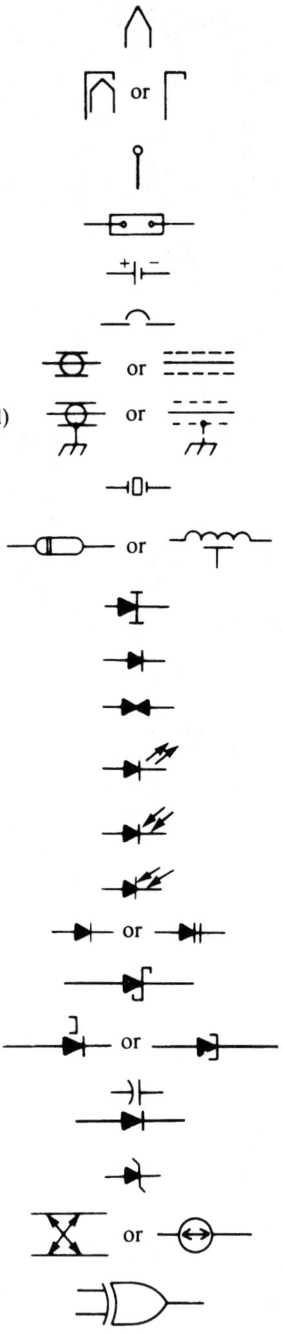

female contact (general)
ferrite bead
fuse
galvanometer
ground (chassis)
ground (earth)
handset
headphone (double)
headphone (single)
headphone (stereo)
inductor (air-core)
inductor (bifilar)
inductor (iron-core)
inductor (tapped)
inductor (variable)
integrated circuit
inverter or inverting amplifier
jack (coaxial or phono)
jack (phone, two-conductor)
jack (phone, two-conductor interrupting)
jack (phone, three-conductor)
jack (phono)
key (telegraph)
lamp (incandescent)
lamp (neon)
male contact (general)
meter (general)

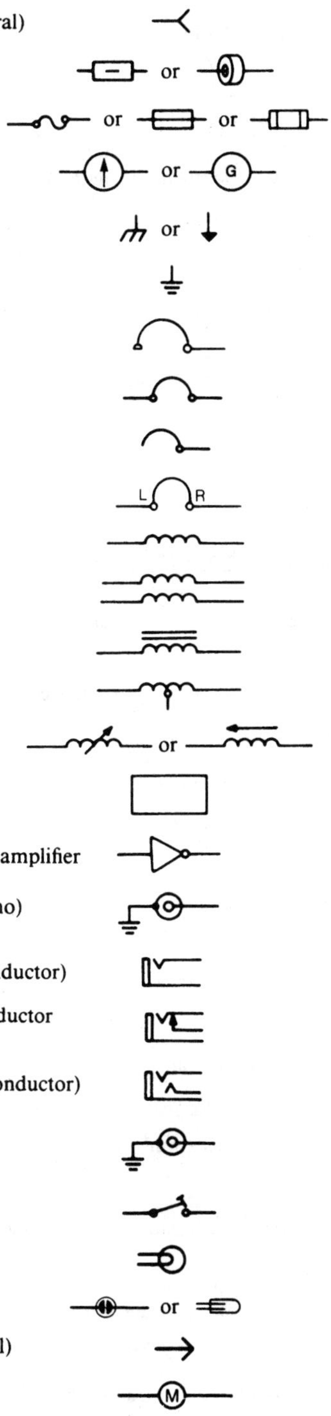

microammeter
microphone
microphone (directional)
milliammeter
NAND gate
negative voltage connection
NOR gate
operational amplifier
OR gate
outlet (nonpolarized)
outlet (polarized)
outlet (utility, 117 V, nonpolarized)
outlet (utility, 234 V)
photocell (tube)
plug (nonpolarized)
plug (polarized)
plug (phone, two-conductor)
plug (phone, three-conductor)
plug (phono)
plug (utility, 117 V)
plug (utility, 234 V)
positive-voltage connection

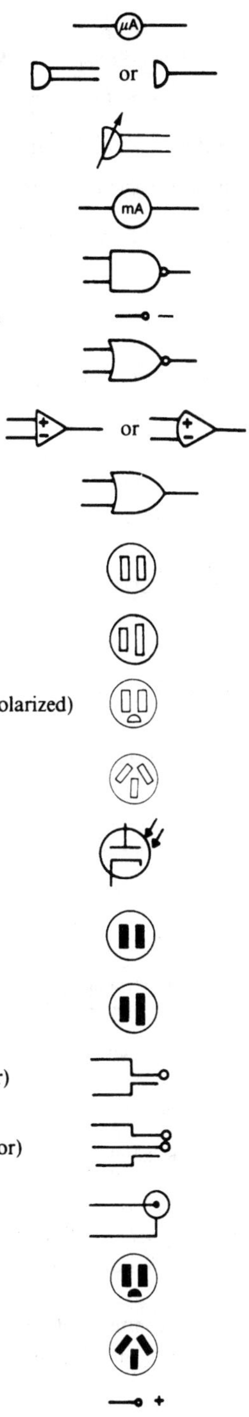

- potentiometer (variable resistor, or rheostat)
- probe (radio-frequency)
- rectifier (semiconductor)
- rectifier (silicon-controlled)
- rectifier (tube-type)
- rectifier (tube-type, gas-filled)
- relay (DPDT)
- relay (DPST)
- relay (SPDT)
- relay (SPST)
- resistor (fixed)
- resistor (preset)
- resistor (tapped)
- resonator
- rheostat (variable resistor, or potentiometer)
- saturable reactor
- shielding
- signal generator
- solar cell
- source (constant-current)
- source (constant-voltage)
- speaker

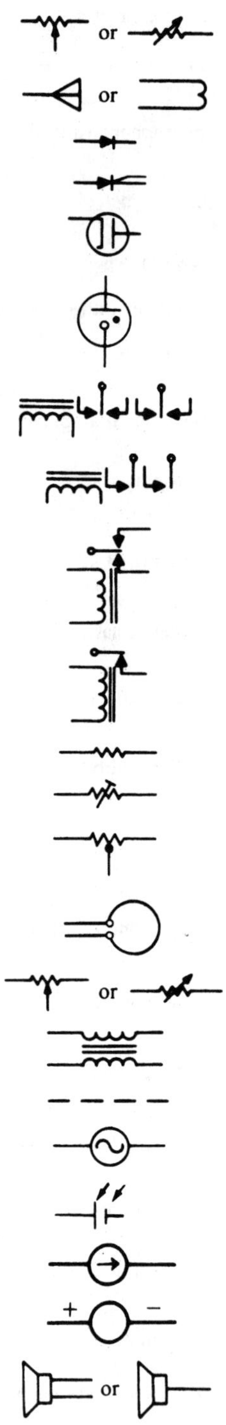

- switch (DPDT)
- switch (DPST)
- switch (momentary-contact)
- switch (rotary)
- switch (silicon-controlled)
- switch (SPDT)
- switch (SPST)
- terminals (general, balanced)
- terminals (general, unbalanced)
- test point
- thermocouple
- thyristor (diac)
- thyristor (triac)
- transformer (air-core)
- transformer (air-core, adjustable)
- transformer (iron-core)
- transformer (iron-core, adjustable)
- transformer (powdered-iron-core)
- transformer (tapped-primary)

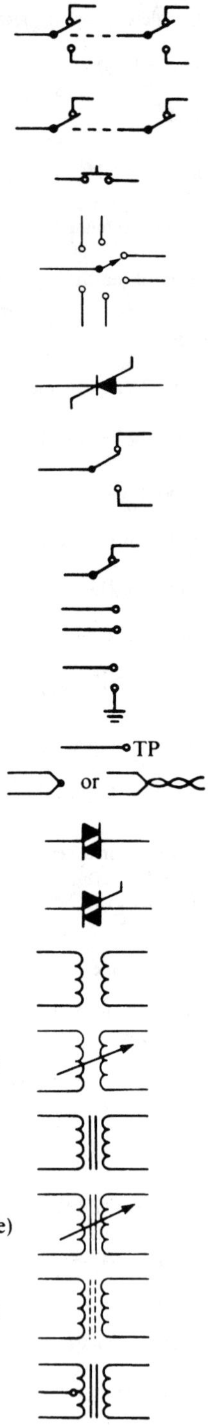

transformer (tapped-secondary) 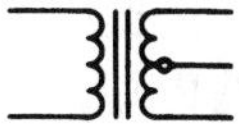

transistor (bipolar, npn) 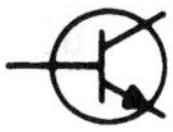

transistor (bipolar, pnp) 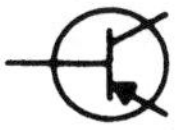

transistor (insulated-gate field-effect, or IGFET) 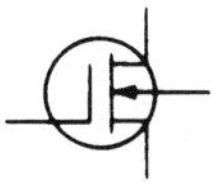

transistor (junction field-effect, or JFET) 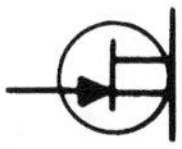

transistor (field-effect, n-channel) 

transistor (field-effect, p-channel) 

transistor (metal-oxide, dual-gate)  or 

transistor (metal-oxide, single-gate)  or 

transistor (photosensitive)  or 

transistor (unijunction)  or 

tube (diode) 

tube (pentode) 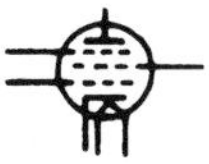

tube (photomultiplier) 

tube (tetrode) 

tube (triode) 

unspecified unit or component 

voltmeter 

wattmeter 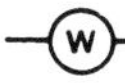

or 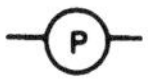

waveguide (circular) 

waveguide (rectangular) 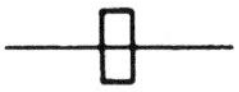

waveguide (flexible) 

waveguide (twisted) 

wires (crossing, connected)  or  or 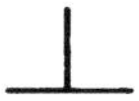

wires (crossing, not connected)  or 

Molded paper capacitor codes (capacitance given in pF)

Color	Digit	Multiplier	Tolerance
Black	0	1	20%
Brown	1	10	
Red	2	100	
Orange	3	1000	
Yellow	4	10000	
Green	5	100000	
Blue	6	1000000	
Violet	7		
Gray	8		
White	9		10%
Gold			5%
Silver			10%
No color			20%

5-dot radial-lead ceramic capacitor

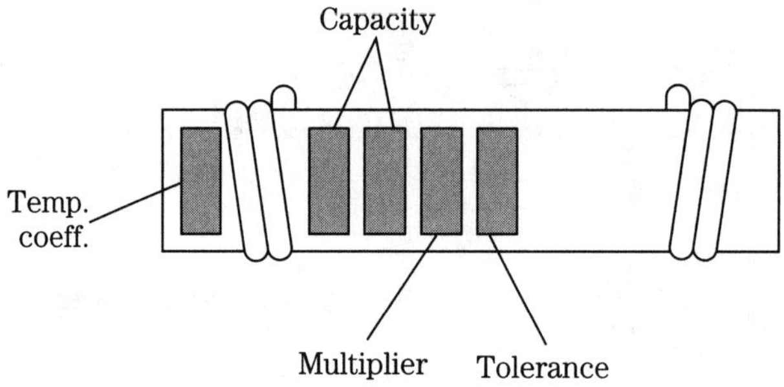

Current standard code

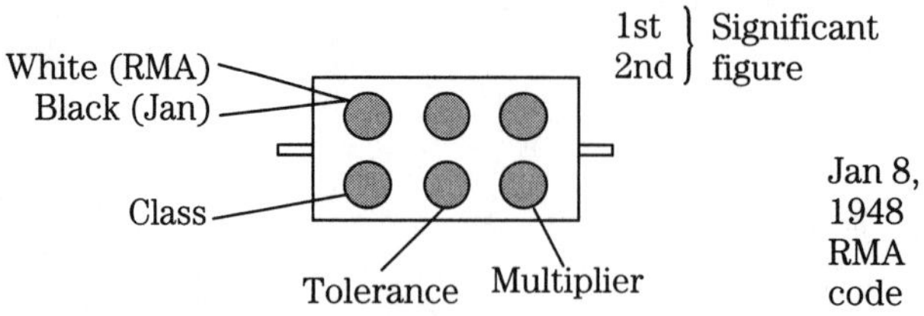

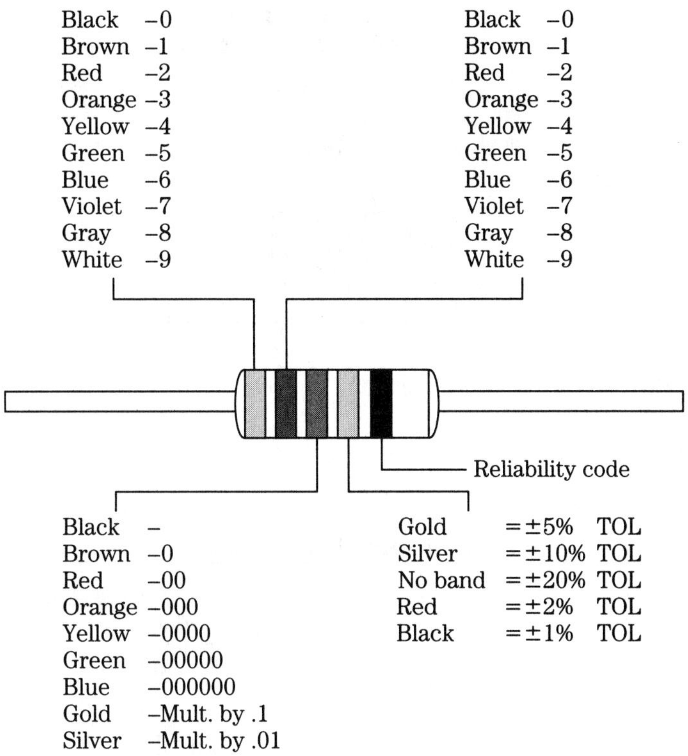

Tubular capacitor color codes

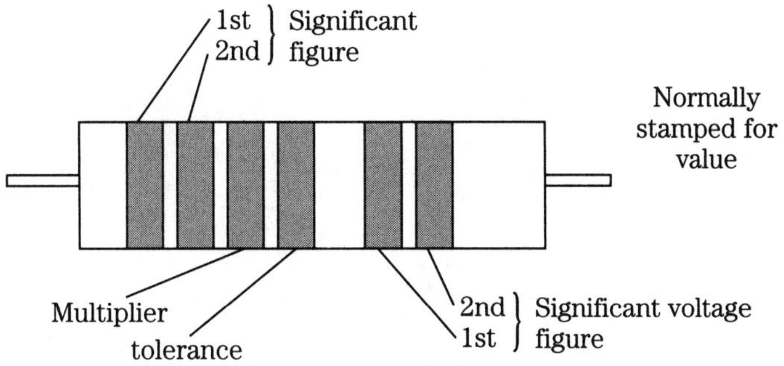

A 2-digit voltage rating indicates more than 900 V.
Add 2 zeros to end of 2-digit number.

Axial lead ceramic capacitor

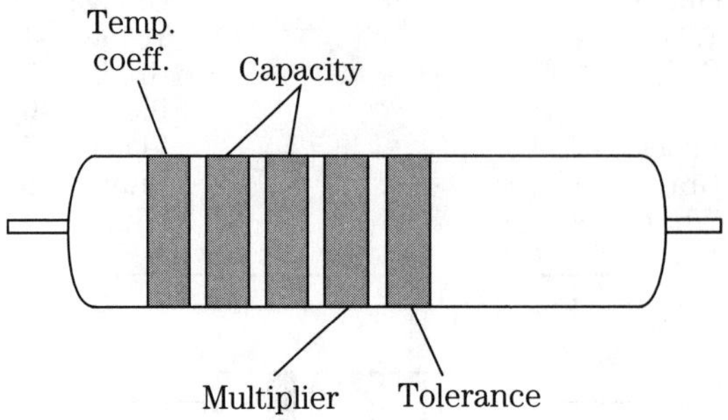

Disc ceramic RMA code

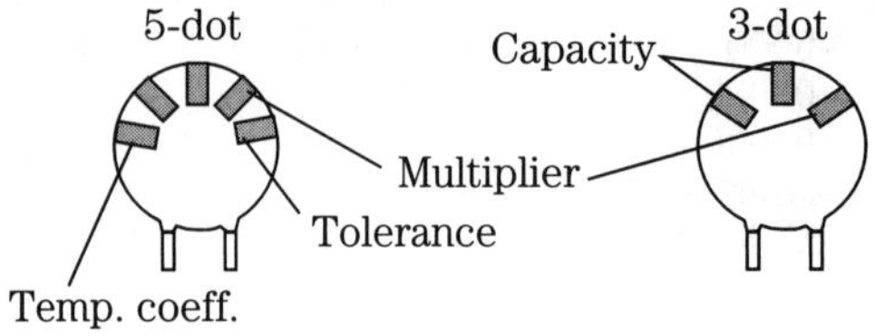

Ceramic capacitor codes (capacitance given in pF)

In reference to RS-198-A of the EIA, ceramic dielectric capacitors have three major classifications:

Class 1, temperature compensating ceramics requiring high Q and capacitance stability.

Class 2, where Q and stability of capacitance are not required.

Class 3, low voltage ceramics, where dielectric losses, high insulation resistance, and capacitance stability are nof of major importance.

				Class 1				Class 2
			*Tolerance**		Temperature coefficient	*Temp.* Significant	*Coeff.*	
Color	**Digit**	**Multiplier**	**10 pF or less**	**Over 10 pF**	**PPM°C**	**figure**	**Multiplier**	**Tolerance***
Black	0	1	±2.0 pF	±20%	0	0.0	–1	±20%
Brown	1	10	±0.1 pF	±1%	–33		–10	
Red	2	100		±2%	–75	1.0	–100	
Orange	3	1000		±3%	–150	1.5	–1000	
Yellow	4	10,000			–220	2.2	–10,000	+100%, –0%
Green	5		±0.5 pF	±5%	–330	3.3	+1	±5%
Blue	6			–470		4.7	+10	
Violet	7			–750		7.5		
Gray	8	.01	±0.25 pF		+150 to –1500		+1000	+80%, –20%
White	9	.1	±1.0 pF	±10%	+100 to –750		+10,000	±10%
Silver								
Gold								

*Tolerance on class-3 ceramic capacitors is indicated by its code, either ±20% (code M) or +80, –20% (code Z).

C

APPENDIX

Parts identification and substitution guide

The following component part numbers, references, and basic specifications might be helpful in choosing replacement parts for projects. Experimenters can use the basic parameters given as the basis for trying new circuit configurations, or in adapting old circuits to new uses. When "scavenging" old equipment for parts, this appendix will be helpful in parts identification, and in determining the suitability of parts for various applications.

For more detailed information, consult component substitution and characteristics reference guides, which are available from many dealers.

Integrated circuits

Converters, counters, displays

Part #	Description	Pins
ICL7106CPL	3½-Digit LCD Single-Chip A/D Converter	40
ICL7107CPL	3½-Digit LED Single-Chip A/D Converter	40
ICL7109CPL	12-Bit Binary A/D Converter	40
ICL7135CPI	Precision 4½-Digit A/D Converter	28
ICM7211IPL	4-Digit LCD Display Driver; Multiplexed Interface	40
ICM7211MIPL	4-Digit LCD Display Driver; Microprocessor Interface	40
ICM7216BIPI	8-Digit Universal Counter	28
ICM7216DIPI	8-Digit Frequency Counter	28
ICM7224IPL	4½-Digit LCD Counter Driver	40

Linear ICs

Part #	Description
OP07CP	Low-offset, Low-drift Op Amp
OP27GP	Low-noise, High-speed, Precision Op Amp
TL061CP	JFET-Input Low-Power Op Amp
TL064CN	JFET-Input Low-Power Op Amp
TL071CP	Low-noise, JFET Op Amp
TL072CP	Low-noise, Dual JFET Op Amp
TL074CN	Low-noise, Quad JFET Op Amp
TL081CP	JFET Input Op Amp
TL082CP	Wide Bandwidth JFET-Input Dual Op Amp
TL084CN	JFET-Input Quad Op Amp
TLC271CP	Programmable Low-Power Op Amp
TLC272CP	Precision Dual Op Amp
TL288CP	JFET-Input Op Amp
LM301N	Improved Op Amp
LM307N	Op Amp
LM308N	Precision Op Amp
LM310J	Noninverting Op Amp/Voltage Follower
LM310N	Noninverting Op Amp/Voltage Follower
LM311H	Hi-Performance Voltage Comparator
LM311N	Voltage Comparator
LM318N	Precision Hi-Speed Op Amp
LM319N	High-Speed Dual Comparator
LM324N	Low-Power Quad Op Amp
LM339N	Quad Comparator
LF347N	Wide-Band Quad JFET-Input Op Amp
LM348N	Quad 741 Op Amp
LF351N	BIFET Op Amp
LF353N	Dual BIFET Op Amp
LF355N	JFET-Input Low-Power Op Amp
LF356N	Wide-Band JFET-Input Op Amp
LF357N	Wide-Band Decompensated JFET-Input Op Amp
LM358N	Low-Power Dual Op Amp
LM380N	2-Watt Audio Power Amp
LM381N	Low-Noise Dual Pre-Amplifier; 9 V to 40 V
LM383T	7-Watt Audio Power Amp; 5 V to 20 V
LM384N	5-Watt Audio Amp
LM386N1	Low-Voltage Audio Amp; 0.250 W at 6 V
LM386N3	Low-Voltage Audio Amp; 0.500 W at 9 V
LM392N	Op Amp and Comparator
LM393N	Dual Comparator
LF411CN	Low-Offset, Low-Drift JFET-Input Op Amp
LF412CN	Low-Offset, Low-Drift JFET-Input Dual Op Amp
LF441CN	JFET-Input Low-Power Op Amp

Part #	Description
LF442CN	Dual JFET-Input Low-Power Op Amp
LF444CN	Quad JFET-Input Low-Power Op Amp
LM709CN	Op Amp
LM710CN	Differential Comparator
LM711N	Dual Differential Comparator
LM741CN	Op Amp
LM741CH	Op Amp
LM747CN	Dual 741 Op Amp
MC1458CP1	Quad 741 Op Amp
LM748CH	Frequency Adjusted Op Amp
LM1458N	Dual Op Amp
LM1875T	20-W Power Audio Amp; 16 V to 60 V
LM2002T	8-W Audio Power Amp
VA2707PJ	Dual Op Amp
LM2877P	Dual 4-W Audio Power Amp; 6 V to 24 V
LM2878P	Dual 4-W Audio Power Amp; 6 V to 32 V
LM2901N	Low-Power, Low-Offset Voltage Quad Comparator
MC3350P	Triple Differential Amp
LM3900N	Quad Op Amp
RC4136N	Quad Op Amp
NE5205	Wide-Band High-Frequency Amp
NE5532	Internally Compensated Dual Low-Noise Op Amp
NE5534	Single and Dual Low-Noise Op Amp
NE5539	UHF Op Amp
LM13600N	Dual Transconductance Amp
LF13741N	Monolithic JFET-Input Op Amp

CA series

CA3080E	Operational Transconductance Amp
CA3081N	Common Emitter NPN Transistor Array
CA3083N	NPN Transistor Array
CA3086N	Transistor Array
CA3130E	Super CMOS Op Amp
CA3140E	BI-MOS Op Amp
CA3146E	High-Voltage Transistor Array
CA3161E	BCD to 7-Segment Decoder Driver
CA3162E	Dual Slope and Rate A/D Converter
CA3306CE	CMOS 6-Bit Flash A/D Converter

Other converters

ADC0804LCN	8-Bit A/D Converter
ADC0809CCN	8-Bit A/D Converter with 8-Channel Multiplexer
ADC0817CCN	8-Bit A/D Converter w/16-Channel Multiplexer
ADC0820CCN	8-Bit, Hi-Speed A/D Converter, w/Track/Hold

Part #	Description
DAC0800LCN	8-Bit D/A Converter; Same as MC1408-6
DAC0807LCN	8-Bit D/A Converter; Same as MC1408-7
DAC0808LCN	8-Bit D/A Converter; Same as MC1408-8
DAC0830LCN	8-Bit Up D/A Converter; 0.05%
MB40576	6-Bit Ultra-High Speed A/D Converter
MB87020	16-Bit CMOS A/D or D/A Selectable Converter
TLC549IP	8-Bit A/D Converter; 0.05%
ZN427E8	8-Bit Successive Approximation A/D Converter
ZN428E8	8-Bit Latched Input D/A Converter
ZN429E8	8-Bit D/A Converter
ZN449E	8-Bit MPU Compatible A/D Converter

Timers

LM322N	Precision Timer
LM555CH	Timer
LM555CN	Timer
LM556N	Dual 555 Timer; Same as MC3456P
NE558N	Quad Timer
LM3905N	Precision Timer
XRL555	Micropower 555 Timer
XR2240	Programmable Timer/Counter
MC1455P	Timer
XR2242CP	Long-Range Timer

Transistor arrays

TBA331	General-Purpose Array
ULN2002A	Darlington Array
ULN2003A	High-Voltage, High-Current Darlington Array
ULN2004A	High-Voltage, High-Current Darlington Array
ULN2803A	High-Voltage, High-Current Darlington Array
MC3346P	General-Purpose Transistor Array

Various specialty devices

LM34CZ	Precision Fahrenheit Temperature Sensor
LM35CZ	Precision Centigrade Temperature Sensor
LM335CZ	Precision Temperature Sensor
LM3911N	Temperature Measurement And/Or Controller
MC3340	Electronic Attenuator
TDA2030A	18-W Hi-Fi Amp
TDA2040	20-W Hi-Fi Amp
ULN3330Y	Optoelectronic Switch
ULN3113	Hall-Effect Switch
ULN2429	Fluid/Moisture Detector
ZN414	AM Radio Circuit: 150 kHz to 3 MHz

Part #	Description
ZN416	ZN414 With 18-dB Audio Stage
MC33171	Low-Power Single-Supply Op Amp: 3 V to 44 V
ULN3755W	Dual-Power Op Amp

Voltage regulators

LM304H	Precision Negative Voltage Regulator; 0 to 40 V
C305C	Adjustable, Positive 4.5 V to 40 V
LM305H	Adjustable, Positive 4.5 V to 40 V
LM309H	5-V Regulator
LM309K	5-V 1-A Regulator
LM317K	Adjustable, Positive: 1.5 A at 1.2 V to 37 V
LM317LZ	Adjustable, Positive: 100 mA at 1.2 V to 37 V
LM317T	Adjustable, Positive: 1.5 A at 1.2 V to 37 V
LM323K	5 V, 3 A Positive
LM330T	5 V Positive: Low In/Out Automotive Regulator
LM337K	Adjustable, Negative: 1.5 A at -1.2 V to -37 V
LM337T	Adjustable, Negative: 1.5 A at -1.2 V to -37 V
LM338K	Adjustable, Positive: 5 A at 1.2 V to 32 V
LM338T	Adjustable, Positive: 5 A at 1.2 V to 32 V
LM345K	5 V, 3 A Negative
LM350T	Adjustable, Positive 3 A
LM396K	Adjustable, Positive 10 A
TL497ACN	Switching Regulator
LM723CH	Adjustable: 2 V to 37 V
LM723CN	Adjustable: 2 V to 37 V
LT1086CT	Adjustable, Positive: 1.5 A at 1 V to 25 V
LT1172CK	1.25-A Hi-Efficiency 100-kHz Switching Reg.
LT1172CN8	1.25-A Hi-Efficiency 100-kHz Switching Reg.
LT1172CT	1.25-A Hi-Efficiency 100-kHz Switching Reg.
78L05ACH	Positive: 5 V at 100 mA
78L05	Positive: 5 V at 100 mA
78L06	Positive: 6 V at 100 mA
78L09	Positive: 9 V at 100 mA
78L12	Positive: 12 V at 100 mA
78L15	Positive: 15 V at 100 mA
78M05	Positive: 5 V at 0.5 A
78M06	Positive: 6 V at 0.5 A
78M09	Positive: 9 V at 0.5 A
78M12	Positive: 12 V at 0.5 A
78M15	Positive: 15 V at 0.5 A
7805K	Positive: 5 V at 1 A
7812	Positive: 12 V at 1 A
7815	Positive: 15 V at 1 A
7805T	Positive: 5 V at 1 A

Part #	Description
7806T	Positive: 6 V at 1 A
7808T	Positive: 8 V at 1 A
7810T	Positive: 10 V at 1 A
7812T	Positive: 12 V at 1 A
7815T	Positive: 15 V at 1 A
7818T	Positive: 18 V at 1 A
7824T	Positive: 24 V at 1 A
78S40	1.5-A Switching Reg., 1.25 A at 1.25 to 40 V
79L05	Negative: 5 V at 100 mA
79L12	Negative: 12 V at 100 mA
79L15	Negative: 15 V at 100 mA
7905T	Negative: 5 V at 1 A
7912T	Negative: 12 V at 1 A
7915T	Negative: 15 V at 1 A
7918T	Negative: 18 V at 1 A
7924T	Negative: 24 V at 1 A

Rectifiers

1-A rectifiers

Part #	Voltage rating
1N4001	50
1N4002	100
1N4003	200
1N4004	400
1N4005	600
1N4006	800
1N4007	1000

3-A rectifiers

1N5400	50
1N5401	100
1N5402	200
1N5404	400
1N5406	600
1N5407	800
1N5408	1000

6-A rectifiers

6A05	50
6A1	100
6A2	200
6A4	400
6A6	600

1-A bridge rectifiers

Part #	Voltage rating
DB101	50
DB102	100
DB103	200
DB104	400
DB105	600

1.5-A bridge rectifiers

RB151	50
RB152	100
RB153	200
RB154	400
RB155	600

2-A bridge rectifiers

BR805D	50
BR81D	100
BR82D	200
BR84D	400
BR86D	600
BR810D	1000

5% 500-mW zener diodes

Part #	Voltage	Part #	Voltage	Part #	Voltage
1N5226B	3.3	1N5237B	8.2	1N5248B	18
1N5227B	3.6	1N5238B	8.7	1N5249B	19
1N5228B	3.9	1N5239B	9.1	1N5250B	20
1N5229B	4.3	1N5240B	10	1N5251B	22
1N5230B	4.7	1N5241B	11	1N5252B	24
1N5231B	5.1	1N5242B	12	1N5253B	25
1N5232B	5.6	1N5243B	13	1N5254B	27
1N5233B	6.0	1N5244B	14	1N5255B	28
1N5234B	6.2	1N5245B	15	1N5256B	30
1N5235B	6.8	1N5246B	16	1N5257B	33
1N5236B	7.5	1N5247B	17	1N5258B	36

1N746 = 1N5226B	1N753 = 1N5234B	1N757 = 1N5239B
1N749 = 1N5229B	1N754 = 1N5235B	1N758 = 1N5240B
1N751 = 1N5231B	1N756 = 1N5237B	1N759 = 1N5242B

5% 1-W zener diodes

Part #	Voltage	Part #	Voltage	Part #	Voltage
1N4728A	3.3	1N4737A	7.5	1N4746A	18
1N4729A	3.6	1N4738A	8.2	1N4747A	20
1N4730A	3.9	1N4739A	9.1	1N4748A	22
1N4731A	4.3	1N4740A	10	1N4749A	24
1N4732A	4.7	1N4741A	11	1N4750A	27
1N4733A	5.1	1N4742A	12	1N4751A	30
1N4734A	5.6	1N4743A	13	1N4752A	33
1N4735A	6.2	1N4744A	15	1N4753A	36
1N4736A	6.8	1N4745A	16		

Light sources

Referring to the emission wavelengths of the various types of LEDs, it becomes apparent that various applications can be affected by the reception characteristics of apparatus. Examples would be in the interferometer, when used as a measuring instrument, and the solar resistance measurements of the pyrheliometer.

Two of the more available other kinds of light sources are the IREDs, (infrared-emitting diodes), and the type of laser diodes used in the optoelectronics industry: in CD players, videodisc players, laser copiers and printers, and in certain measuring instruments.

Briefly, IREDs have peak emission wavelengths of from 870 to about 950 nm. This is significantly higher than the emission wavelengths of the red visible LED, which has an emission wavelength of up to 700 nm.

Laser light (coherent, monochromatic light produced by low-power diodes in consumer products) has an average emission wavelength between that of the visible and the infrared LEDs, generally from about 775 to 825 nm. Typical power outputs are from 3 to 5 mW, but some laser diodes produce 10 to 15 mW of power output.

Another related item concerns the solar cells used when constructing the solarimeter and the pyrheliometer: silicon cells can usually deliver more power per area than selenium cells. That is, in part, caused by the fact that silicon cells convert not only visible light, but also some of the invisible spectrum into electricity. Selenium cells have a response similar to that of the human eye.

Bulbs, lamps, LEDs, and pilot lights

Dial lamps

Type	Volts	Amps	Base type
14	2.47	0.30	Screw
40	6.3	0.15	Screw
41	2.5	0.50	Screw
42	3.2	0.35	Screw
43	2.5	0.50	Bayonet
44	6.3	0.25	Bayonet
45	2.5	0.35	Bayonet
46	6.3	0.25	Screw
47	6.3	0.15	Bayonet
48	2.0	0.06	Screw
49	2.0	0.06	Bayonet
50	7.5	0.22	Screw
51	7.5	0.22	Bayonet
53	14.4	0.12	Bayonet
55	7.0	0.41	Bayonet
57	14.0	0.24	Bayonet
67	13.5	0.60	Screw/Bayonet
382	14.0	0.08	Flanged
511	2.0	0.25	Screw
1021 (Lantern)	4.5	1.25	Screw
24-68M	24.0	0.68	Bayonet

Pilot lamps

Part #	Type	Volts	Amps	Hours
NE2A-E	Neon Glow	110/220[a]	0.30	25,000
NE4	Neon Glow	110/220[a]	0.50	15,000
NE45	Neon Glow	105/125[b]	0.002	——
NE51B[d]	Standard	105/125[b]		15,000
NE51HB[d]	High Brightness	105/125[c]		15,000
NE51S[e]	Standard	105/125[b]		25,000
NE51HS[e]	High Brightness	105/125[c]		25,000

a. An external resistor connected in series with the lead of these lamps to adjust the lamp to its working voltage; for 110-V operation, use a 100-kΩ resistor. For 220-V operation, use a 200-kΩ resistor.
b. For these lamps, a 200-kΩ resistor connected in series with one of the leads will be required to operate the lamp on its listed working voltage.
c. For these lamps, a 47-kΩ resistor connected in series with one of the leads will be required to operate the lamp on its working voltage.
d. These lamps are bayonet based.
e. These lamps are screw based.

Light-emitting diodes

Part #	Color	Lens type	Wavelength	Intensity
SE5011	Red	Jumbo Diffused	655	1.0
SE5021	Green	Jumbo Diffused	565	6.0
SE5031	Yellow	Jumbo Diffused	585	3.5
SE5041	Orange	Jumbo Diffused	635	5.2
SEU50151H	Red	Jumbo Diffused	660 (Hi-Brite)	200
SEU50153H	Red	Jumbo Clear	660 (Hi-Brite)	650
SEU50154H	Red	Jumbo Clear	660 (Hi-Brite)	650
SE3011	Red	Mini-Diffused	655	1.0
SE3021	Green	Mini-Diffused	565	4.5
SE3031	Yellow	Mini-Diffused	585	2.3
SE3041	Orange	Mini-Diffused	635	5.0
SE5013D	Red	Jumbo Clear	655	10
SE5023D	Green	Jumbo Clear	565	35
SE5033D	Yellow	Jumbo Clear	585	35
SE5043D	Orange	Jumbo Clear	635	32
SE10101[a]	Red	Rectangular Diffused	695	1.6
SE10121[a]	Green	Rectangular Diffused	565	5.0
SE6501DK	Red	Rectangular Diffused	695	1.0
SE6521DK	Green	Rectangular Diffused	565	2.5

Wavelength is “peak” (in nm) and *intensity* is at “average” power output.
a. The LEDs have the cathode lead as the longer lead. All others listed have the cathode as the shorter lead.

Transistors

Small-signal transistors

Part #	Description
2N697	NPN Metal: 40 V at 150 mA
2N706	NPN Metal: 20 V at 100 mA
2N1711	NPN Metal: 50 V at 500 mA
2N2218	NPN Metal: 30 V at 500 mA
2N2219A	NPN Metal: 40 V at 500 mA

Part #	Description
2N2222A	NPN Metal: 30 V at 500 mA
PN2222A	NPN Plastic
2N2369	NPN Metal: 15 V at 100 mA
2N2857	NPN Metal: UHF and Microwave
2N2904	PNP Metal: 40 V at 500 mA
2N2905	PNP Metal: 60 V at 500 mA
2N2907	PNP Metal: 40 V at 500 mA
PN2907	PNP Plastic
2N2907A	PNP Metal: 60 V at 500 mA
2N3053	NPN Metal: 40 V at 150 mA
2N3440	NPN Metal: 250 V at 1 A, 1 W
2N3643	NPN Plastic
2N3903	NPN General Purpose
2N3904	NPN Plastic, General Purpose (GP)
2N3906	PNP Plastic
2N4123	NPN: GP
2N4400	NPN: Plastic
2N4401	NPN: Plastic
2N4402	PNP: Plastic
2N4403	PNP: Plastic
2N5087	PNP Plastic: Low-noise, Hi-Gain 50 to 100 mA
2N5089	NPN Plastic: Low-noise, Hi-Gain-to 100 mA
2N5109	NPN: Wideband A
2N5179	NPN Metal: 900 MHz to 2 GHz
2N5400	PNP Plastic
2N5401	PNP Plastic
2N5551	NPN
BFR90	NPN Plastic: to 5 GHz
MPQ2222	Small-Signal, Quad NPN GP, in 14-Pin IC
MPQ2907	Small-Signal, Quad PNP GP, in 14-Pin IC
MPS918	NPN Plastic: to 900 MHz
MPS2924	NPN Plastic
MPS2925	NPN Plastic
MPS2926	NPN Plastic
MPS3638A	PNP Plastic
MPS3646	NPN Plastic
MPS4123	NPN Plastic
MPS4124	NPN Plastic
MPS4125	PNP Plastic
MPS4126	PNP Plastic
MPS6507	NPN Plastic:VHF
MPSA05	NPN Plastic
MPSA06	NPN Plastic
MPSA12	NPN Plastic
MPSA13	NPN Plastic: Darlington

Part #	Description
MPSA14	NPN Plastic: Darlington
MPSA18	NPN Plastic
MPSA20	NPN Plastic
MPSA55	PNP Plastic
MPSA56	PNP Plastic
MPSA64	PNP Plastic
MPSA70	PNP Plastic
MPSH11	NPN Plastic:VHF/UHF

Field-effect transistors

2N3819	JFET: 25 V
2N3823	JFET: VHF A
2N4351	Small-Signal MOSFET
2N4416A	JFET: N-Channel GP Amp, also MPF103
2N5458	JFET: N-Channel GP Amp, also MPF104
2N5459	JFET: N-Channel GP Amp, also MPF105
2N5484	JFET: N-Channel, VHF/UHF A
2N5485	JFET: N-Channel, VHF/UHF A
2N5486	JFET: N-Channel, VHF/UHF Amp, also MPF107
2N5951	JFET: N-Channel, VHF/UHF A
J308-J310	JFET: N-Channel, VHF/UHF Amp/Osc./Mixers
MPF102	JFET: N-Channel, VHF A

Small Signal MOSFETs (Dual Gate)

3N140	VHF Amp/Mixer: NF 5 dB at 200 MHz (NTE221)
3N200	VHF Amp/Mixer: NF 6 dB at 200 MHz (NTE222)
3N201	UHF/ VHF Preamp: NF 4.5 dB at 200(NTE454)
3N204	Same as 3N200
3N213	Same as above
40673	Same as above
MFE121	Same as above
MFE140	Same as 3N140
NTE455	UHF RF A to 900 MHz

Power transistors

Power MOSFETs

Part #	Description
IRF511	80 V at 5.6 A
IRF520	100 V at 9 A
IRF521	80 V at 9 A
IRF530	100 V at 14 A
IRF531	80 V at 14 A
IRF540	100 V at 28 A
IRF541	80 V at 28 A
IRF620	220 V at 5 A

Part #	Description
IRF720	400 V at 3.3 A
IRF740	400 V at 10 A
IRFZ32	50 V at 25 A
IRFZ40	50 V at 50 A

Bipolar power transistors (TO-3 case style)

2N3055	NPN: 60 V, 15 A, 115 W
2N3715	NPN: 60 V, 10 A, 150 W
2N3716	NPN: 80 V, 10 A, 150 W
2N3771	NPN: 40 V, 30 A, 150 W
2N3772	NPN: 60 V, 30 A, 150 W
2N6282	NPN: 60 V, 20 A: Darlington
2N6285	PNP: 60 V, 20 A: Darlington
MJ413	NPN: 400 V, 10 A, 125 W
MJ423	NPN: 400 V, 10 A, 125 W
MJ802	NPN: 100 V, 30 A: Also MJ4502
MJ900	PNP: 60 V, 8 A: Also MJ1000
MJ11015	PNP: 140 V, 30 A, 200 W
MJ11016	NPN: 140 V, 30 A, 200 W
MJ15001	NPN: 140 V, 15 A
MJ15002	PNP: 140 V, 15 A

TIP-series power transistors[a]

Part#	Description
TIP29	NPN: 1 A at 30 W
TIP30	PNP: 1 A at 30 W
TIP31	NPN: 3 A at 40 W
TIP32	PNP: 3 A at 40 W
TIP33	NPN: 10 A
TIP34	PNP: 10 A
TIP35	NPN: 25 A
TIP36	PNP: 25 A
TIP41	NPN: 6 A
TIP42	PNP: 6 A
TIP102	NPN: Power Darlington: 8 A at 100 V
TIP106	PNP: Power Darlington: 8 A at 80 V
TIP120	NPN: Power Darlington: * A at 60 V
TIP121	NPN: Power Darlington: 80 V
TIP122	NPN: Power Darlington: 100 V
TIP125	PNP: Power Darlington: 8 A at 60 V
TIP126	PNP: Power Darlington: 80 V
TIP127	PNP: Power Darlington: 100 V
TIP142	NPN: Power Darlington: 10 A at 100 V
TIP147	PNP: Power Darlington: 10 A at 100 V

a. Suffix on the TIP Series for transistor numbers from 29 to 42 as follows: No suffix = 40 V rating. A suffix = 60-V rating. B suffix = 80-V rating. C suffix = 100-V rating.

International transistors

Part#	Polarity NPN/PNP	Volts Emit-coll	Current Max. coll	Watts Peak
BC308	PNP	45	0.1 A	1
BC309	PNP	25	0.1	1
BC327	PNP	45	0.8	0.8
BC328	PNP	25	0.8	0.8
BC547	NPN	45	0.1	0.5
BC548	NPN	30	0.1	0.5
BC549	NPN	30	0.1	0.5
BC549C	NPN	30	0.1	0.5
BC557	PNP	45	0.1	0.3
BC558	PNP	25	0.1	0.3
BC559	PNP	30	0.1	1.5
BC560	PNP	45	0.1	1.5
BC639	NPN	80	1.0	1
BC640	PNP	80	1.0	1
BD139	NPN	60	1.0	8
BD140	PNP	80	1.0	8
BD681	NPN	100	4.0	40
BD682	PNP	100	4.0	40
BF199	NPN	25	0.25	0.23
BF469	NPN	250	0.05	1.8
BF470	PNP	250	0.05	1.8

Early transistors

Part#	Description
2N107	PNP: 9 V for Audio, IF and Switching
2N111	PNP: 9 V for IF, Converter, and Audio
2N139	PNP: 9 V for Audio, IF, and Switching
2N218	PNP: 9 V for Audio, IF, and Switching
CK722	PNP: 9 V for Audio, IF, and Switching
2N170	NPN: 9 V for IF, RF, Converter, and Audio
2N193	NPN: For Mixer, Converter, and Oscillator
2N292	NPN: 9 V for IF, RF, Converter, and Audio
2N388	NPN: For Mixer, Converter, and Oscillator
2N1302	NPN: For Mixer, Converter, and Oscillator
ECG121	30-W Audio Output Transistor

D
APPENDIX

Parts suppliers

Antique Electronics Supply
6221 S. Maple Ave.
Tempe, AZ 85283

All Electronics Corp.
P.O. Box 567
Van Nuys, CA 91408

DC Electronics
P.O. Box 3203
Scottsdale, AZ 85271

Edmund Scientific Co.
101 E. Gloucester Pike
Barrington, NJ 08007

The Electronic Goldmine
P.O. Box 5408
Scottsdale, AZ 85261

Fair Radio Sales Co.
P.O. Box 1105
Lima, OH 45802

Steinmetz Electronics
7519 Maplewood Ave. R.E.
Hammond, IN 46324

Universal Radio, Inc.
6830 Americana Pky.
Reynoldsburg, OH 43068

E
APPENDIX
Selected reading list

Here is a selected reading list of books and periodicals that can be of help with principles, procedures, theories, and observations associated with parts of this book.

Books

ARRL Amateur Radio Handbook by the ARRL, 1993 (70th edition), ARRL.
The ultimate radio amateur's "bible," containing theory, reference material, and circuitry for the radio amateur and the electronics hobbyist.

The World Below 500 Kilohertz by Peter Carron, Jr., 1985, Universal Radio Research.
A book dedicated to aspects of longwave listening, covering ELF, VLF, Omega stations, receiver and antenna information, and more.

Weather Radio by Anthony R. Curtis, 1992, Tiare Publications.
Weather Radio is a frequency guide of worldwide channels of weather information, from 60 kHz through satellite frequencies.

G. E. Tubes: Essential Characteristics, by G.E. Corp, 1973 (14th edition), GE.
This book contains all G.E. tube characteristics, foreign to American interchangeability, industrial/military cross-reference, and some semiconductor information.

How to Repair Old-Time Radios by Clayton L. Hallmark, 1979, TAB Books.
A classic! *How to Repair Old-Time Radios* contains radio theory, circuit descriptions, repair procedures, and receiving tube data. It's a must-have book for the radio restorer, and for the builder for testing parts and circuits.

The Shortwave Propagation Handbook by Jacobs and Cohen, 1982, CQ Publishing.

Principles and theories of radio propagation are discussed. Information about the ionosphere, sunspots, and unusual propagation characteristics is also included.

The 1934 Official Shortwave Radio Manual by Hugo Gernsbeck, 1934 and 1987, Lindsay Publications.

High-quality reprint of Hugo Gernsbeck's 1934 publication, which contains radio history, set building, and servicing guides.

RCA Receiving Tube Manual by RCA Corporation, 1965, RCA.

This book contains tube data for the RCA-manufactured tube types. It also contains some circuits and related information.

Tube Substitution Handbook, 1980, Howard W. Sams & Co., Inc.

Tube basing and interchangeability charts.

Periodicals

Earthquakes and Volcanoes, U. S. Geologic Survey.

This bimonthly publication by the Geologic Survey provides information on earthquakes, seismology, and volcanoes. It runs about a year behind current activity. It is available by subscription from:

Superintendent of Documents
U.S. Government Printing Office
Washington, DC 20402

Preliminary Determination of Epicenters, U.S. Geological Survey.

This publication runs nearly one year behind current date. It includes detailed reports and adjustments of earthquakes in the United States, which are very useful for plotting historical data.

Storm Data, NOAA.

This is another government publication that runs about one year behind current date. The listings are state by state with significant weather and storm data, including many weather spotter reports. This monthly publication is available by subscription from:

National Climate Data Center
Federal Building
37 Battery Ave.
Asheville, NC 28801-2733

Mariners Weather Log, NOAA.

This quarterly publication contains marine weather information, weather radio station information, typhoon and hurricane pictures and data, and feature stories. It is available by subscription from:

User Services Branch
National Oceanographic Data Center
NOAA
Washington, DC 20235

F
APPENDIX

Solar, weather, and seismic data
Figures, observations, and significant events

This listing contains observations made from a permanent installation near Fresno, California, from equipment described in this book, and from radio station WWV.

Many more events occurred; the following compilation includes the most significant events recorded. See appendix E for publications that will have more detailed information when they are written and released.

Geomagnetic, seismic, and solar activity Significant events March through December 1993

March

2	Solar flare at 21:49Z
4	Satellite observed Proton event at 15:05Z
6	Solar flare at 20:49Z
8	Geomagnetic storm began at 21:39Z
11	Solar flare at 22:02Z
11–14	Geomagnetic storm conditions
12	Solar flare at 18:16Z
25	Earthquake in Portland, Oregon

April

4–6 Geomagnetic storm conditions
8–22 Polar cap disturbance
13 Geomagnetic storm
18 Solar flare at 19:15Z

May

8 Polar cap disturbance began at 22:00Z
8–10 Geomagnetic storm conditions
27 5.0 Earthquake at Frazier Park, California at 21:47L

June

4 Geomagnetic storm began at 14:00Z
7 Solar flare at 14:25Z
14 Earthquake in Yosemite, California; 3.0 magnitude
24 Solar flare at 07:35Z
25 Solar flare at 03:22Z

July

2 Geomagnetic storm

August

11 Earthquake near San Jose, California; 5.0 at 15:34L
15–18 Geomagnetic storm conditions

September

13 Geomagnetic storm
20 Series of earthquakes near Kalamath Falls, Oregon; 5.7 at 08:39, 5.5 at 10:30, 4.8 at 11:48L
22 Aftershocks to the Kalamath Falls series, topped by a magnitude 4.3 at 22:30L
28 Earthquake 4 miles south of Redlands, California; 3.0 at @ 22:00L
28 10 moderate earthquakes in southeastern Mexico
29 Major 6.4 earthquake; 250 miles southeast of Bombay, India

October

1 Geomagnetic storm conditions
9 Geomagnetic storm conditions
11 Earthquakes: 4.1 at 00:30L; 2½ miles south of Hollister, California, on the San Andreas Fault. A 6.7 earthquake also occurred near Tokyo, Japan

November

4 Geomagnetic storm conditions
14 Earthquake near Parkfield, California; 4.8 at 04:28L
24 Earthquake; 9 miles west of San Francisco, CA, 3.0 at 07:18L

December

2 Geomagnetic storm conditions
4 Earthquake; 5.4 near Kalamath Falls, Oregon, at 07:18L
7 Geomagnetic storm began at 15:00Z

10 Earthquakes in California: 3.0 in Ridgecrest, 3.0 in Mojave
25 Beginning of stratospheric warming episode
28 Earthquake; 4.7 30 miles west of Jackson, Wyoming
31 Earthquake; 4.0 at 13:30L at Mammoth Lakes, California
31 Earthquake; 5.2 at Gren Menem, Venezuela

All Times: L = Local; Z = GMT

Solar flux values March 1993 through February 1994

Date	MAR	APR	MAY	JUN	JLY	AUG	SEP	OCT	NOV	DEC	JAN	FEB
1	132	124	103	137	108	100	96	117	91	109	148	94
2	144	121	105	139	110	101	81	127	93	104	146	96
3	151	117	107	136	111	98	80	125	94	106	133	98
4	167	116	111	130	107	96	79	121	96	105	129	96
5	165	119	118	134	101	94	80	124	96	101	137	93
6	167	133	120	128	95	84	79	118	97	103	132	95
7	153	130	131	112	93	91	78	117	95	106	126	96
8	146	143	129	112	93	91	79	113	92	105	123	95
9	143	136	129	115	90	96	79	110	90	98	117	101
10	146	126	134	113	86	102	79	107	90	96	110	94
11	150	116	133	102	83	108	79	103	89	93	101	93
12	161	103	126	93	86	101	79	98	91	91	98	98
13	142	97	120	87	87	97	80	95	94	88	95	98
14	136	92	112	84	93	92	82	91	94	88	90	101
15	131	88	104	82	93	91	85	94	104	85	96	104
16	122	90	99	82	97	90	88	91	100	85	100	105
17	124	96	96	83	99	93	85	90	100	84	105	106
18	111	106	91	85	100	92	85	83	103	85	101	106
19	127	111	91	86	101	92	84	89	101	87	103	108
20	128	119	91	91	102	92	80	94	100	91	105	108
21	131	119	91	91	105	95	80	95	97	93	111	105
22	128	117	92	97	108	94	79	92	99	100	113	107
23	121	122	95	109	110	93	78	92	100	105	118	107
24	115	129	98	123	106	91	70	92	100	111	129	105
25	117	125	100	119	102	89	96	87	97	119	132	97
26	117	123	108	122	104	88	106	89	93	125	128	94
27	123	118	120	128	103	88	104	87	92	140	120	95
28	126	114	126	124	102	88	111	88	93	134	119	93
29	129	107	129	123	100	90	116	92	94	129	102	
30	129	106	140	116	98	90	121	91	104	143	99	
31	125		138		96	89		91		141	98	

Geomagnetic A index values March 1993 through February 1994

Date	MAR	APR	MAY	JUN	JLY	AUG	SEP	OCT	NOV	DEC	JAN	FEB
1	14	8	7	5	16	1	1	25	7	15	14	5
2	15	5	5	5	23	1	2	6	4	32	13	12

Date	MAR	APR	MAY	JUN	JLY	AUG	SEP	OCT	NOV	DEC	JAN	FEB
3	19	1	10	15	15	3	25	4	3	28	11	7
4	20	26	6	35	10	16	22	5	44	4	6	8
5	5	64	7	38	2	18	12	5	17	8	1	20
6	8	10	10	15	2	18	10	10	21	4	8	16
7	11	13	13	16	9	12	10	2	17	16	2	50
8	19	18	37	10	14	10	5	12	13	41	4	48
9	51	19	28	6	12	11	3	30	7	2	0	34
10	21	20	44	10	10	7	4	21	10	5	0	29
11	29	16	3	9	24	3	5	21	4	5	14	36
12	20	10	13	14	9	6	8	15	4	5	18	29
13	23	25	5	12	8	5	43	13	7	2	22	24
14	16	17	15	11	2	1	24	6	9	1	14	23
15	38	26	8	7	3	8	18	2	8	7	17	19
16	28	12	13	2	1	34	5	3	9	18	14	14
17	14	8	16	2	1	33	4	5	5	15	17	9
18	14	16	4	3	3	19	3	7	16	13	20	5
19	14	8	11	3	2	13	2	6	20	9	23	16
20	9	23	11	3	12	9	14	3	7	10	12	14
21	19	25	2	1	14	5	7	6	5	12	5	45
22	19	15	1	1	11	4	6	12	4	10	5	58
23	9	9	1	12	6	3	6	12	5	7	3	7
24	27	7	2	23	2	3	15	6	2	8	2	4
25	14	13	1	16	2	2	12	27	3	6	3	11
26	8	4	2	17	3	5	12	15	9	3	18	3
27	10	3	14	7	8	15	2	34	0	6	17	6
28	11	4	18	4	5	7	2	13	2	1	15	5
29	13	10	9	10	10	9	3	8	8	0	6	
30	18	12	3	7	5	2	3	2	2	2	6	
31	7		5		3	2		4		13	5	

Flux and A index values for March and April 1994

	March		April	
Date	Flux	A	Flux	A
1	94	10	82	5
2	98	12	79	20
3	101	23	77	44
4	98	1	77	30
5	96	3	77	25
6	96	6	73	30
7	91	36	73	27
8	91	40	73	31
9	90	33	73	34
10	88	24	75	25
11	88	24	74	31

	March		April	
Date	**Flux**	**A**	**Flux**	**A**
12	93	33	74	17
13	90	21	74	22
14	87	25	79	19
15	87	37	80	15
16	86	16	82	23
17	85	27	82	62
18	87	18	84	17
19	89	13	85	16
20	89	9	86	11
21	91	20	87	9
22	91	13	85	7
23	92	13	85	5
24	93	16	83	9
25	91	14	83	6
26	89	8	81	7
27	88	5	78	4
28	88	8	77	2
29	86	6	78	5
30	86	11	75	3
31	85	7		

Explanation of figures contained in appendix F

Figure F-1 is the key to the weather map symbols. Figure F-2 shows storm tracks in the eastern north Pacific at the end of the tropical storm season, as the 1991-92-93 ENSO event ends. Figure F-3 shows storm tracks that begin to signal drought conditions for central and northern California, as storms are directed to the arctic in the north and southward along the West Coast to south and east of California.

Figure F-4 shows a cold, northerly flow over California. Central California had several frosts beginning on January 7, with that morning's low temperature of 30°F being 13° colder than on January 5. Figure F-5 depicts a Pacific air-flow pattern that shows the opening of the "Siberian Express" to the Midwest and East.

Figures F-6 and F-7 show storm tracks and jet-streams trying to get into California, with the West Coast finally getting some rain. Figures F-8 and F-9 show the jet streams and storm tracks going north again; and severe weather moving through California on March 5 through 6, 1994 is depicted in Figs. F-10 and F-11.

Figures F-12 and F-13 are maps of the storm tracks and severe weather locations of March 22 through 24, 1994. Figure F-13 shows the storm that dropped southward along the West Coast, moving in along with the jet stream in extreme southern California, spawning severe thunderstorms and tornadoes. This pattern is significant because it led to heavy rains and tornadoes in the Midwest and heavy rains in the Southeast.

Figures F-14 and F-15 are graphs of the Solar Flux and Geomagnetic A Index for March and April 1994, and Figs. F-16 and F-17 are graphs of the 25-kHz and 300-Hz radio propagation activity for those months.

Wind direction is shown by arrows that fly "with the wind."
Feathers and flags indicate wind speed.

Symbol	MPH	Knots	Symbol	MPH	Knots
	Calm	Calm		44 to 49	38 to 42
	1 to 2	1 to 2		50 to 54	43 to 47
	3 to 8	3 to 7		55 to 60	48 to 52
	9 to 14	8 to 12		61 to 66	53 to 57
	15 to 20	13 to 17		67 to 71	58 to 62
	21 to 25	18 to 22		72 to 77	63 to 67
	26 to 31	23 to 27		78 to 83	68 to 72
	32 to 37	28 to 32		84 to 89	73 to 77
	38 to 43	33 to 37		119 to 123	103 to 107

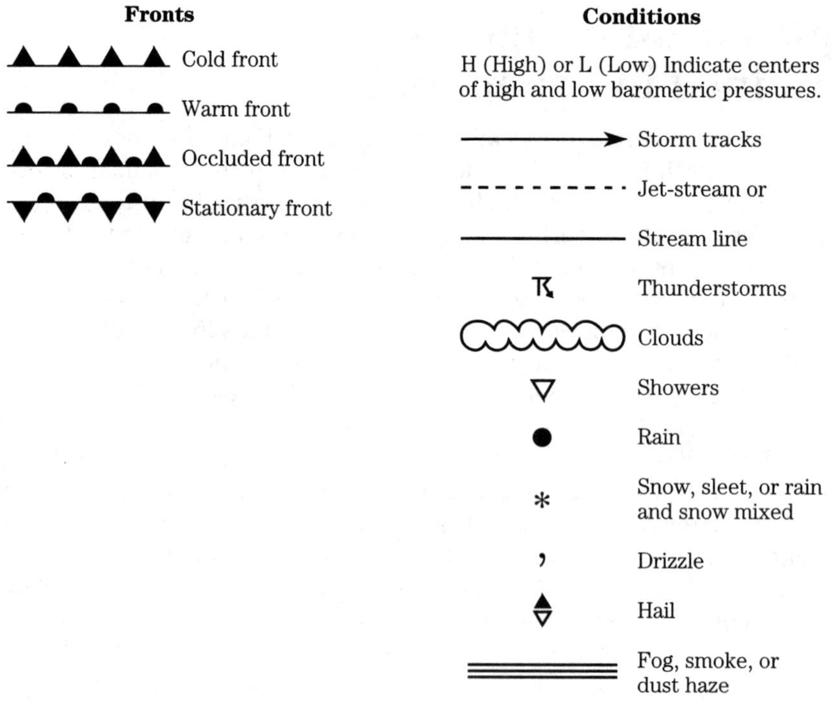

F-1 Key to weather maps.

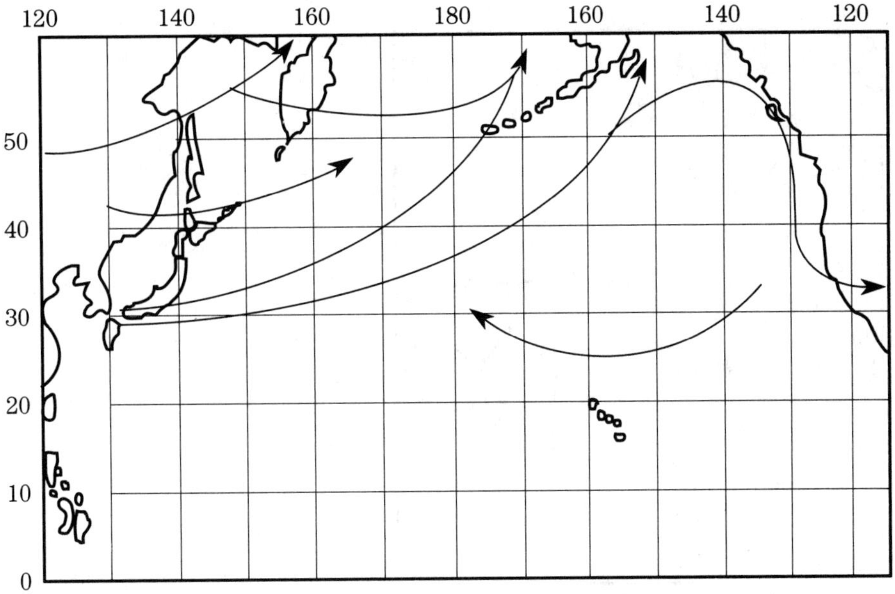

F-2 Storm tracks, jet stream from November 5 through 15, 1993.

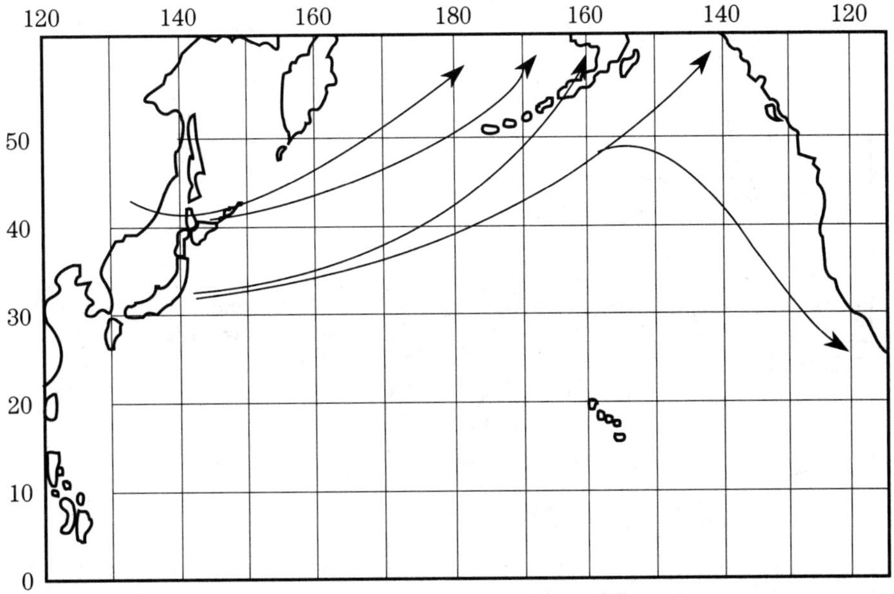

F-3 Storm tracks, jet stream from December 10 through 20, 1993.

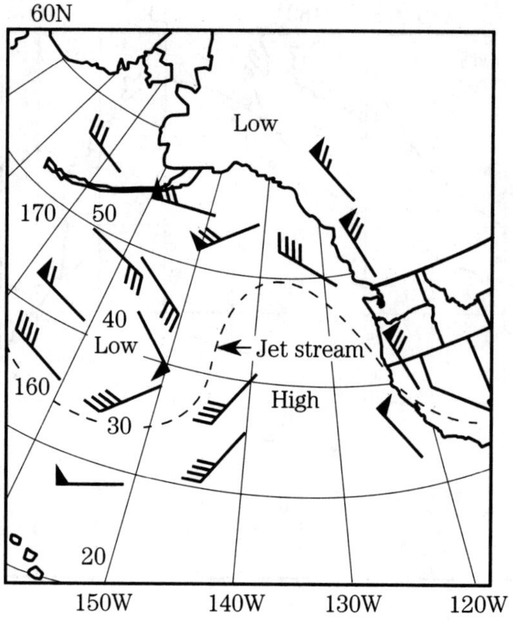

F-4
500-MB winds and jet stream from January 6, 1994.

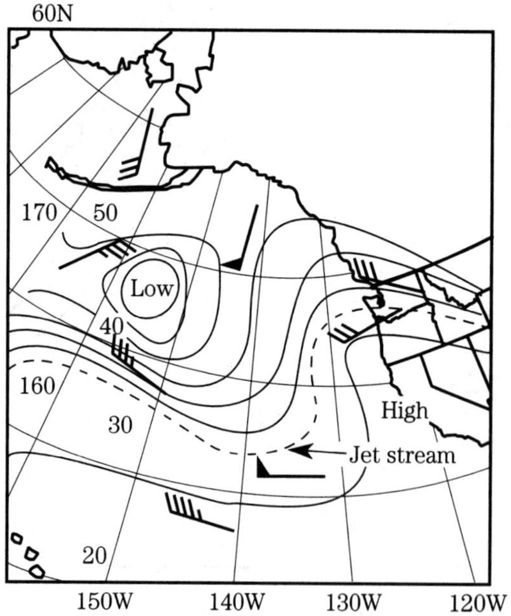

F-5
500-MB winds, jet stream, and pressures from January 15 through 18, 1994.

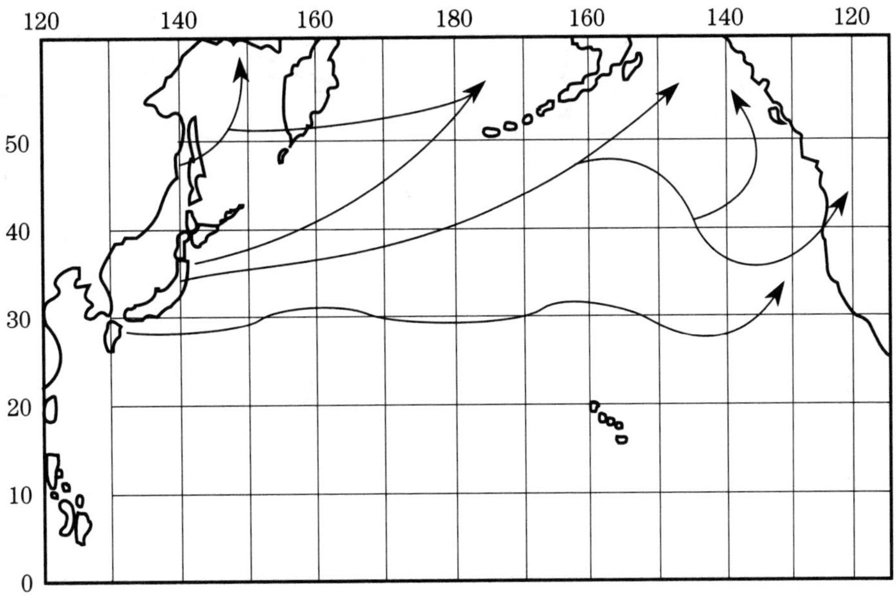

F-6 Storm tracks and jet stream January 15 through 25, 1994.

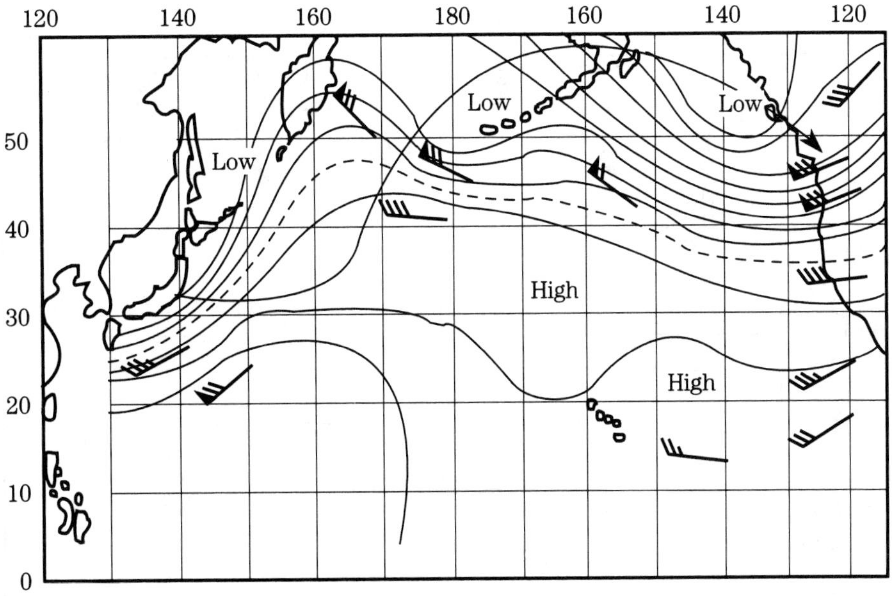

F-7 Storm tracks February 10 through 20, 1994, with 500-MB jet stream and winds February 10 through 15, 1994.

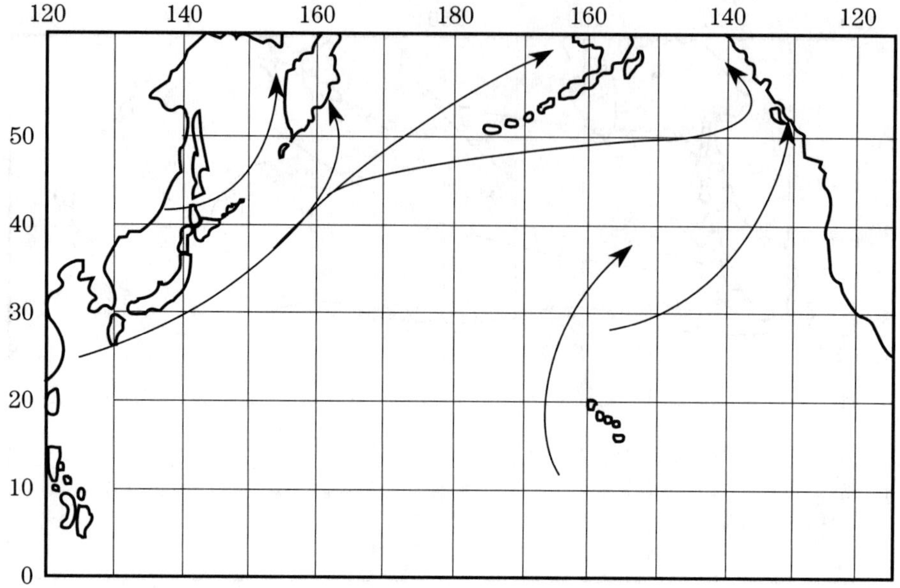

F-8 Storm tracks and jet stream February 15 through 25, 1994.

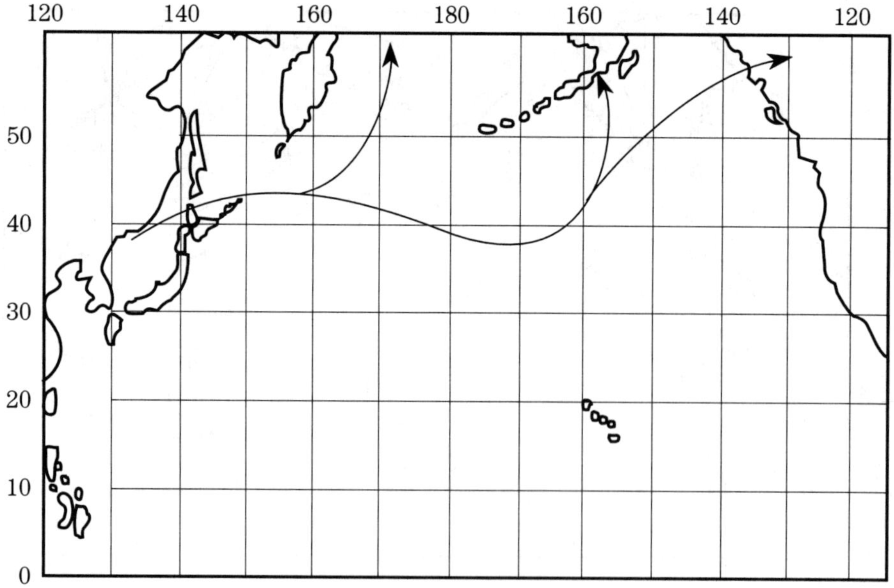

F-9 Storm tracks and jet stream March 5 through 15, 1994.

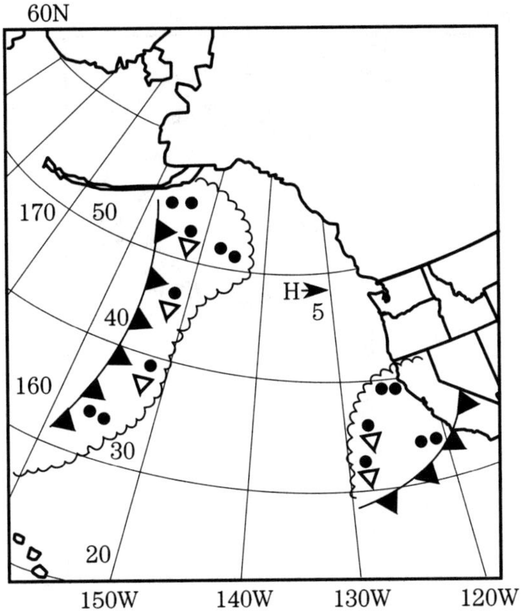

F-10
Significant weather March 5 through 15, 1994.

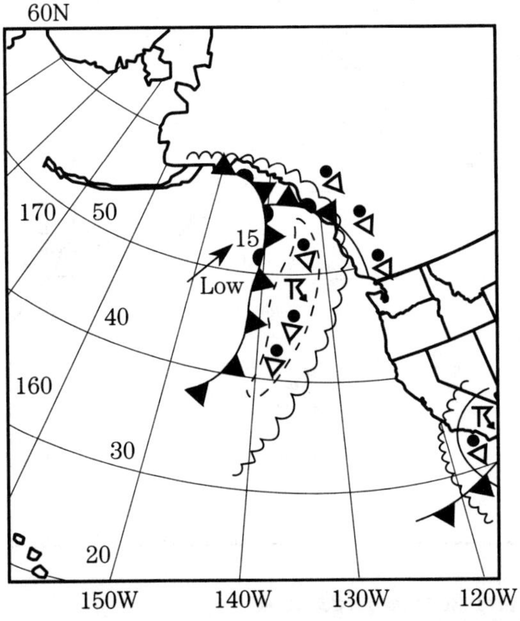

F-11
Significant weather March 6 and 7, 1994.

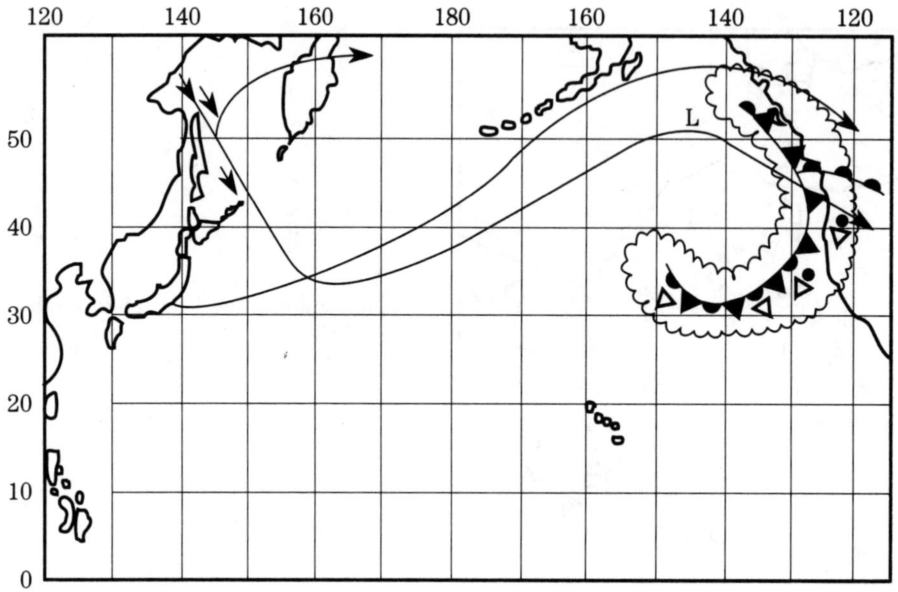

F-12 Significant weather March 20 and 21, with storm tracks March 15 through 25, 1994.

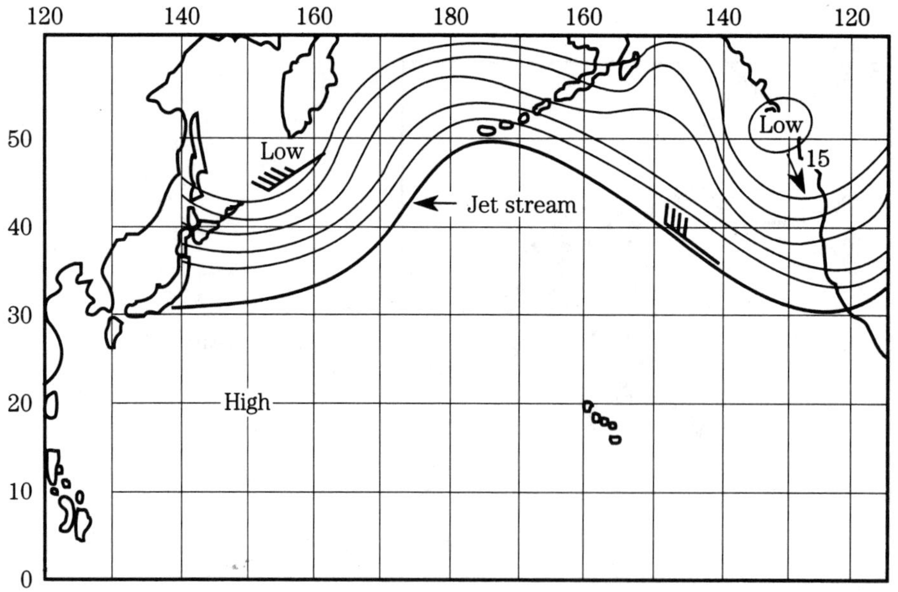

F-13 500-MB jet stream and winds March 20 through 22, 1994.

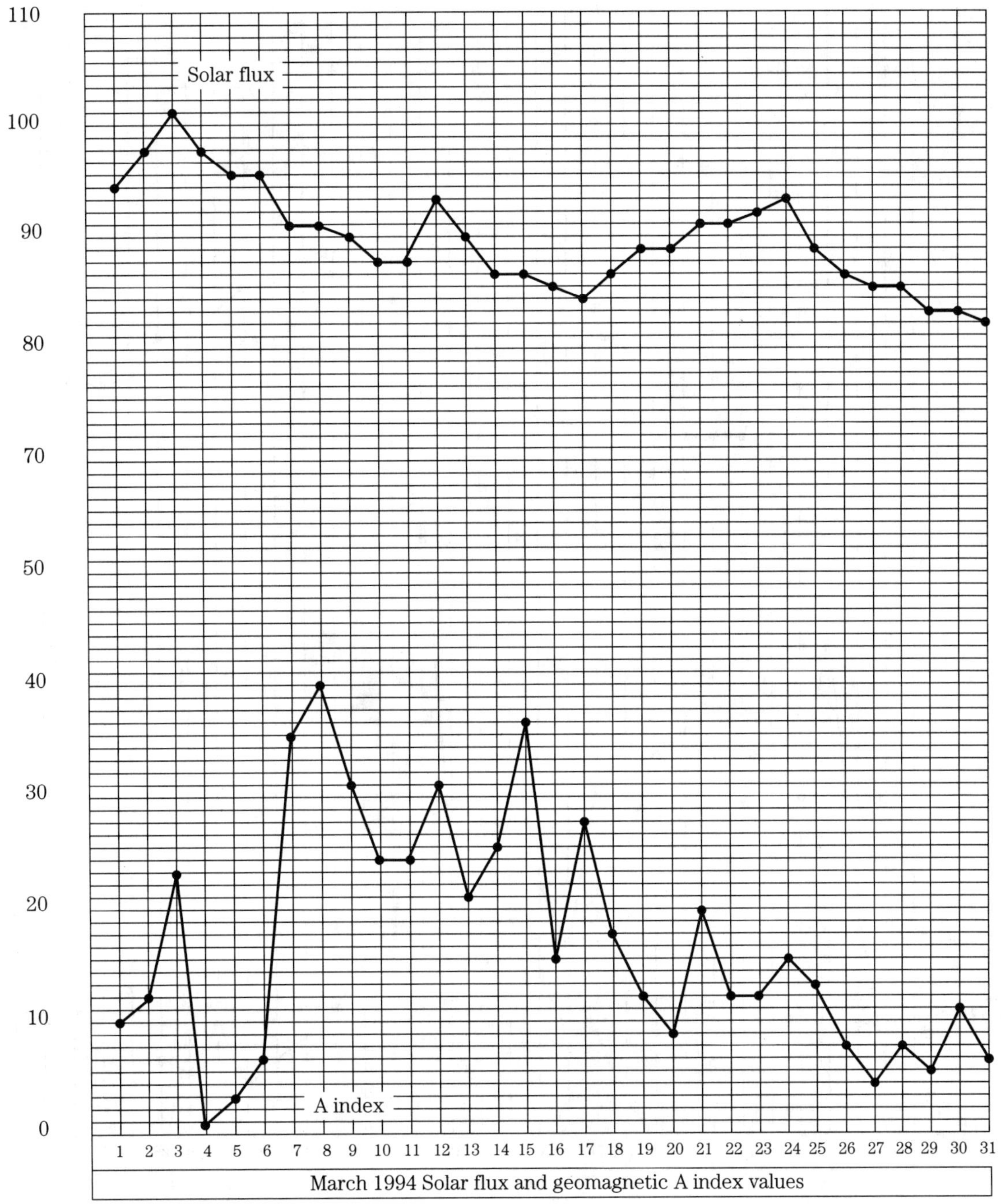

F-14 Solar flux and geomagnetic A index graphs for March 1994.

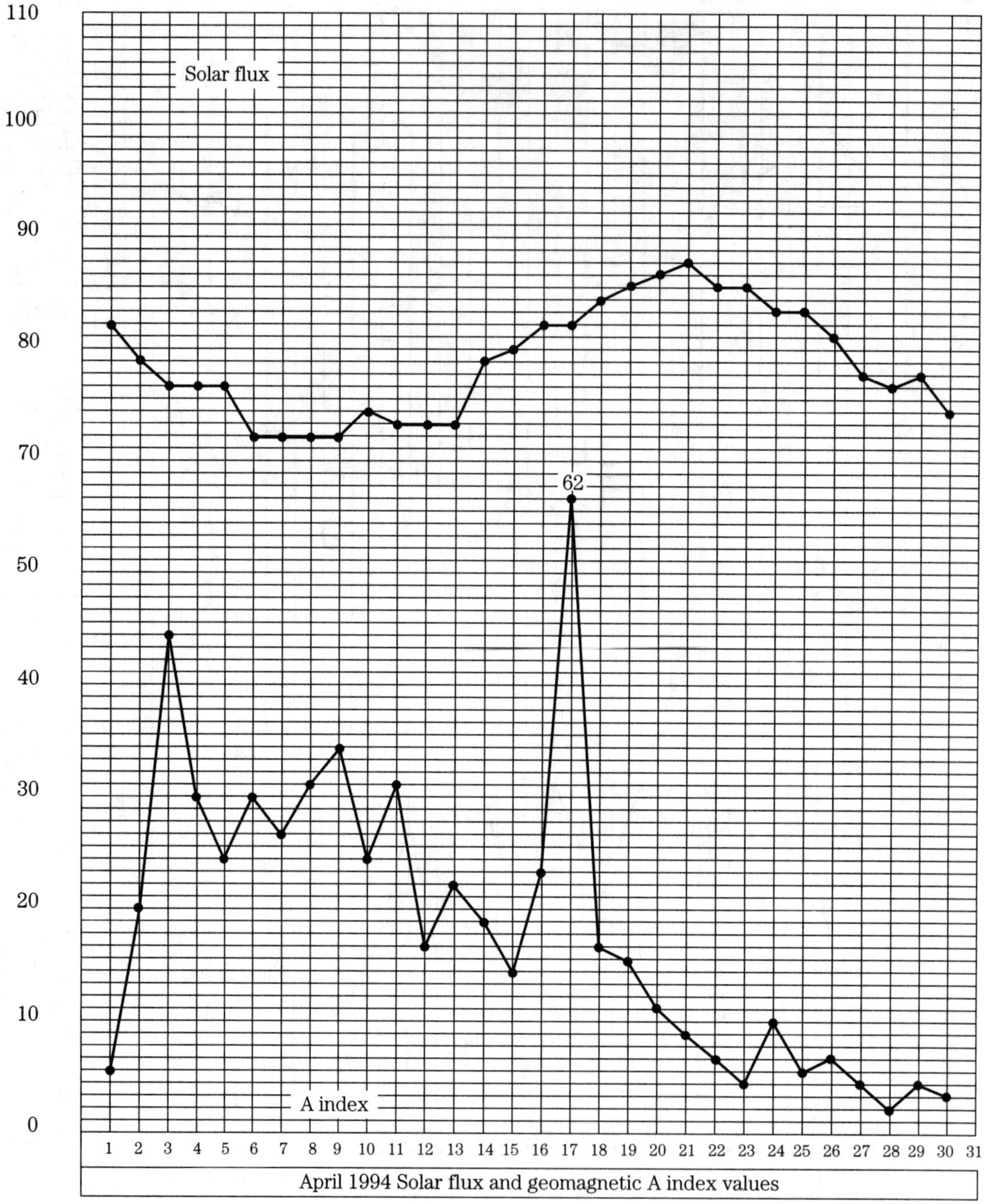

F-15 Solar flux and geomagnetic A index graphs for April 1994.

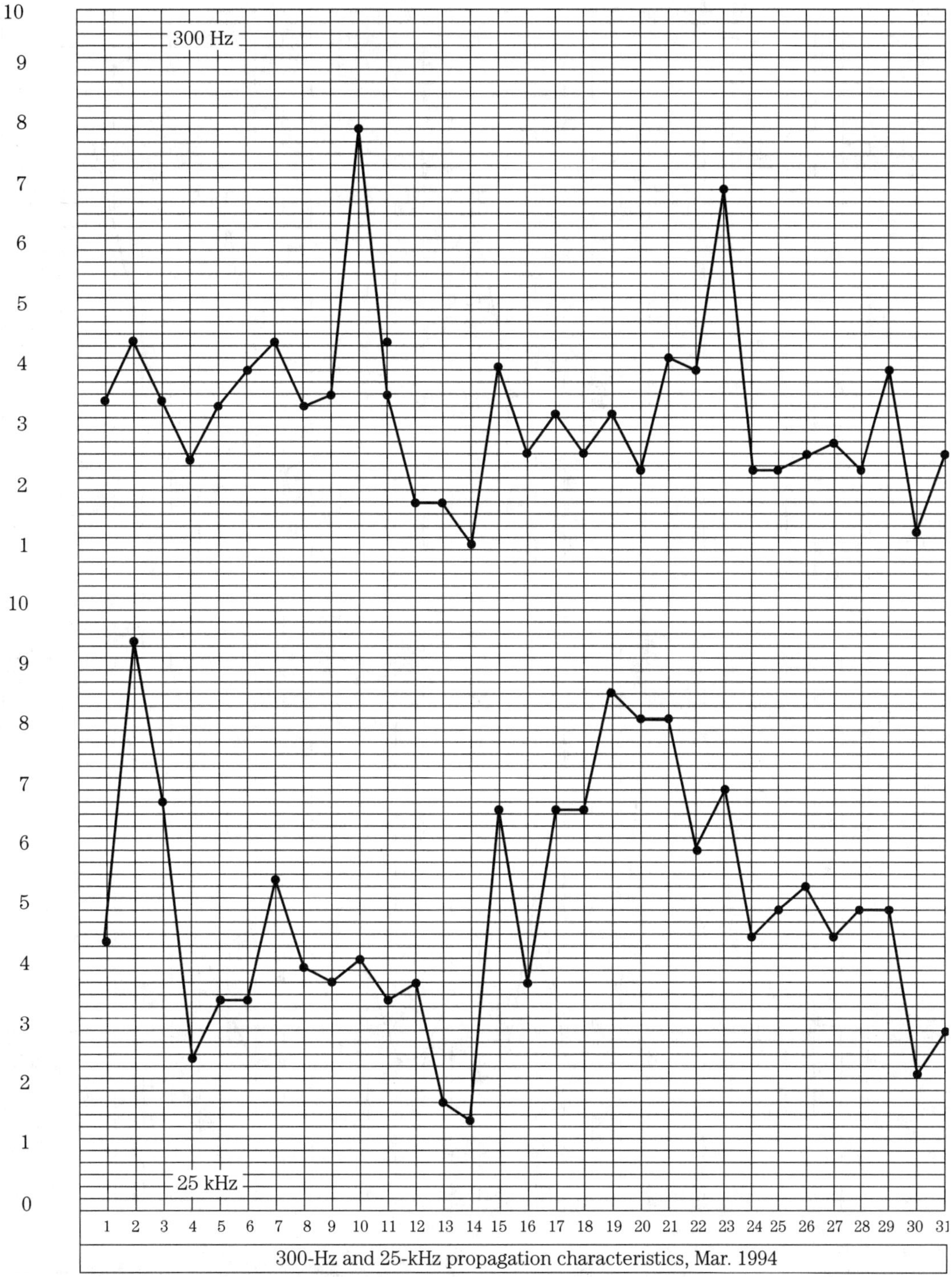

F-16 25-kHz and 300-Hz propagation graphs for March 1994.

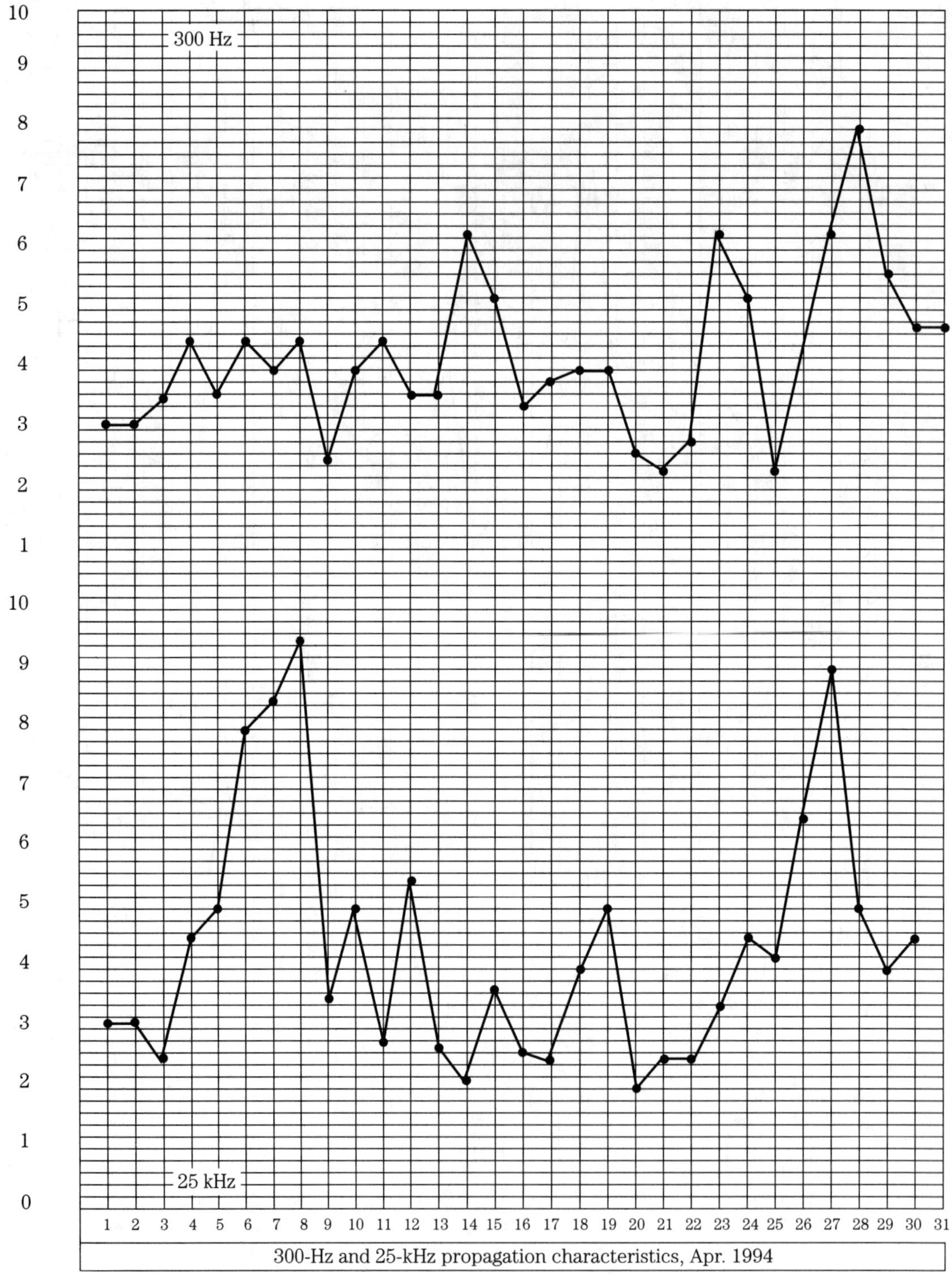

F-17 25-kHz and 300-Hz propagation graphs for April 1994.

Glossary

absorption The conversion of radiant energy into energy of a different form.

alpha rays High-speed, positively charged helium nuclei emitted by radioactive elements.

amplitude The peak value of a wave; the amount of displacement in any oscillatory motion.

anode The positive electrode of an electrochemical apparatus.

appendage An elongated structure that projects from a body.

attenuation A loss of current or power in a circuit, usually measured in decibels.

AWG (American Wire Gauge) The adopted standard wire sizes; the larger the number of the wire, the smaller the wire.

B+ supply The power supply used to supply plate voltage to a vacuum tube.

batholith A very large igneous rock intrusion that has no visible bottom, but extends deeply toward the center of the earth.

beta rays Negatively charged electrons emanating from radium and other radioactive substances.

bifilar An apparatus fitted with two filaments or threads.

caldera A large crater that is believed to be the result of the sinking in of the top of the cone because of withdrawal of lava from beneath.

capacitor A condenser; a device for storing electricity that consists of two metal plates separated by air or some other insulating material.

cathode The electrode in a cell (voltaic or primary) that attracts the positive ions and repels the negative ions; the negative pole.

center tap A connection made to the center of a transformer winding or to the center of a resistor.

choke coil A coil of low ohmic resistance and comparatively high impedance to alternating current; an inductor.

chromosphere A deep layer of incandescent gas outside the photosphere of the sun.

circadian Biological rhythm of about a 24-hour period.

convection The transmission of heat from one place to another by fluid circulation between the spots of different temperature.

convection current The movement of warm air upward and of colder air downward, which promotes the uniformity of temperature.

corona In meteorology, colored rings observed around the moon or sun, which are caused by light diffraction by water droplets.

cosmic rays Strongly penetrating rays, discovered by Miliken in 1925, coming from beyond the earth's atmosphere and containing particles that move at extremely high speed.

damping A decrease in the amplitude of an oscillation or wave motion with time; the checking of a motion because of friction or a similar cause.

detector 1. A device, usually a triode tube, for rectifying current oscillations so that they can affect a speaker; 2. a simple galvanometer that indicates the presence of current in a circuit.

earthquake Shaking of a portion of the earth's crust, often caused by the slipping of rock strata.

electric induction The effect of a charged body on a conductor placed next to it; the conductor becomes charged because of the redistribution of electricity.

El Nino An annual Christmastime appearance of warm water off the coastline of Ecuador and Peru that disrupts local fishing.

ENSO Abbreviation for the El Nino southern oscillation; the most intense events of the El Nino phenomenon are capable of affecting climatic patterns around the globe.

epicenter The point of the maximum shock of an earthquake, it lies directly above the earthquake's focus.

fault A break in the mass of rock on one side of which the rock moves up or down, and no level connects at the same level with rock on the other side.

focal depth The depth of the maximum displacement and intensity of an earthquake.

focus The point of the maximum displacement and intensity of an earthquake.

galvanometer An instrument to indicate and measure small currents.

gamma rays A highly penetrating radiation, traveling with the speed of light, from certain radioactive substances.

Geiger counter A device to detect and measure radiation emissions by the ionization produced in a gas-filled metal tube with a needle electrode.

geomagnetic field The field of the earth's magnetism.

geomagnetic storm The activity that is associated with ionization of atmospheric particles from the sun, which disrupts and distorts readings of the earth's magnetic field.

GOES Geostationary Orbiting Earth Satellite; U.S. meteorological satellites that remain fixed in an orbital point over the earth.

graticule Also *reticle*; a network of fine lines, wires, squares, etc., placed in the focus of an optical device.

ground An electrical connection to the earth, usually a pipe or metal plates driven into the earth.

HF High frequency; in the radio spectrum usually denoted as 3 MHz to 30 MHz.

halo A colored ring (or rings) that encircle the sun or the moon as a result of the dispersion of light by ice crystals in cirrus clouds.

high-pressure cell In meteorology; an anti-cyclonic, clockwise flow of air.

hydrology The study of water, its properties and occurrence on the earth's surface, underground, and in the atmosphere.

hydrometer An instrument to measure the specific gravity of liquids.

ITCZ Intertropical Convergence zone; the area of the earth north and south of the equator, where convective activity, thunderstorms, and related activity is most pronounced.

inductance The property of an electrical circuit by which its variations induce an electromotive force in that circuit or in a nearby circuit.

integrator An instrument to indicate the total amount of a mean value.

interferometer An instrument to compare unknown wavelengths with a standard wave of light with the aid of interference; used to analyze a narrow spectral region (optical).

ionization An electrically neutral atom or molecule that acquires a charge.

ionosphere A region of ionized air beginning about 65 miles above the earth; it reflects radio waves.

jet stream A narrow stream of air at a high altitude moving at speeds of up to 200 to 300 miles per hour.

laser Light amplification by stimulated emission of radiation; a microwave amplifier operating in the visible spectrum; also *maser*.

LED Light-emitting diode.

LF Low frequency; in the radio spectrum, usually for frequencies below the broadcast band, below 530 kHz.

lightning A discharge of electricity between two or more clouds or between the clouds and the earth.

lithosphere The outer solid rock layer of the earth; the earth's crust.

L-wave A type of seismic wave; longitudinal waves.

LW (longwaves) In the radio-frequency spectrum, usually below 500 kHz.

low-pressure cell In meteorology; a cyclonic, counterclockwise flow of air.

magma Molten igneous rock.

magnetic storms Disturbances in the earth's magnetic field usually caused by solar flares. They can generate currents that interfere with radio communications.

magnitude In seismology; the measure of the strength of an earthquake.

Mercalli scale A scale that is used to determine the intensity of an earthquake; it is used by observing the earthquake's effects on people and surroundings.

microseism A weak, recurrent earth tremor, usually caused by earthquakes.

MW Medium wave; in the radio frequency spectrum, *MW* usually refers to the broadcast band, 540 to 1700 kHz.

neap tides The lowest high tide and the highest low tide among the spring tides; they occur when the tidal influences of the sun and moon are acting at right angles with relation to the earth.

oscillation 1. The to and fro motion of a pendulum; 2. the back-and-forth motion of electrons in a circuit.

oscillator A device or circuit that emits sonic, supersonic, or electromagnetic waves.

oscillograph An apparatus that records electrical oscillations.

oscilloscope A device in which a controlled beam of electrons from a cathode-ray tube traces a wave form on a fluorescent screen.

pendulum A body suspended so that it can swing to and fro on a horizontal axis.

pentode An electron tube that contains a filament, plate, and three grids.

pH A value, based on the hydrogen-ion concentration, that is used to express acidity or alkalinity on a scale running from 1 to 14. 7.0 is neutral; values under 7.0 indicate acidity, with the strength increasing with decreasing values; values above 7.0 indicate alkalinity, with strength increasing with increasing values.

pH meter A device that determines the hydrogen-ion concentration.

photoelectric cell Photocell; a device in which electrons are released under the effect of light.

photoemissive detector An apparatus that detects and measures radiant energy by emitting electrons from a photocathode.

photovoltaic cell Photochemical cell; a photoelectric cell that consists of two electrodes of the same metal immersed in an electrolyte; the illumination of one of the electrodes gives rise to a potential difference.

photosensitivity A sensitivity to visible or invisible light.

plate tectonics Theory of the historical movements and changes of continental land masses over geologic time.

plumb line A string to which a weight is attached; it is used to determine the true vertical direction.

polarization 1. A gas accumulation on the electrodes of primary cells, which reduces the voltage and current; 2. changing ordinary light to waves whose vibrations are only in one plane; 3. the separation of the positive and negative charges in a molecule.

potentiometer An instrument that is used to compare a voltage with a constant standard; also, a device for potential adjustment; a variable resistor.

p-waves A kind of seismic wave: Primary waves.

rainbow A curved band of colored light in the sky, caused by drops of water acting as prisms refracting and internally reflecting the sunlight.

Rayleigh waves Seismic waves, usually caused by earthquakes.

rectifier A device for changing alternating current into direct current.

refraction The change in the direction of a light ray or a radio wave when it passes from one medium to another.

regenerative circuit A radio circuit in which the current in the grid circuit is increased by the inductive action of the anode circuit.

resistor A device designed to provide resistance; it may consist of carbon or metal wire and may have a fixed resistance of a specified number of ohms, or it may be variable.

resonance The conditions under which the capacitive reactance and inductance in a tuned circuit are equal.

Richter scale A method of determining the strength of an earthquake using quantitative measurements.

ring-of-fire A geologic feature around the rim of the Pacific Ocean, characterized by seismic, tectonic, and volcanic activity.

saturation (1) The condition of air when it has all the water vapor it can hold; (2) the condition of a magnetic material when it has reached its highest flux density; (3) the state of purity and maximum brilliance of color; (4) a condition in a vacuum tube when all the electrons have been taken up by the anode and/or the other electrodes; (5) dissolving the largest possible amount of a substance in a solvent to obtain a saturated solution.

scattering The diffuse reflection of wave energy when a ray encounters a body or passes through a heterogeneously ionized region.

seism An earthquake.

seismic detector A microphone connected to a recording oscillograph and located in a strategic position on the earth's surface to detect sound waves transmitted through the earth.

seismogram The recording of an earthquake, made on a seismograph.

seismograph An extremely sensitive instrument used to record vibrations of the earth's crust and to detect the location of earthquakes.

seismology The study of earthquakes.

sensor A device capable of receiving and responding to a stimulus.

shear waves A type of seismic wave, usually made by an earthquake.

solar constant The rate at which the earth is receiving energy from the sun, corrected for loss through absorption by the atmosphere; about 1.8 horsepower falls on each square meter of the earth each second; approximately 1.34 × 106 ergs per square centimeter per second.

solar flare Bright eruptions that occur on the sun's surface, emitting intense radiation.

solar radiation The radiation from the sun which covers a wide range of wave lengths, from infrared to ultraviolet.

solarimeter A pyrheliometer for direct reading of solar radiation.

southern oscillation In meteorology; a massive see-sawing of air pressure gradients between the southeastern and the western tropical areas over the Pacific Ocean.

specific gravity The ratio of a mass of a given volume of a substance to that of an equal volume of water.

spectroheliograph An instrument for photographing the sun by monochromatic light; it consists of a spectroscope with a second slit that can be adjusted in such a way that it lets light of only a desired wavelength pass.

stratosphere The layer of the atmosphere, about six to ten miles from the earth's surface; it has an even, low temperature and no convection, but is subject to warming during periods of ionization from charged particles emitted during periods of intense solar activity.

suspension line The device or mechanism by which an instrument is suspended. Examples; the wire holding the coil of a galvanometer; the plumb line of sensors.

SW Short Wave; in the radio frequency spectrum, usually involving frequencies from 1.7 to 30 MHz.

S-waves A type of seismic wave.

temblor An earthquake.

transducer A device inserted between electrical, acoustical, and mechanical elements of a communications system for transformation of the power or impulse into another medium.

tsunami Japanese for the giant sea wave often associated with large earthquakes occurring at sea.

turbidity The cloudiness of water or air due to the presence of finely divided particles in suspension.

ULF Ultra-low-frequency; in the radio frequency spectrum, usually applied to frequencies below 3 kHz.

VLF Very-low-frequency; in the radio frequency spectrum, usually applied to frequencies below 300 kHz.

whiteout A condition occurring in arctic regions when approximately equal intensities of light coming from above and reflected from below cause clouds and snow-covered surfaces to blend so that no object casts a shadow, and only very dark objects are visible.

x-rays A very short wave-length radiation having the power of penetrating matter which is opaque to ordinary light.

zero-potential Optimum ground condition/connection, where connection between a ground rod and to the earth produces a "zero-ohms" reading on an ohmmeter.

Index

T